CW01263139

Mapping Hazardous Terrain using Remote Sensing

The Geological Society of London
Books Editorial Committee

Chief Editor
BOB PANKHURST (UK)

Society Books Editors
JOHN GREGORY (UK)
JIM GRIFFITHS (UK)
JOHN HOWE (UK)
PHIL LEAT (UK)
NICK ROBINS (UK)
JONATHAN TURNER (UK)

Society Books Advisors
MIKE BROWN (USA)
ERIC BUFFETAUT (France)
RETO GIERÉ (Germany)
JON GLUYAS (UK)
DOUG STEAD (Canada)
RANDELL STEPHENSON (The Netherlands)

Geological Society books refereeing procedures

The Society makes every effort to ensure that the scientific and production quality of its books matches that of its journals. Since 1997, all book proposals have been refereed by specialist reviewers as well as by the Society's Books Editorial Committee. If the referees identify weaknesses in the proposal, these must be addressed before the proposal is accepted.

Once the book is accepted, the Society Book Editors ensure that the volume editors follow strict guidelines on refereeing and quality control. We insist that individual papers can only be accepted after satisfactory review by two independent referees. The questions on the review forms are similar to those for *Journal of the Geological Society*. The referees' forms and comments must be available to the Society's Book Editors on request.

Although many of the books result from meetings, the editors are expected to commission papers that were not presented at the meeting to ensure that the book provides a balanced coverage of the subject. Being accepted for presentation at the meeting does not guarantee inclusion in the book.

More information about submitting a proposal and producing a book for the Society can be found on its web site: www.geolsoc.org.uk.

It is recommended that reference to all or part of this book should be made in one of the following ways:

TEEUW, R. M. (ed.) 2007. *Mapping Hazardous Terrain using Remote Sensing*. Geological Society, London, Special Publications, **283**.

FERRIER, G., RUMSBY, B. & POPE, R. 2007. Application of hyperspectral remote sensing data in the monitoring of the environmental impact of hazardous waste derived from abandoned mine sites. *In*: TEEUW, R. M. (ed.) *Mapping Hazardous Terrain using Remote Sensing*. Geological Society, London, Special Publications, **283**, 107–116.

GEOLOGICAL SOCIETY SPECIAL PUBLICATION NO. 283

Mapping Hazardous Terrain using Remote Sensing

EDITED BY

R. M. TEEUW
University of Portsmouth, UK

2007
Published by
The Geological Society
London

THE GEOLOGICAL SOCIETY

The Geological Society of London (GSL) was founded in 1807. It is the oldest national geological society in the world and the largest in Europe. It was incorporated under Royal Charter in 1825 and is Registered Charity 210161.

The Society is the UK national learned and professional society for geology with a worldwide Fellowship (FGS) of over 9000. The Society has the power to confer Chartered status on suitably qualified Fellows, and about 2000 of the Fellowship carry the title (CGeol). Chartered Geologists may also obtain the equivalent European title, European Geologist (EurGeol). One fifth of the Society's fellowship resides outside the UK. To find out more about the Society, log on to www.geolsoc.org.uk.

The Geological Society Publishing House (Bath, UK) produces the Society's international journals and books, and acts as European distributor for selected publications of the American Association of Petroleum Geologists (AAPG), the Indonesian Petroleum Association (IPA), the Geological Society of America (GSA), the Society for Sedimentary Geology (SEPM) and the Geologists' Association (GA). Joint marketing agreements ensure that GSL Fellows may purchase these societies' publications at a discount. The Society's online bookshop (accessible from www.geolsoc.org.uk) offers secure book purchasing with your credit or debit card.

To find out about joining the Society and benefiting from substantial discounts on publications of GSL and other societies worldwide, consult www.geolsoc.org.uk, or contact the Fellowship Department at: The Geological Society, Burlington House, Piccadilly, London W1J 0BG: Tel. +44 (0)20 7434 9944; Fax +44 (0)20 7439 8975; E-mail: enquiries@geolsoc.org.uk.

For information about the Society's meetings, consult *Events* on www.geolsoc.org.uk. To find out more about the Society's Corporate Affiliates Scheme, write to enquiries@geolsoc.org.uk.

Published by The Geological Society from:
The Geological Society Publishing House, Unit 7, Brassmill Enterprise Centre, Brassmill Lane, Bath BA1 3JN, UK

(*Orders*): Tel. +44 (0)1225 445046, Fax +44 (0)1225 442836)
Online bookshop: www.geolsoc.org.uk/bookshop

The publishers make no representation, express or implied, with regard to the accuracy of the information contained in this book and cannot accept any legal responsibility for any errors or omissions that may be made.

© The Geological Society of London 2007. All rights reserved. No reproduction, copy or transmission of this publication may be made without written permission. No paragraph of this publication may be reproduced, copied or transmitted save with the provisions of the Copyright Licensing Agency, 90 Tottenham Court Road, London W1P 9HE. Users registered with the Copyright Clearance Center, 27 Congress Street, Salem, MA 01970, USA: the item-fee code for this publication is 0305-8719/07/$15.00.

British Library Cataloguing in Publication Data

A catalogue record for this book is available from the British Library.

ISBN: 978-1-86239-229-8

Typeset by Techset Composition Ltd, Salisbury, UK

Printed by MPG Books, Bodmin, UK

Distributors

North America
For trade and institutional orders:
The Geological Society, c/o AIDC, 82 Winter Sport Lane, Williston, VT 05495, USA
Orders: Tel. +1 800-972-9892
　　　　　Fax +1 802-864-7626
　　　　　E-mail: gsl.orders@aidcvt.com

For individual and corporate orders:
AAPG Bookstore, PO Box 979, Tulsa, OK 74101-0979, USA
Orders: Tel. +1 918-584-2555
　　　　　Fax +1 918-560-2652
　　　　　E-mail: bookstore@aapg.org
　　　　　Website http://bookstore.aapg.org

India
Affiliated East-West Press Private Ltd, Marketing Division, G-1/16 Ansari Road, Darya Ganj, New Delhi 110 002, India
Orders: Tel. +91 11 2327-9113/2326-4180
　　　　　Fax +91 11 2326-0538
　　　　　E-mail: affiliat@vsnl.com

Contents

Preface	vii
TEEUW, R. M. Introducing the remote sensing of hazardous terrain	1
KERVYN, M., KERVYN, F., GOOSSENS, R., ROWLAND, S. K. & ERNST, G. G. J. Mapping volcanic terrain using high-resolution and 3D satellite remote sensing	5
STEWART, S. F., PINKERTON, H., BLACKBURN, G. A. & GUÐMUNDSSON, M. T. Comparison and validation of Airborne Thematic Mapper thermal imagery using ground-based temperature data for Grímsvötn caldera, Vatnajökull, Iceland	31
RIEDMANN, M. & HAYNES, M. Developments in synthetic aperture radar interferometry for monitoring geohazards	45
WALSTRA, J., CHANDLER, J. H., DIXON, N. & DIJKSTRA, T. A. Aerial photography and digital photogrammetry for landslide monitoring	53
TREFOIS, P., MOEYERSONS, J., LAVREAU, J., ALIMASI, D., BADRYIO, I., MITIMA, B., MUNDALA, M., MUNGANGA, D. O. & NAHIMANA, L. Geomorphology and urban geology of Bukavu (R.D. Congo): interaction between slope instability and human settlement	65
MILLER, S., HARRIS, N., WILLIAMS, L. & BHALAI, S. Landslide susceptibility assessment for St. Thomas, Jamaica, using geographical information system and remote sensing methods	77
TEEUW, R. M. Applications of remote sensing for geohazard mapping in coastal and riverine environments	93
FERRIER, G., RUMSBY, B. & POPE, R. Application of hyperspectral remote sensing data in the monitoring of the environmental impact of hazardous waste derived from abandoned mine sites	107
BOURGUIGNON, A., DELPONT, G., CHEVREL, S. & CHABRILLAT, S. Detection and mapping of shrink–swell clays in SW France, using ASTER imagery	117
VAN DER WERFF, H. M. A., NOOMEN, M. F., VAN DER MEIJDE, M. & VAN DER MEER, F. D. Remote sensing of onshore hydrocarbon seepage: problems and solutions	125
MANNING, J. Remote sensing for terrain analysis of linear infrastructure projects	135
DOWMAN, I. & BALAN, P. Mapping remote areas using SRTM and ASTER digital elevation model data: a solution to orientation problems	143
ANDREWS DELLER, M. E. Space technology for disaster management: data access and its place in the community	149
Index	165

Preface

This book has its origins in a conference with the same theme that was run by the Geological Remote Sensing Group (GRSG) in December 2004. The GRSG is a Special Interest Group of the Geological Society of London and the Remote Sensing and Photogrammetric Society. A key objective of the GRSG is to increase the awareness of geoscientists, and the general public, about ways in which remote sensing can be used to map and monitor the Earth's surface. To that end the GRSG holds at least one conference each year, produces a quarterly newsletter and runs a website (http://www.grsg.org).

Thanks go to the many people have been involved with the production of this book, notably the GRSG committee who arranged the 2004 conference, the authors who produced the scientific papers, and the experts who reviewed the papers. The reviewers included Roy Alexander, Peter Beaumont, Peter Collier, John Diggens, Gerald Ernst, Claire Fleming, Jim Griffiths, Mike Hall, Norman Kerle, Matthieu Kervyn, Bruce King, Alex Koh, Geoff Lawrence, Philippa Mason, John McMahon Moore, Nick McWilliam, Todd Rubin, Richard Sanders, Geoff Wadge, Nick Walton, Martin Whiteside, Tim Wright and Paul Zukowskyj.

This book features numerous colour images – without which it would be difficult to illustrate many remote sensing applications – which has inevitably led to above-average printing costs. Thanks to Arup, ER Mapper, ITT Visual Information Solutions (ENVI), the Geohazard Research Centre at the University of Portsmouth and the GRSG for generous contributions towards colour printing costs.

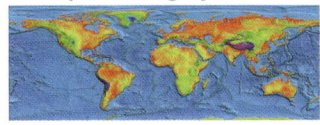

RICHARD TEEUW
Chair, Geological Remote Sensing Group

Introducing the remote sensing of hazardous terrain

R. M. TEEUW

*Geohazard Research Centre, School of Earth & Environmental Sciences,
University of Portsmouth, Portsmouth (e-mail: richard.teeuw@port.ac.uk)*

With human population and associated environmental degradation continuing to increase, it is inevitable that more and more people will be living in zones of hazardous terrain, and therefore the risk of disaster will increase. Add to this the growing evidence for global warming and the increased severity of geohazards related to larger and more frequent storms (i.e. storm surges, coastal erosion, landslides, fluvial flooding, erosion) and both the frequency and severity of disasters is set to increase (e.g. Kesavan & Swaminathan 2006; Smolka 2006; van Aalst 2006).

Remote sensing (measuring and mapping the Earth's surface from aircraft or satellites) can help us to rapidly assess, and therefore better manage, geohazards. Remote sensing and hazardous terrain mapping can play key roles in the management and mitigation of natural disasters (e.g. Cutter 2003; Zeil 2003), with applications grouped into three main stages.

(1) *Pre-disaster.* Maps showing the distribution of geohazards and their relative severities can be used by key decision makers in government, the insurance industry and the local population, to minimize the danger to people and infrastructure. For this stage, geomorphological mapping based on stereoscopic aerial photography has been widely used for many years (e.g. Verstappen & Van Zuidan 1968; Doornkamp *et al.* 1979; Mantovani *et al.* 1996). A new development is mapping based on spectral responses and digital elevation models (DEMs), either regionally using satellite sensors (e.g. Liu *et al.* 2004; Andrews Deller 2006; Theilen-Willige 2006), or in detail using airborne hyperspectral sensors and laser altimetry (e.g. Brown 2004; Deronde *et al.* 2004). Furthermore, early warnings of disasters caused by slope instability, seismic activity or volcanic eruption should be feasible within a few years, using constellations of radar satellites to detect ground deformation (e.g. Kerle & Oppenheimer 2002; Bruno *et al.* 2005; Singhroy 2006).

(2) *Event crisis.* With the onset of a disaster, medium-resolution (15–250 m pixel) satellite imagery (e.g. Landsat or MODIS) is useful for assessing the regional extent and relative severities of the various impacts. With rapid-onset disasters, such as earthquakes, explosive volcanic eruptions or landslides, aerial photography or very high resolution (0.5–2 m pixel) satellite imagery (e.g. Ikonos, Quickbird) is needed to assist rapid search and rescue operations. A useful summary of remote sensing applications in the emergency response to a major disaster, the 2004 Sumatra earthquake and Indian Ocean tsunami, has been given by Kelmelis *et al.* (2006).

(3) *Post disaster.* Geomorphological and geo-ecological mapping, based on the interpretation of aerial photography or spectral and DEM data from satellites, can assist disaster recovery by highlighting locations with essential resources (e.g. water, wood fuel) and materials for reconstruction, such as timber and sand–gravel or clay deposits. Earth resource maps of Banda Aceh, produced by the British Geological Survey in the 1970s, were very useful in the reconstruction of that region after the 2004 earthquake and tsunami (M. Culshaw, pers. comm. 2005).

The hazardous terrains examined in this publication include landslides, flooding, contaminated land, shrink–swell clays, subsidence, fault zones and volcanic landforms (Table 1). Featured remote sensing systems include aerial photography, thermal scanning, hyperspectral sensors, laser altimetry, radar interferometry and multispectral satellites, notably Landsat and ASTER. Related techniques, such as the processing of DEMs and data analysis using geographical information systems (GISs), are also discussed.

The coverage starts with tectonic geohazards, then moves on to ground instability and flooding, before finishing with linear engineering projects and disaster management. **Kervyn *et al.*** provide a comprehensive review of optical and radar approaches to the satellite remote sensing of volcanic terrain, with case studies from Hawaii and the Rungwe rift valley of Tanzania. Next, the thermal properties of Iceland's Grímsvötn volcano are examined by **Stewart *et al.*** New developments in satellite radar interferometry, which can detect millimetre-scale ground deformation, are presented by **Reidmann & Haynes**, who give examples from Turkey's 1999 Izmit earthquake and subsidence in Russia. **Dowman & Balan** provide a new approach for correcting errors that occur when linking together DEMs produced by the Shuttle Radar

From: TEEUW, R. M. (ed.) *Mapping Hazardous Terrain using Remote Sensing*. Geological Society, London, Special Publications, **283**, 1–3.
DOI: 10.1144/SP283.1   0305-8719/03/$15.00   © The Geological Society 2007.

Table 1. *Summary of topics covered in this book*

Type of remote sensing	Seismic or faults	Volcanic	Subsidence; clay types	Landslides	Flooding; erosion or deposition	Pollution	Others: routes, refugees and general mapping
Space							
Landsat, SPOT	Kervyn	Kervyn			Teeuw; Manning		Andrews Deller; Manning
ASTER	Kervyn; Andrews Deller	Kervyn; Andrews Deller	Bourguignon; Andrews Deller	Andrews Deller	Andrews Deller; Bourguignon Teeuw	Andrews Deller	Dowman
ERS radar: InSAR, DifSAR	Kervyn; Reidmann	Kervyn	Riedmann; Teeuw				
Space Shuttle DEM: SRTM	Kervyn; Andrews Deller	Kervyn; Andrews Deller		Andrews Deller	Teeuw; Andrews Deller		Dowman; Manning
Ikonos, Quickbird	Trefois	Kervyn; Manning		Trefois; Manning			Manning
Corona spy-satellite photographs		Kervyn					Manning
Aerial							
Airphoto interpretation, photogrammetry	Trefois	Kervyn	Manning	Walstra; Miller; Trefois Teeuw	Teeuw	Werff	Manning
Multispectral, including thermal		Stewart			Teeuw		
Hyperspectral, including laboratory spectra			Bourguignon; Ferrier		Teeuw; Ferrier	Ferrier; Werff	
SAR imagery, InSAR				Teeuw	Teeuw		
Laser altimetry (LiDAR)					Teeuw		Manning
Other							
Linked GIS analysis				Miller; Trefois			

Names refer to the lead authors of relevant chapters.

Topographic Mission, using the tectonically active Zagros Mountains of Iran as their study area.

The papers covering landslide mapping (**Walstra et al.**, **Trefois et al.** and **Miller et al.**) make extensive use of aerial photography, for both hazard identification and change analysis. The latter two papers also illustrate some of the benefits of using a GIS to integrate datasets, analyse spatial distributions, assess factors contributing to slope instability and more effectively manage hazardous terrain. A different approach was used by **Bourguignon et al.** to map the regional distribution of shrink–swell clays in SW France: they identified the spectral properties of relevant clay minerals and then detected distinctive spectral responses from various clay types using the 14 spectral bands of ASTER satellite imagery.

Uses of remote sensing to map geohazards in riverine and coastal settings are reviewed by **Teeuw**, with the focus on flooding, erosion–sedimentation and ground instability. Airborne hyperspectral remote sensing and associated spectral analyses of pollutants have been used by **Ferrier et al.** and **van der Werff et al.** to map the extent of pollution associated with toxic mine waste and hydrocarbon seepages. Remote sensing applications, both spaceborne and airborne, associated with geohazard mapping along linear civil engineering projects are summarized by **Manning**. The final paper (**Andrews Deller**) examines the uses of free, or low-cost, satellite imagery to map geohazards and raise public awareness in developing countries. This is a fitting conclusion, as **Andrews Deller** illustrates how affordable satellite imagery can be used to map many of the hazardous terrain features illustrated in this volume.

References

ANDREWS DELLER, M. E. 2006. Facies discrimination in laterites using Landsat Thematic Mapper, ASTER and ALI data—examples from Eritrea and Arabia. *International Journal of Remote Sensing*, **27**, 2389–2409.

BROWN, K. 2004. Increasing classification accuracy of coastal habitats using integrated airborne remote sensing. *EARSeL eProceedings*, **3**, 34–42.

BRUNO, D., HOBBS, S. & OTTAVIANELLI, G. 2005. Active and passive geosynchronous SAR systems: concept design and possible applications to monitor geohazards and climate change. *In:* TEEUW, R. M., WHITWORTH, M. Z. & LAUGHTON, K. (eds) *Measuring, Mapping and Managing a Hazardous World. Proceedings RSPSoc2005*. RSPSoc, Nottingham, on CD.

CUTTER, S. L. 2003 GI science, disasters and emergency management. *Transactions in GIS*, **7**, 439–445.

DERONDE, B., HOUTHUYS, R., STERCKX, S., DEBRUYN, W. & FRANSAER, D. 2004. Sand dynamics along the Belgian coast, based on airborne hyperspectral data and LIDAR data. *EARSeL eProceedings*, **3**, 26–33.

DOORNKAMP, J. C., BRUNSDEN, D., JONES, D. K. C., COOKE, R. U. & BUSH, P. R. 1979. Rapid geomorphological assessments for engineers. *Quarterly Journal of Engineering Geology*, **12**, 189–204.

KELMELIS, J. A., SCHWARTZ, L., CHRISTIAN, C., CRAWFORD, M. & KING, D. 2006. Use of geographic information in response to the Sumatra–Andaman earthquake and Indian Ocean tsunami of December 26, 2004. *Photogrammetry Engineering & Remote Sensing*, August, 862–876.

KERLE, N. & OPPENHEIMER, C. 2002. Satellite remote sensing as a tool in lahar disaster management. *Disasters*, **26**, 140–160.

KESAVAN, P. C. & SWAMINATHAN, M. S. 2006. Managing extreme natural disasters in coastal areas. *Philosophical Transactions of the Royal Society of London, Series A*, **364**, 2191–2216.

LIU, J. G., MASON, P. J., CLERICI, N. ET AL. 2004. Landslide hazard assessment in the Three Gorges area of the Yangtze river, using ASTER imagery. *Geomorphology*, **61**, 171–187.

MANTOVANI, R., SOETERS, R. & VAN WESTERN, C. J. 1996. Remote sensing techniques for landslide studies and hazard zonation in Europe. *Geomorphology*, **15**, 213–225.

SINGHROY, V. 2006. Applications of radar remote sensing. *In:* TEEUW, R. M. (ed.) *Remote Sensing of Earth Resources: Exploration, Extraction and Environmental Impacts*. Proceedings GRSG2006, http://www.grsg.org/GRSG2006_details.doc.

SMOLKA, A. 2006. Natural disasters and the challenge of extreme events: risk management from an insurance perspective. *Philosophical Transactions of the Royal Society of London, Series A*, **364**, 2147–2165.

THEILEN-WILLIGE, B. 2006. Tsunami risk site selection in Greece, based on remote sensing and GIS methods. *Science of Tsunami Hazards*, **24**, 35–48.

VAN AALST, M. K. 2006. The impacts of climate change on the risk of natural disasters. *Disasters*, **30**, 5.

VERSTAPPEN, H. T. & VAN ZUIDAN, R. A. 1968. *Photo-interpretation*. ITC, Enschede.

ZEIL, P. 2003. Management and prevention of natural disasters—what are the requirements for the effective application of remote sensing? *International Archives of the Photogrammetry, Remote Sensing and Spatial Information Sciences*, **34**, 54–56.

Mapping volcanic terrain using high-resolution and 3D satellite remote sensing

M. KERVYN[1], F. KERVYN[2], R. GOOSSENS[3], S. K. ROWLAND[4] & G. G. J. ERNST[1]

[1]*Mercator & Ortelius Research Centre for Eruption Dynamics, Department of Geology and Soil Sciences, Ghent University, Krijgslaan 281/S8, B-9000 Gent, Belgium (e-mail: Matthieu.KervynDeMeerendre@UGent.be)*

[2]*Cartography & Photo-Interpretation Section, Geology and Mineralogy Department, Royal Museum for Central Africa, Leuvensesteenweg 13, B-3080 Tervuren, Belgium*

[3]*Remote Sensing & Photogrammetry, Department of Geography, Ghent University, Krijgslaan 281/S8, B-9000 Gent, Belgium*

[4]*Department of Geology & Geophysics, University of Hawai'i at Mānoa, 1680 East–West Road, Honolulu, HI 96822, USA*

Abstract: Most of the hazardous volcanoes, especially those in developing countries, have not been studied or regularly monitored. Moderate-to-high spatial resolution and 3D satellite remote sensing offers a low-cost route to mapping and assessing hazards at volcanoes worldwide. The capabilities of remote sensing techniques are reviewed and an update of recent developments is provided, with emphasis on low-cost data, including optical (Landsat, ASTER, SPOT, CORONA), topographic (3D ASTER, SRTM) and synthetic aperture radar data. Applications developed here illustrate capabilities of relevant remote sensing data to map hazardous volcanic terrain and derive quantitative data, focusing on mapping and monitoring of volcanic morphology. Limitations of the methods, assessment of errors and planned new sensors are also discussed.

The volcanology community has long recognized the advantages of multispectral and synthetic aperture radar (SAR) remote sensing (RS) techniques. The multispectral nature of the data and the repeated coverage of extensive volcanic terrains are major advantages. For over two decades, satellite data have been used to study volcanic activity and map volcanic terrains (e.g. Francis & Baker 1978; Mouginis-Mark *et al.* 2000; Ramsey & Dean 2004; Ramsey & Flynn 2004). Spaceborne sensors permit observation of volcanoes that are remote or difficult to access for political reasons. Sensors allow information retrieval from ongoing eruptive activity for which field data collection is too hazardous. Satellite data allow study of diverse volcanic phenomena (e.g. Francis *et al.* 1996), including to:

(1) detect hot areas, their temporal and spatial patterns (e.g. Harris *et al.* 2000);
(2) monitor volcanic clouds (e.g. Rose *et al.* 2000);
(3) map recently erupted volcanic (e.g. Patrick *et al.* 2003; Rowland *et al.* 2003) or volcano-related deposits (e.g. lahars, debris avalanches; Kerle *et al.* 2003);
(4) discriminate fresh volcanic rock surfaces in terms of mineralogical, textural and compositional differences (e.g. Gaddis 1992; Ramsey & Fink 1999; Byrnes *et al.* 2004);
(5) distinguish weathered volcanic surfaces and assess the terrains' relative age (e.g. Kahle *et al.* 1988), or identify mechanically weak sectors in a volcanic edifice (Crowley & Zimbelman 1997);
(6) characterize volcano morphology and study its changes (e.g. Amelung *et al.* 2000; Rowland & Garbeil 2000; Lu *et al.* 2003);
(7) assess ground deformation using interferometric synthetic aperture radar (InSAR; e.g. Zebker *et al.* 2000).

The potential of high temporal resolution satellite imagery for monitoring volcanic activity and mitigating hazards has been reviewed by Oppenheimer (1998) and Harris *et al.* (2000). Thanks to increased RS data accessibility and reduced time between data acquisition and availability to users (Ramsey & Flynn 2004), low-to-medium spatial resolution (>1 km to 250 m pixel) satellite data are now routinely used to monitor volcanoes; for example, to detect thermal anomalies (Wright *et al.* 2004), or to detect and track ash- and gas-rich volcanic clouds (e.g. Rose *et al.* 2000; Ramsey & Dean 2004). It is not the purpose of this paper to review RS capabilities to monitor

From: TEEUW, R. M. (ed.) *Mapping Hazardous Terrain using Remote Sensing*. Geological Society, London, Special Publications, **283**, 5–30.
DOI: 10.1144/SP283.2   0305-8719/07/$15.00   © The Geological Society 2007.

eruptive events (for that, see Mouginis-Mark et al. 2000), but instead to focus on assessing future hazards by mapping the distribution of existing eruption-related features.

Until recently, research focused on monitoring active volcanoes or on studying *a posteriori* volcanic events and their deposits. Repose periods between eruptions can be decades to thousands of years long. Some of the largest and most devastating historical eruptions occurred at volcanoes that had been considered as dormant or inactive (e.g. 1991 Mt Pinatubo; Newhall & Punongbayan 1996). The majority of the *c.* 560 on-land historically active volcanoes are poorly known (Simkin & Siebert 1994). Fewer than a quarter of them have detailed hazards maps, and fewer still are regularly monitored. Many of these potentially hazardous volcanoes are located in developing countries, where local systems struggle to assess and mitigate volcanic hazards.

For hazard assessment and risk modelling, there is an obvious need to map and characterize the range of deposit types at poorly known or remote volcanoes, to infer the range of activity that they might exhibit. Prior to geological mapping, one needs a good topographic map, also essential for modelling volcanic processes and risks, such as pyroclastic, lava and mud flows (e.g. Iverson et al. 1998; Stevens et al. 2002; Sheridan et al. 2004). Sources of spatial data, such as aerial photography and good quality topographic maps (e.g. 1:25 000 or better), either do not exist or are very difficult to obtain for many volcanically active regions. The goal of this paper is to illustrate how medium-to-high spatial resolution (*c.* 10 to 100 m pixel) satellite imagery provided by multispectral (e.g. Landsat, ASTER, SPOT, CORONA) and SAR sensors (e.g. ERS-1, ERS-2, SRTM) is helpful both for deriving accurate topographic and geological maps and for assessing volcanic hazards.

A key development is that the availability of medium-resolution (*c.* 30 m pixel) satellite images over the Internet has increased dramatically, together with a rapid decrease in cost (Table 1). Satellite RS techniques (InSAR, digital stereophotogrammetry) now provide some of the best digital topographic datasets. These improvements will help the systematic assessment of geohazards at poorly studied volcanoes. However, there is a need to assess the accuracy of the quantitative data retrieved from satellite imagery.

In this paper, we assess the capabilities of satellite data to map and quantitatively study the structure, morphology and texture of volcanic terrains at different scales, to gain insights into processes and assess hazards. We review previous studies, provide an update on advances and present new case studies to illustrate satellite data use for mapping volcanoes. We review the ability of multispectral and SAR sensors to map volcanic deposits, as well as InSAR's capability to generate digital elevation data, aiding morphological studies. A first case study illustrates the potential of ASTER and CORONA data to provide high-resolution and up-to-date topographic maps. The approach, illustrated for a non-volcanic region of Morocco, can be readily applied to volcanic regions. The second case study (Mauna Kea, Hawai'i) illustrates use of multispectral imagery and SRTM DEM data to map small-scale volcanic features. The goal is to examine the advantages and limitations of RS data types available for developing countries, using a volcanic test area where data cross-validation allows accuracy assessments. With the third case study, we then explore to what extent the RS data can be used for the rapid mapping of geohazards and terrain types, including the monitoring of ground deformation, at the Rungwe Volcanic Province (RVP), SW Tanzania. Finally, we highlight some future perspectives for systematic risk assessment at volcanoes, especially in less developed countries, through integrated RS studies and through the increased capabilities that planned new sensors will provide.

Multispectral and digital elevation model data

An ever-increasing number of spaceborne sensors continue to provide a wealth of high-quality multispectral data. This study focuses on the moderate-to-high spatial resolution (*c.* 10 to 100 m pixel) sensors needed to study volcanoes and that are easily accessible at low cost (Table 1). Very high spatial resolution (*c.* 1 m pixel) data provided by commercial sensors (e.g. Ikonos, Quickbird) are not yet a low-cost alternative to assess volcanic hazards (Table 1), even though such sensors prove helpful to produce up-to-date orthomaps for some active volcanoes (e.g. Nisyros caldera, Greece; Vassilopoulou et al. 2002). Airborne RS, although very useful, is not discussed here because it is comparatively expensive, with coverage restricted to relatively few volcanoes.

Multispectral images are a record of reflected or emitted electromagnetic energy. The data range from visible (VIS), near-infrared (NIR), and mid-infrared (MIR) wavelengths (solar energy reflected by the surface) to thermal-infrared (TIR; solar energy absorbed and then re-emitted by the surface). Multispectral data are considered here from Landsat, SPOT, ASTER and CORONA sensors (properties summarized in Table 1). Digital Elevation Model's (DEM) of volcanic terrains

Table 1. Characteristics and sources of RS data discussed in the text

Sensor	Number of bands and spatial resolution	Scene size	Cost	Data distributor	Distributor website
Medium- to high-resolution optical sensors					
Landsat ETM+	PAN: 1 band 15 m VNIR + SWIR: 6 bands 30 m TIR: 1 band 60 m	180 × 180 km^2	Free	Global Land Cover Facility, University of Maryland	http://glcfapp.umiacs.umd.edu/index.shtml
ASTER	VNIR: 3 bands 15 m SWIR: 6 bands 30 m TIR: 5 bands 90 m	61.5 × 63 km^2	Free US$ 55 Free	Land Processes Distributed Active Archive Center (LPDAAC) NASA Earth Observing System Data Gateway Global Land Cover Facility, University of Maryland	http://edcdaac.usgs.gov/datapool/datapool.asp http://redhook.gsfc.nasa.gov/~imswww/pub/imswelcome http://glcfapp.umiacs.umd.edu/index.shtml
SPOT 5	PAN: 1 band 5 m VNIR: 3 bands 10 m SWIR: 1 band 20 m	60 × 60 km^2	€100–800* €1200–6200	Spotimage	www.spotimage.fr
CORONA	PAN: 2–30 ft (c. 0.6–9 m)	c. 14 × 200 km^2 (KH 1–4)	US$ 24–45	USGS Earth Explorer	http://edcsns17.cr.usgs.gov/EarthExplorer/
Very high-resolution optical sensors					
IKONOS	PAN: 1 band 1 m VNIR: 4 bands 4 m	>49 km^2 11 km swath width	US$ 7–56 km^{-2}	Space Imaging	www.spaceimaging.com
Quickbird	PAN: 1 band 0.6 m VNIR: 4 bands 2.4 m	>25 km^2 16.5 km swath width	€18–42 km^{-2}	DigitalGlobe	www.digitalglobe.com
Synthetic aperture radar					
ERS-1 &-2	C-band 25 m	100 × 110 km^2	€400–1400	Eurimage	www.eurimage.com
RADARSAT	C-band 30 m	100 × 100 km^2	US$2750–4250	Radarsat International	www.rsi.ca
JERS SAR	L-band 18 m	75 × 75 km^2	€950	Eurimage	www.eurimage.com
Digital elevation models					
SRTM C-band	C-band: 30 m	1° lat–long	Free; US only	Seamless Data Distribution System	http://seamless.usgs.gov
SRTM C-band	C-band: 90 m	1° lat–long	Free	Seamless Data Distribution System	http://seamless.usgs.gov
SRTM X-band	X-band: 25 m	0.25° lat–long	€400	EOWEB, German Aerospace Agency	www.eoweb.dlr.de
SPOT HRS	PAN: 10 m	>3000 km^2; 120 km swath width	€2.3 km^{-2}	Spotimage	www.spotimage.fr

*Only for European scientific users.
Range of price is indicative, varying with processing level, data type (archived or newly acquired data), and delivery time. Data distributors are given as examples of where those data can be acquired.

can be obtained by: (i) digital photogrammetry based on stereoscopic pair of aircraft of satellite images; (ii) digitalization and interpolation of topographic maps; (iii) radar interferometry; (iv) laser scanning or (v) field survey. Advantages and limitations of the different DEM-generation techniques have been received by Baldi *et al.* (2002) and Stevens *et al.* (e.g. 2002, 2004).

Landsat

The Enhanced Thematic Mapper + (ETM+) instrument onboard the Landsat 7 satellite acquires data in six bands at VIS to MIR wavelengths at 30 m spatial resolution. It also provides a panchromatic band (i.e. black and white) spanning the visible spectrum at 15 m spatial resolution, and two TIR bands (at the same wavelength but with different gain settings) at 60 m spatial resolution (Williams 2003). An ETM+ scene (180×180 km^2) covers a much larger area than that of an airphoto; it does not suffer from radial distortion (after systematic data pre-processing), and it is cheaper than airphoto coverage for the same area. The ETM+ image discussed here was downloaded, for free, from the Global Land Cover Facilities website (GLCF 2005; Table 1). Similar archive images are now freely available for most regions. The Landsat generation of satellites has acquired data since 1972, resulting in a large volume of available archived data. On request, a new image can be acquired, over any given region, every 16 days. However, the cost of on-request data remains high (up to €600; Eurimage website 2005). Regions for which no specific data acquisition requests have been made might not be covered by cloud-free data for several years.

Data acquired by the Thematic Mapper (TM) and ETM+ instruments onboard the Landsat 5 and 7 satellites, respectively, have been used to monitor volcanic activity as these were the first sensors to provide sufficiently fine spatial resolution to map the extent, and characterize the spatial evolution, of hot volcanic areas (e.g. lava flows or domes; e.g. Harris *et al.* 2004). The fine spatial resolution of TM data was also exploited to identify potentially active volcanoes in the Andes (Francis & De Silva 1989) based on crater morphology, lava flow texture and evidence of post-glacial eruptive activity. TM's spectral capability also proved useful for identifying and mapping the distribution of distinct lithologies within volcanic debris avalanches (e.g. Francis & Wells 1988; Wadge *et al.* 1995), and for identifying basaltic scoria cones with specific iron oxidation features in dry environments (e.g. Chagarlamudi & Moufti 1991). Landsat data are now often used in morphological studies of volcanoes or *a posteriori* descriptions of a volcanic event because they easily provide a synoptic view of a volcano across multiple wavelengths (e.g. Patrick *et al.* 2003). On the other hand, the temporal resolution (16 days at best), time lag between image acquisition and availability to users (1–14 days), and high cost of on-request data over specific targets prevent true real-time monitoring of eruptive activity with Landsat. The small number of spectral bands, as well as their broad width in the shortwave IR (SWIR) and TIR also limit capabilities to discriminate volcanic lithologies.

ASTER

The Advanced Spaceborne Thermal Emission and Reflection Radiometer (ASTER) is a medium-to-high spatial resolution, multispectral imaging system flying aboard TERRA, a satellite launched in December 1999. Volcano hazards monitoring, geology and soil mapping are specific applications for which ASTER was developed (e.g. Pieri & Abrams 2004). An ASTER scene, covering 61.5 km $\times 63$ km, contains data from 14 spectral bands ranging from the VIS and NIR (VNIR; three bands at 15 m resolution), SWIR (six bands at 30 m resolution), to TIR (five bands at 90 m resolution; Abrams & Hook 2003).

A key advantage for volcanic hazard assessment is that ASTER acquires stereoscopic images at 15 m spatial resolution for deriving digital elevation models (DEMs). Specifically, in the VNIR, one nadir-looking (band 3N) and one backward-looking telescope (band 3B, $27.7°$ off-nadir with an effective viewing angle at the Earth's surface of $30°$) provide the stereo-pair images (Hirano *et al.* 2003). The advantage of this along-track mode of stereo-image acquisition is that the stereo-images are acquired only a few minutes apart, under uniform environmental and lighting conditions, whereas scenes from across-track data acquisition (e.g. SPOT) are several days apart (Hirano *et al.* 2003; Stevens *et al.* 2004). DEM accuracy depends on the availability, spatial distribution and accuracy of high-quality ground control points (GCPs) and on the contrast within the image (see Mauna Kea test case). Hirano *et al.* (2003) suggested that the root mean square error (RMSE) in ASTER DEM elevations ranges from ± 7 m to ± 15 m, depending on GCPs and image quality.

Processed ASTER DEMs can be acquired at the same cost as an ASTER scene (Land Processes Distributed Active Archive Center website (LPDAAC 2005); Table 1). The routine procedure utilizes the ephemeris and attitude data derived from both the ASTER instrument and the TERRA spacecraft platform to compute a relative DEM. Absolute DEMs are based on GCPs specified by the user. ASTER DEMs are derived at 30 m resolution, in an attempt to minimize errors in the matching process. The RMSE of the relative DEM in x, y and z is expected to range between 10 m and 30 m. An RMSE of 7–50 m is expected for absolute DEMs if at least four GCPs are provided (LPDAAC 2005). The capability of retrieving accurate relative DEMs without having to collect GCPs in the field is of great interest, especially for remote volcanoes. The exact accuracy of these DEMs still needs to be assessed, especially for high-relief terrains such as volcanoes.

The easy availability, low cost and unique combination of multispectral and 3D capabilities at 15 m resolution are the main advantages of ASTER for volcano studies. Although Stevens et al. (2004) demonstrated that 3D ASTER allows for accurate topographic mapping of volcanoes and highlighted the potential for volcano terrain deformation analysis by repeat DEM generation combined with spectral change analyses, the full capability of these data has not yet been fully explored. Ramsey & Fink (1999) highlighted capabilities of ASTER multispectral data in the TIR for estimating surface vesicularity contrast in volcanic rocks. ASTER multispectral data were used by Byrnes et al. (2004) to discriminate between different lava flow morphologies based on the spectral reflectance in the VNIR, and on the emissivity in the TIR. The main limitation of ASTER data, as with other optical RS data, for regular monitoring is their sensitivity to cloud cover, a frequent problem with high-relief subtropical volcanoes.

SPOT

The SPOT satellites (Satellites Pour l'Observation de la Terre) combine high-resolution multispectral bands with stereoscopic capabilities. SPOT acquires data in three bands in the VNIR at 20 m spatial resolution (10 m for the most recent satellite, SPOT 5), and one panchromatic band at 10 m spatial resolution (5 or 2.5 m for SPOT 5). SPOT 4 and 5 also provide one SWIR band (20 m resolution). Despite having fewer spectral bands than Landsat or ASTER, SPOT has the advantages of higher spatial resolutions and variable viewing angles. This increases the potential temporal resolution and the chances of acquiring cloud-free images. With three SPOT satellites currently operational, it is possible to observe almost the entire planet in a single day.

SPOT can acquire stereo-images pairs, but in cross-track rather than along-track orientation (compared with ASTER). The two stereoscopic scenes can be acquired in tandem mode on the same day by using two of the three satellites. However, the stereo-pair images are not acquired under identical illumination or atmospheric conditions. This can affect the accuracy of the resulting DEM (e.g. vertical accuracy between ± 5 and ± 20 m; Hirano et al. 2003). SPOT 5, launched in May 2002, carries the High Resolution Stereoscopic (HRS) instrument, which can acquire simultaneous stereo images with 10 m spatial resolution. The simultaneity of acquisition for the stereo-pair images increases the quality of the derived DEMs (Kormus et al. 2004). A current problem is that the two images used to produce the DEM are not made available to users. Only the processed DEMs can be acquired. The cost of these DEMs has so far limited their applicability (Table 1).

Even though SPOT satellites have been acquiring data since 1986, applications for volcanic studies have been less frequent than those using Landsat data. The lower number of spectral bands and higher cost (e.g. a minimum of €100 per scene for European scientists, and much higher price for non-Europeans) in comparison with Landsat are the main factors limiting its current applicability. SPOT images were used to map lava flows and monitor eruptive activity at Sabancaya volcano (Chile; e.g. Legeley-Padovani et al. 1997), to map and assess structure and morphology of lahar deposits (see Kerle et al. 2003, for a review) and to map volcanic vents and lava flows, in combination with other RS and field data, at Galápagos volcanoes (e.g. Rowland 1996; Rowland et al. 2003). At Galápagos volcanoes, panchromatic and multispectral SPOT images allow identification of flow boundaries based on contrast in surface albedo and on surface colour change associated with flow ageing. All volcanic cones higher than a few metres were also readily identified (Fig. 1a; Rowland 1996).

Declassified satellite images

The CORONA programme was conceived by the US Air Force and Central Intelligence Agency, to take pictures from space of the Soviet Bloc countries and other parts of the world. In 1995, a first set of images, acquired by the CORONA and ARGON systems, was declassified and made available to the public at low cost via the US Geological Survey (USGS Earth Explorer website 2005; Table 1). Between 1960 and 1972, in the operational phase, panchromatic images were recorded by a panoramic camera at flight height of c. 150 km. Depending on the sensor used, the best ground resolution varied from c. 7.3 m to 1.8 m. Satellites used on CORONA missions (sensors KH-1 to KH-4B) carried two cameras, which recorded stereo images of the Earth's surface (McDonald 1995). The principal coverage areas are Asia, Eastern Europe and Northern Africa. The filmstrips suffer from typical panoramic distortion. Time-consuming processing allows derivation

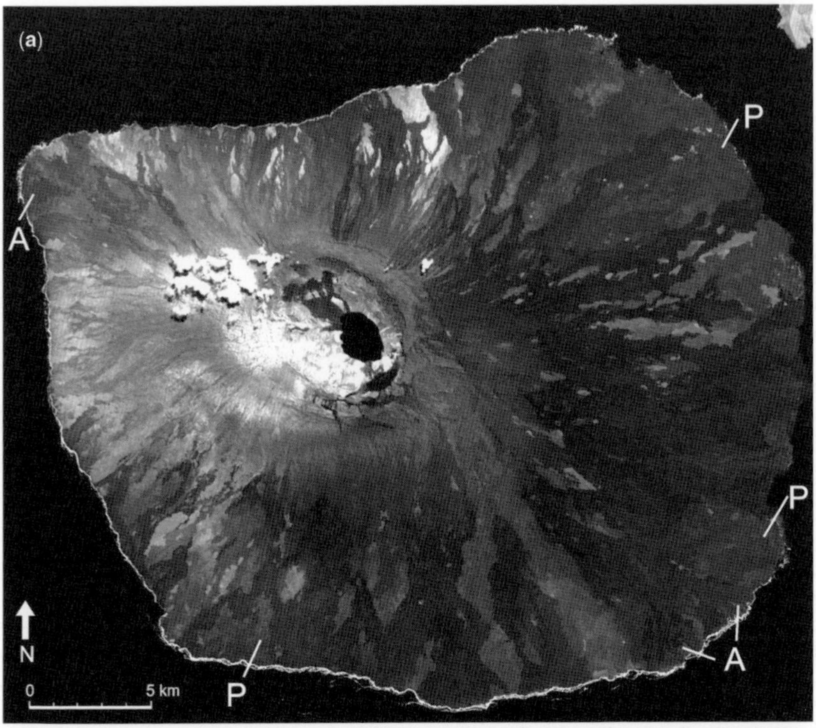
(a)
P
A
P
N
A
0
5 km
P

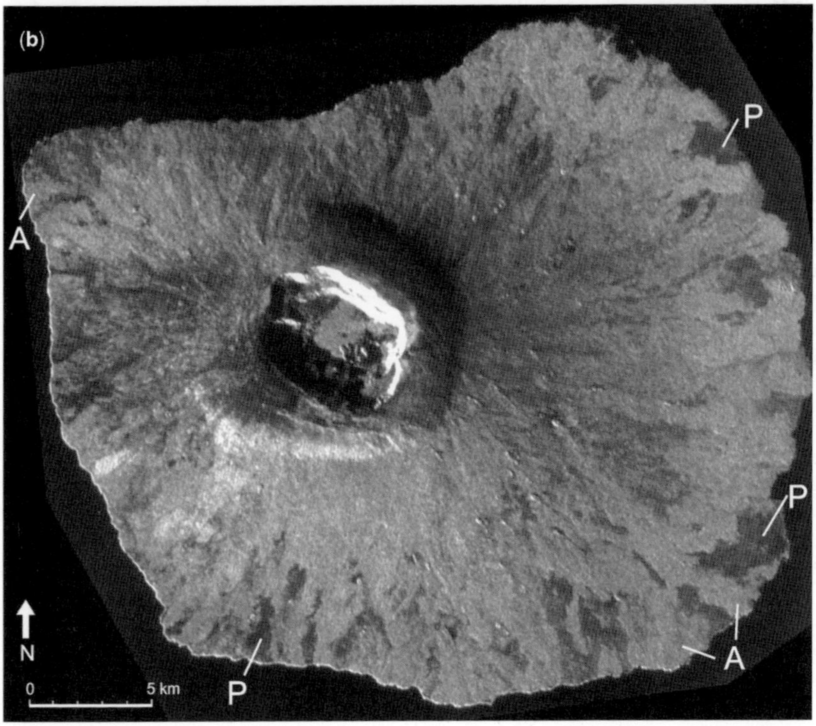
(b)
P
A
P
N
A
0
5 km
P

of a DEM and ortho-image from the original image (see Altaimer & Christoph 2002).

A second dataset from systems GAMBIT (KH-7 surveillance system; 1963–1967) and HEXAGON (KH-9 mapping system; 1973–1980) was declassified in 2002. The KH-7 system does not offer stereocapabilities but acquired very high resolution data (0.6–1.3 m pixel) for many specific regions including volcanic terrains of Kamchatka and Indonesia. KH-9 acquired panchromatic images over much more extensive areas with a spatial resolution of 6.1–9.2 m. When acquired in stereo-mode, successively acquired images overlap by 60%, allowing DEM derivation with standard photogrammetry techniques. In contrast to CORONA, HEXAGON images do not suffer from panoramic distortions. Because of their recent release, the capabilities of the HEXAGON data for DEM generation remain to be assessed. Although cloud cover is a limitation, a rapid search for declassified images returned potentially useful imagery for tens of active or dormant volcanoes in subtropical LDCs especially from KH4 (A and B) and KH-9 missions (USGS Earth Explorer website 2005).

The declassified satellite images provide a wealth of panchromatic, very high resolution data for many regions, including some lacking airphoto coverage. These images may offer a low-cost route to high-resolution DEM generation; for example, this option is about 3000 times cheaper than that using IKONOS data (<1 cent km^{-2} for CORONA v. c. \$25km^{-2} for IKONOS; Altaimer & Christoph 2002). As with airphotos, CORONA data require intensive processing to correct for geometric distortions. Good quality GCPs are needed to obtain absolute elevations. Images acquired several decades ago might be of interest for assessing a volcano's morphology prior to its most recent eruptions or for studying its morphological changes.

Satellite radar systems

Synthetic aperture radar (SAR)

Radar sensors provide terrain information by recording the amplitude and phase of the backscattered signal. Amplitude and phase are influenced by radar system parameters (wavelength, incidence angle, polarization) and by terrain properties such as surface roughness on the scale of the radar wavelength, ground slopes, and dielectrical properties of the surface (e.g. McKay & Mouginis-Mark 1997). Different radar wavelength can be used: X-band (2.8 cm), C-band (5.6 cm; ERS, ENVISAT, RADARSAT sensors) and L-band (23.5 cm; JERS sensor). The SAR sensors (ERS-1 until 2000, ERS-2, JERS until 1998, ENVISAT, RADARSAT) differ in wavelength, looking angle and other acquisition parameters.

Radar penetrates clouds, a major advantage for the study of volcanic regions, especially in the subtropics. SAR sensors are 'active': they record their own backscattered signal, are not dependent on sun illumination, and so can work day and night; they combine ascending and descending orbit data (i.e. halving the return period from 35 days to 17.5 days in the case of ERS and ENVISAT). Radar image use for mapping is limited by the geometric distortion and shadowing effect caused by oblique viewing and topography.

Backscatter intensity has been used to map distinct pahoehoe and aa lava flow surfaces (Gaddis 1992; see below), and surface textures of lava flows generally (Byrnes et al. 2004). Different polarizations were also used for mapping lava flows (McKay & Mouginis-Mark 1997). When combined with cross-validation field data, this provides insights into lava flow emplacement and constraints for lava flow modelling. Radar was also used on remote volcanoes to map new lava, debris or pyroclastic flows (Rowland et al. 1994; Carn 1999). Figures 1b and 2 illustrate the great capabilities of SAR datasets for volcano mapping. An SIR-C radar image of Volcán Fernandina shows how backscatter intensity varies with contrasted lava flow texture. Pahoehoe flows, which have a 'smooth' texture relative to the radar wavelength (c. 6 cm), returns a low backscatter intensity. The rougher texture of aa flows produces higher backscatter and appears much brighter. This roughness contrast is not observed on the SPOT image (Fig. 1a). SPOT and SAR data served as complementary datasets to map and characterize lava flows (e.g. Fig. 1; Rowland 1996). An ERS-1 scene acquired on Aniakchak volcano in 1992 illustrates the advantages of using SAR for assessing eruptive activity at remote volcanoes that have chronic cloud coverage limiting the applicability of optical RS (e.g. Fig. 2; Rowland et al. 1994).

Fig. 1. (**a**) Panchromatic SPOT image of Fernandina Island, Galápagos, collected in 1988 (10 m resolution). Representative aa (A) and pahoehoe (P) flows are indicated. It should be noted that albedos of aa and pahoehoe flows are similar, making these surfaces difficult to discriminate even on high-resolution satellite images such as this one. (**b**) Shuttle Imaging Radar-C (SIR-C) image of Fernandina Island, Galápagos, collected in 1994. The smooth texture of pahoehoe lava flows (P) produces a low backscatter intensity (dark pixels). Aa (A) flows have a rougher surface, with a roughness scale of same order as the radar wavelength (bright pixels). The caldera structure, intra-caldera and flank cones are visible (see Rowland 1996, for more details).

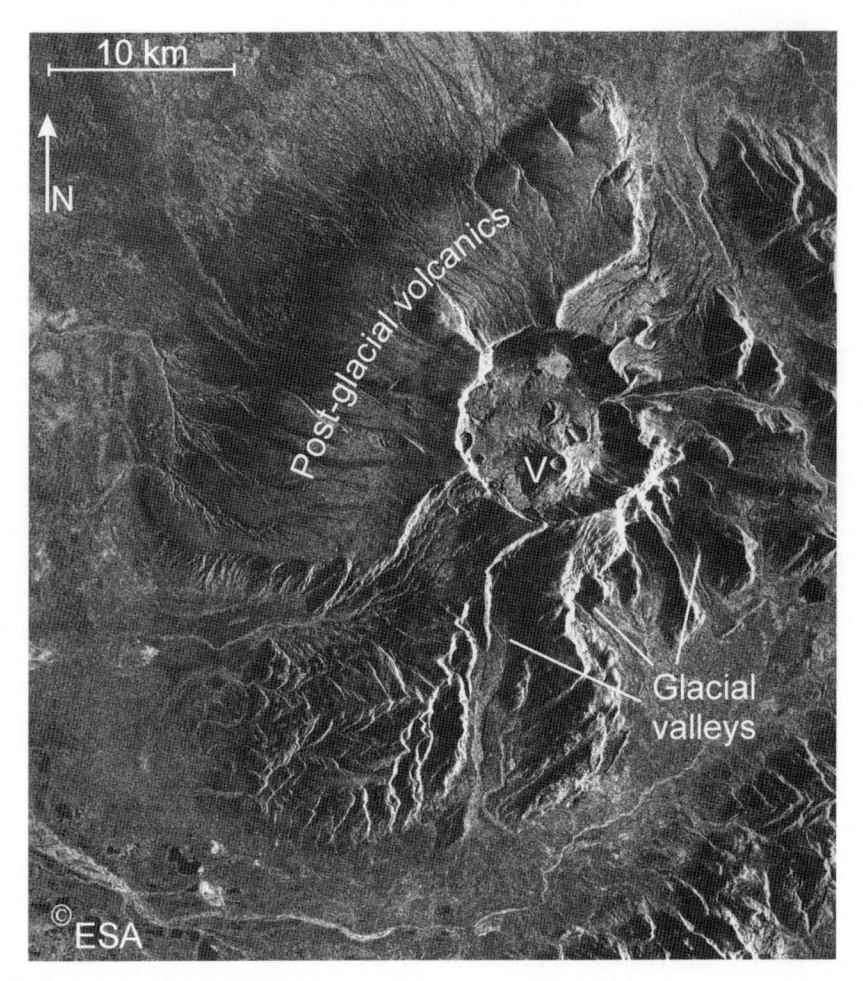

Fig. 2. European Remote-Sensing Satellite-1 (ERS-1) image of Aniakchak volcano, Alaska, collected in 1991. The nearly circular caldera formed c. 3400 years ago, and truncated a volcanic cone that had been glacially eroded. Post-glacial and pre-caldera(?) volcanic deposits have presumably buried glacial valleys on the north and west flanks. Vent Mountain (V) is the tallest of c. 10 intra-caldera vents. It should be noted that topographic features all appear to be steeper on their SE-facing sides. This is the effect of radar foreshortening, whereby summits and ridges are displaced toward the radar, which in this case was looking from SE to NW. (See Rowland et al. (1994) for additional discussion of this image.)

Careful analysis of backscatter intensity allows for recognition of post-glacial emplacement of low-mobility pyroclastic flow on the north and west flanks of the volcano. The break in slope that marks the outer limit of the flow deposit, about 10 km from the caldera rim, is highlighted by the low incidence angle of the radar beam (23°).

SAR interferometry and SRTM

InSAR has emerged as a powerful technique to derive high-resolution DEMs and study ground deformation. InSAR involves comparing the phase of the backscatter signal for each corresponding pixel of two radar images acquired from different positions (Zebker et al. 2000), either at the same time by two antennas separated by a fixed baseline (e.g. SRTM or airborne), or during successive passes of a single antenna (e.g. ERS, JERS, RADARSAT, ENVISAT). InSAR gives exceptional results, although it has limitations. Preservation of the phase coherence (i.e. the level of correlation between the images of an interferometric couple) is the most limiting factor. Coherence is controlled by the system geometry (Zebker & Villasenor 1992) and by changes in surface conditions at the scale of the radar wavelength (c. 1–20 cm) during the time interval separating acquisition of the two images. Dense vegetation causes rapid coherence loss for data acquired in C-band (Kervyn 2001) but not L-band (Stevens & Wadge 2004). Decorrelation (i.e. loss of coherence) can also be used positively to map new lava flow areas in regions with an overall good coherence conservation (Lu *et al.* 2003).

As far as the surface 'stability' is concerned, the shorter the period between the two acquisitions, the higher the coherence will be. This is the main motivation for single-pass interferometry, such as in the recent SRTM mission (Rabus *et al.* 2003). On the other hand, a repeat-pass configuration offers more flexibility, particularly regarding the geometry of the acquired data. Depending on the application, one can look for a pair of images offering the optimum geometric baseline (i.e. spatial distance between the two points in space from which the two images were obtained, projected perpendicularly to the line of sight), incidence angle, orbit mode and revisiting rate. If the goal is to derive topography, a long geometric baseline is preferable; the phase difference will be associated with a smaller difference in elevation (e.g. Zebker *et al.* 1994). A shorter geometric baseline is more suitable for ground deformation studies. In this case, a phase difference is also produced by the displacement of the surface; deformation studies require that the 'topographic phase' is filtered out.

Although radar can penetrate clouds, a phase delay can still occur that must be taken into account for data interpretation. Differential interferometry (DifSAR) has to discriminate between the various phase difference origins, such as geomorphology, atmospheric conditions and ground deformation (Massonnet & Feigl 1995).

InSAR monitoring

The contribution of InSAR to monitor volcanic activity is particularly important in areas where cloud coverage is common and where deformations are recorded (e.g Zebker *et al.* 2000; Stevens & Wadge 2004). However, monitoring is inherently time-related and coherence loss can prevent the use of 'conventional' DifSAR in the long term or for areas with dense vegetation. Recent developments make use of the persistent scatterers (PS) technique. Isolated targets in a highly decorrelated pair may still preserve phase coherence and be used to retrieve ground deformation, provided the PS density is high enough (Ferretti *et al.* 2001). The PS method, which requires a large number of scenes (>15, ideally >30), has proved highly successful and accurate in urban areas where hundreds PS km^{-2} can be identified (Ferretti *et al.* 2004). Hooper *et al.* (2004) proposed an adaptation of the method for natural areas devoid of such urban-type permanent scatterers.

Airborne InSAR (e.g. AirSAR; AirSAR Jet Propulsion Laboratory website 2005) has been used to produce repeated high-resolution DEMs over volcanic terrains, to study volcano morphology, assess morphological changes or to constrain the volume of newly erupted material (Rowland *et al.* 1999, 2003; Lu *et al.* 2003). So far, these applications have, however, been limited to volcanoes in Hawai'i, Alaska, the Aleutians and New Zealand (Stevens *et al.* 2002).

SRTM and InSAR topography

The Shuttle Radar Topography Mission (SRTM) flew on the Space Shuttle Endeavour in February 2000. Using two radar antennas separated by a 60 m long mast, it collected single-pass DEM data over nearly 80% of Earth's land surfaces (i.e. between 60°N and 56°S), using both a C- and X-band radar. The 11 day mission generated the most complete high-resolution digital topographic database for Earth. For the USA, the processed C-band DEM data have been released at 30 m spatial resolution (1 arc second). For the rest of the world, SRTM C-band DEMs are available only at 90 m spatial resolution (3 arc second). The nominal absolute horizontal and vertical accuracies are ±20 m and ±16 m, respectively (SRTM mission, JPL website 2005), although within a single scene the relative accuracies are considerably higher (Rabus *et al.* 2003). Table 2 shows the proportion of historically active and Holocene volcanoes for which SRTM DEMs at 30 or 90 m resolution are now available.

Table 2. *Illustration of how different fractional areas of the Earth with historically active and Holocene volcanoes are best covered by near-global digital elevation models*

	Historical	Holocene	Total
SRTM 30 m (%)	7.8	10.4	9.4
SRTM 90 m (%)	76.6	82.6	80.2
GTOPO30 1 km (%)	5.3	4.2	4.6
No data (%)	10.2	2.9	5.8
Number of volcanoes	561	840	1401

C-band SRTM 30 m data are available for the USA (including Hawaii and Alaska). SRTM 90 m data are available for all land surfaces between 60°N and 56°S. The 1 km resolution GTOPO data are available for all land surfaces (Global Topographic Data). No data: submarine and subglacial volcanoes for which there is no global coverage data at high resolution. Data on volcano location and activity are from Simkin & Siebert (1994; updated list available from Global Volcanism Program—Smithsonian Institution 2005).

The potential of gaining understanding about how volcanoes work, hazards or what controls volcano morphology and growth with SRTM data is tremendous. What controls the diversity, variability and complexity of volcano morphology has not been studied systematically. The only textbook on the subject dates from the 1940s (Cotton 1944). A considerable amount of novel work is now expected, exploring what can be learned from the shape, size and vent distributions of volcanoes.

Because of their shorter wavelength, the X-band DEMs have a higher relative height accuracy (i.e. within a scene) by almost a factor of two. The X-band dataset has been processed and is available at a 25 m spatial resolution. The X-band DEMs do not cover the entire globe (i.e. c. 25% of C-band SRTM data coverage is not covered by the X-band SRTM; the areas not covered are evenly distributed worldwide). Another limitation is the €400 per scene cost of the X-band DEMs (X-SAR SRTM website 2005).

Mapping topography with ASTER

Topographic maps at 1:50 000 scale or larger are not available for many regions, especially in developing countries. In the framework of a large mapping project of the Drâa valley (Morocco; Drey et al. 2004), Goossens et al. (2003) took advantage of the high-resolution and 3D capability of ASTER and CORONA data to derive a topographic map for a region where the best available maps were at 1:100 000 scale. This case study, albeit in a non-volcanic terrain, is summarized here to illustrate the capabilities of the method.

Using a limited number of high-quality GCPs (10 and 7, respectively, for the two scenes used) collected with a differential GPS (TRIMBLE Pathfinder Pro XRS), ortho-photomaps were produced for two areas based on two ASTER scenes. The RMSE on the GCPs in the x, y directions was $c.$ 1 pixel ($c.$ 15 m), whereas the vertical RMSE was 5 m and 10 m, respectively, for the two scenes. The field campaign illustrated that high-quality GCPs that can be accurately located on the satellite image are difficult to find in regions with little human infrastructure and a monotonous landscape (e.g. most volcano landscapes). Comparison between the new DEM and existing topographic map shows that the ASTER-derived contour lines closely match available information on the 1:100 000 topographic map. More detailed contour lines with a smaller elevation interval were generated (20 m instead of 50 m intervals; Fig. 3a and b). The ortho-photomap (Fig. 3a and b) has a 15 m resolution and provides more information than the existing map. ASTER has the advantage that, with a limited number of GCPs, a DEM covering a zone of 60×60 km^2 can be readily generated.

Figure 3c and d illustrates the orthomap with contours obtained by processing a pair of CORONA stereo-images for the same region at a spatial resolution of 4 m. Based on 15 GCPs collected with a Leica 300 Differential global positioning system (accuracy of 5 cm) the resulting topographic map has a 1:10 000 scale. The DEM vertical accuracy is $c.$ 20 m (Schmidt et al. 2002). These products are of much higher resolution than the 1:100 000 scale topographic maps that were the only ones available in this area. These two methods are thus especially suitable for mapping purposes in areas where no accurate topographic maps exist. This is the case for many volcanic regions. Collection of high-quality GCPs with a homogeneous spatial distribution on the satellite image is a key factor for the systematic production of orthophotomaps over hazardous volcanic terrains.

Case study: Mauna Kea, Hawai'i

To assess and compare capabilities of the above-mentioned RS data types, data were collected for Mauna Kea volcano, Hawai'i. The goal was to compare information retrieved from satellite datasets with analyses of airphotos and topographic maps. We focus on the fine-scale capabilities of the different datasets to provide quantitative morphological data on the scoria cones on Mauna Kea's flanks.

Pyroclastic constructs are the most common on-land volcanic landforms on Earth (Riedel et al. 2003). They range in size from spatter mounds only a few tens of metres wide, formed by low-energy eruption, to scoria cones several kilometres in diameter and hundreds of metres high. Cone-building eruptions (Riedel et al. 2003) are explosive, can be associated with high-rising ash clouds and widely dispersed ash, and pose a threat where cities are built on cone fields (e.g. Mexico City). Recent cone-building eruptions led to evacuations of settlements, building collapses (e.g. Heimaey, Iceland, 1973), or airport closures (e.g. Catania, July 2001 Etna eruption).

The morphology of small-scale pyroclastic constructs can be described by average values of crater width W_{cr} and construct height H_{co} relative to basal cone diameter W_{co}. Previous studies, based on a limited number of cone fields, found typical W_{cr}/W_{co} ratios of 0.40 and H_{co}/W_{co} ratios of 0.18 for pristine cones. However, published datasets (e.g. Wood 1980a) and preliminary results of the present study suggest a wide variability in these ratios. Riedel et al. (2003) gained insights into cone-forming eruptions and hazards by developing

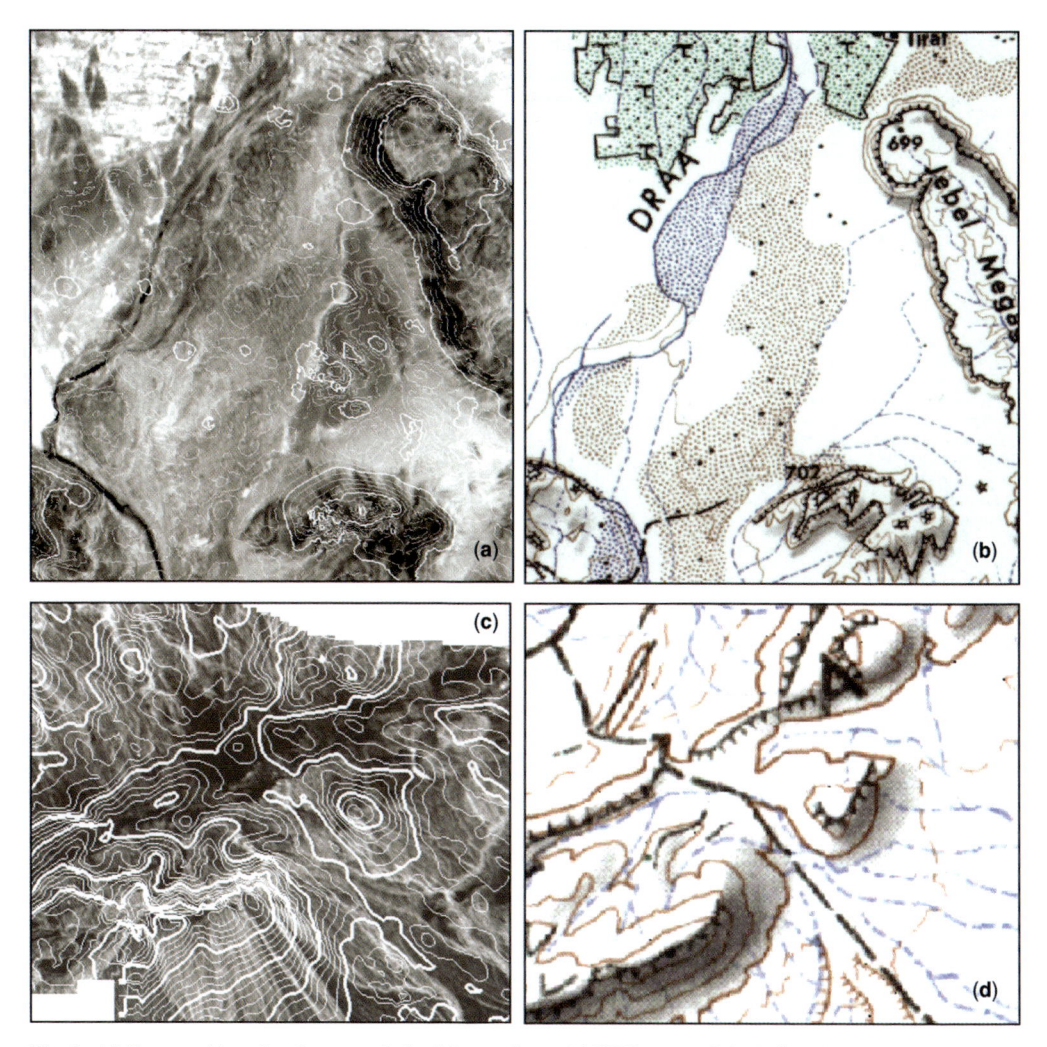

Fig. 3. (**a**) Topographic ortho-photomap derived from subset of ASTER scene of the Drâa valley (Morocco). (**b**) Subset of 1:100 000 topographic map for same region (adapted from Goossens *et al.* 2003). (**c**) Topographic ortho-photomap derived from CORONA data of the Drâa valley (Morocco). (**d**) Subset of 1:100 000 topographic map for same region (adapted from Schmidt *et al.* 2002).

a capability of inverting volcanic construct shape and size data to derive key eruption parameters.

Some previous studies focused on identifying pyroclastic constructs from airphotos or satellite imagery to document volcanic vent spatial distribution and assess tectonic controls (e.g. Chagarlamudi & Moufti 1991; Connor & Conway 2000). Davis *et al.* (1987) used the specific spectra of iron-oxidized scoria to discriminate basaltic pyroclastics from lava flows. All the above-mentioned studies concluded that RS data, usually Landsat imagery, allow to identify 60–80% of the pyroclastic constructs within a cone field. Rowland (1996) identified most of the pyroclastic cones on Volcán Fernandina (Galapagos Island) by combining a TOPSAR-derived DEM and a panchromatic SPOT image. The latter analysis failed to identify only very small arcuate vent constructs (1–3 m high).

Mauna Kea lies in the northern half of the Big Island of Hawai'i (Fig. 4). Activity in the last 130 ka or so has been marked by numerous small gas-rich eruptions that formed *c.* 300 scoria cones (Porter 1972). Mauna Kea most recently erupted *c.* 4000 years ago (Wolfe *et al.* 1997). It was chosen as a study site because of the high density of pyroclastic constructs, the morphologies of which had already been partially documented by

16   M. KERVYN ET AL.

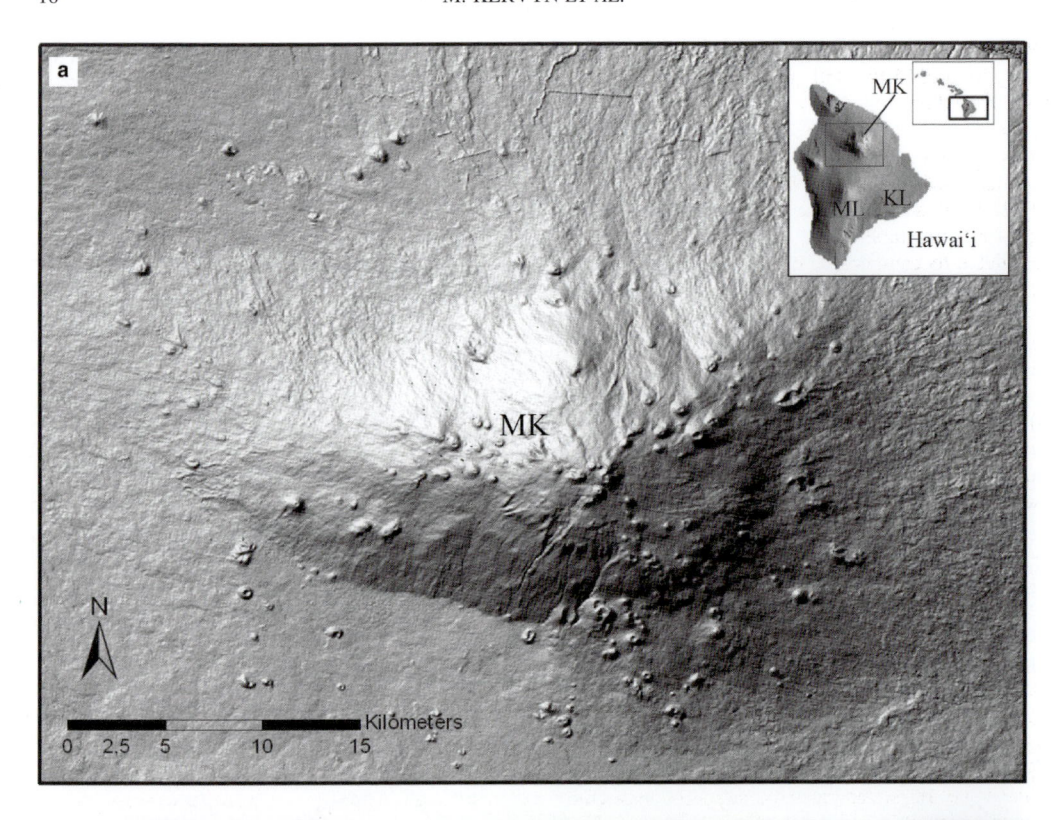

Porter (1972). The cones have a wide size range (W_{co} from c. 15 to 1450 m) and small average size (median diameter 450 m) compared with other cone fields (Wood 1980a), which makes them ideal to evaluate the limitations arising from the spatial resolution of the diverse RS data sources. Traditional sources of spatial data (aerial photographs, topographic and geological maps) were available to compare with our observations and to cross-evaluate RS quantitative measurements. The Mauna Kea cone field, extending from 750 to 4208 m above sea level (a.s.l.), a c. 3500 m elevation range, spans a wide range of climatic conditions and vegetation zones. The contrast in conditions and construct ages (from 150 ka to 4 ka) allows us to assess the general applicability of the approach to volcanic fields and the limitations of several recently developed RS mapping methods.

Data sources and visual interpretation

A Landsat ETM+ image, processed to level L1G (radiometric and systematic geometric corrections applied; Williams 2003) was freely downloaded from the Global Land Cover Facilities website (2005; Table 1). The data were acquired by the satellite on 5 February 2000, and cover the NW part of the Big Island of Hawai'i (Fig. 4b). An ASTER-level 1B image (i.e. registered radiance at sensor with automatic radiometric and geometric correction applied, similar to Landsat L1G; Abrams & Hook 2003) acquired on 5 December 2000 was freely downloaded from the Land Processes Distributed Active Archive Center website (LPDAAC 2005; Table 1). Clouds in the ASTER scene required that we exclude c. 50 cones of the c. 300 Mauna Kea cones (i.e. to have a cloud-free sub-scene for digital photogrammetry processing). SWIR bands of the ASTER scene contained significant 'linear' noise. This is the commonly observed ASTER SWIR crosstalk problem (see ASTERGDS website 2005). C-band SRTM DEMs at both 30 m and 90 m resolution were freely downloaded from the Seamless Data Distribution System (2005, Table 1). The 90 m resolution is used here to exemplify what is achievable for volcanic regions outside the USA.

The ASTER scene (bands 3 N and 3B) was processed using Virtuozo digital photogrammetry software, producing a DEM with 11 GCPs extracted from 1:24 000 scale topographic maps. The processing provided georeferenced and orthorectified images in the three VNIR bands. Several DEMs were produced by varying the DEM spatial resolution and the level of details in the matching process. The user needed to specify the spatial interval (i.e. number of pixels) between two pixels that are automatically matched by the software. Elevations were estimated only for matched pixels and extrapolated in the specified interval. Setting a finer-matching resolution results in more matching errors. In contrast, setting a coarser-matching resolution produces a smoother DEM.

The vertical RMSE of the 11 GCPs was c. 4 m, whereas the horizontal RMSE for these points was c. 20 m. The few points for which exact elevations were provided on the topographic maps typically corresponded to points (e.g. summit of cones) that could not be exactly located on RS images (error of 1 pixel unavoidable). A trade-off had to be found between the high resolution needed in the matching process to render small topographic features (e.g. scoria cone craters) and the increasing editing work needed to correct for matching errors. The ASTER DEM generation procedure takes a few hours, but several days can be spent on fine matching–editing (i.e. manually matching pixels for areas where the automated process generated errors). Here, matching at too fine a level generated troughs or peaks (i.e. a cluster of pixels with abnormally low or high elevations compared with their surroundings) in regions with low contrast. Optimal results were obtained for a matching interval of 3 pixels (45 m) and 30 m spatial resolution. No manual editing of the matching process was undertaken. To reduce errors for regions where the resulting DEM computed slope angles above 45° (i.e. the upper limit expected on Mauna Kea), the pixel elevation was replaced by that from a DEM processed with a matching interval of 7 pixels.

The accuracy of the SRTM and ASTER DEMs was evaluated using a 10 m resolution DEM interpolated from digitized contour lines of the topographic maps. The preliminary comparison of the ASTER and SRTM DEMs led to slope aspect-controlled errors, later attributed to errors with georeferencing of the SRTM DEM. This systematic error was only identified and corrected thanks to the DEM derived from the topographic map. Subtracting the topographic map-derived DEM from the ASTER one, we obtained a mean absolute

---

Fig. 4. (**a**) Shaded relief of Mauna Kea cone field derived from the 30 m resolution SRTM DEM (sun azimuth 315°; sun elevation 80°; elevation exaggerated by factor of three). Inset shows the Big Island of Hawai'i and locations of Mauna Kea (MK), Mauna Loa (ML) and Kilauea (KL). (**b**) True colour composite of the Mauna Kea cone field from the Landsat ETM+ image (R: band 3, 0.63–0.69 μm; G: band 2, 0.53–0.61 μm; B: band 1, 0.45–0.52 μm). White patches on the upper right are clouds, and a small snow patch covers the summit cones. Lava flows to the SW are recent Mauna Loa flows. The arcuate fault-like volcano-scale feature extending from SSW to NNE and 'passing through' MK summit should be noted.

Table 3. *Results of the error analysis of the ASTER and SRTM DEMs (30 m resolution) compared with the DEM derived from topographic map for Mauna Kea volcano*

	DEMs		
	ASTER; 11 GCPs	ASTER; 13 GCPs	SRTM 30
Mean absolute difference (m)	17	11	6
Vertical RMSE (m)	21	13	9

The mean absolute difference is obtained by subtracting DEMs. The vertical RMSE is assessed using a large number of randomly selected points. Original ASTER DEM processing was performed using 11 GCPs. The addition of two GCPs improved the accuracy of the DEM.

difference of 11 m (90% of the errors ranging between −18 and +29 m). Based on 218 control points selected randomly within the study area and taking the topographic map–DEM as a reference, we obtained a vertical RMSE of 13 m. This error is in agreement with results from previous studies (Hirano *et al.* 2003). Errors are attributed to the large altitude range, low spectral contrast in band 3 (near-IR) for the Mauna Kea summit region, and especially to a lack of precise GCPs there. It should be noted that a mean absolute difference of 17 m and an RMSE of 21 m were obtained in a first attempt. Large errors were located in the northern part of Mauna Kea. Significantly, the accuracy of the overall DEM was drastically improved by adding two GCPs on the northern flank (Table 3). Overall, the ASTER DEM tended to underestimate elevations on the higher region of Mauna Kea and to overestimate them in the vegetated lower region. This is consistent with the fact that topographic maps are providing ground elevation value, whereas the ASTER and SRTM DEMs provide an average elevation for the vegetation canopy.

Based on a sample of 450 control points selected randomly on the topographic map–DEM, we obtained an RMSE of 9 m for the SRTM dataset. The mean absolute difference over the entire cone field is 6 m (90% of the errors ranging between −11 and +18 m; Table 3). We cannot exclude that part of this is error in co-referencing the two DEMs. Most pixels for which the SRTM DEM overestimates elevation by more than 10 m are located in the heavily forested regions east of the summit.

Shaded relief, oblique 3D views, and slope maps were generated from the SRTM and ASTER DEMs

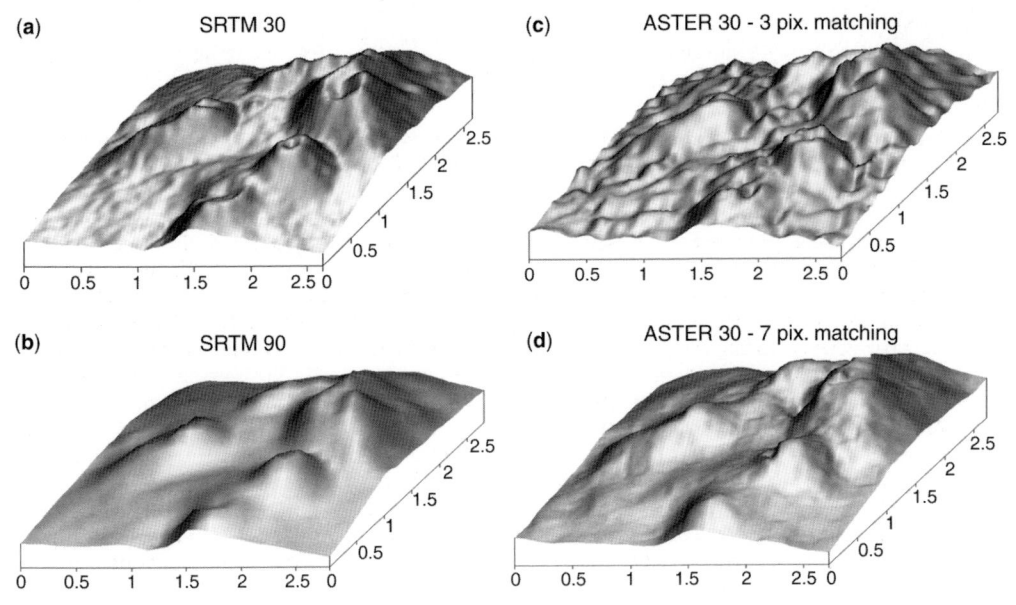

Fig. 5. Oblique view of shaded relief of the Mauna Kea summit displaying five cinder cones. (**a**, **b**) 30 m and 90 m resolution SRTM; (**c**, **d**) 30 m resolution ASTER DEM, the stereo-pair of images being matched every 3 and 7 pixels, respectively.

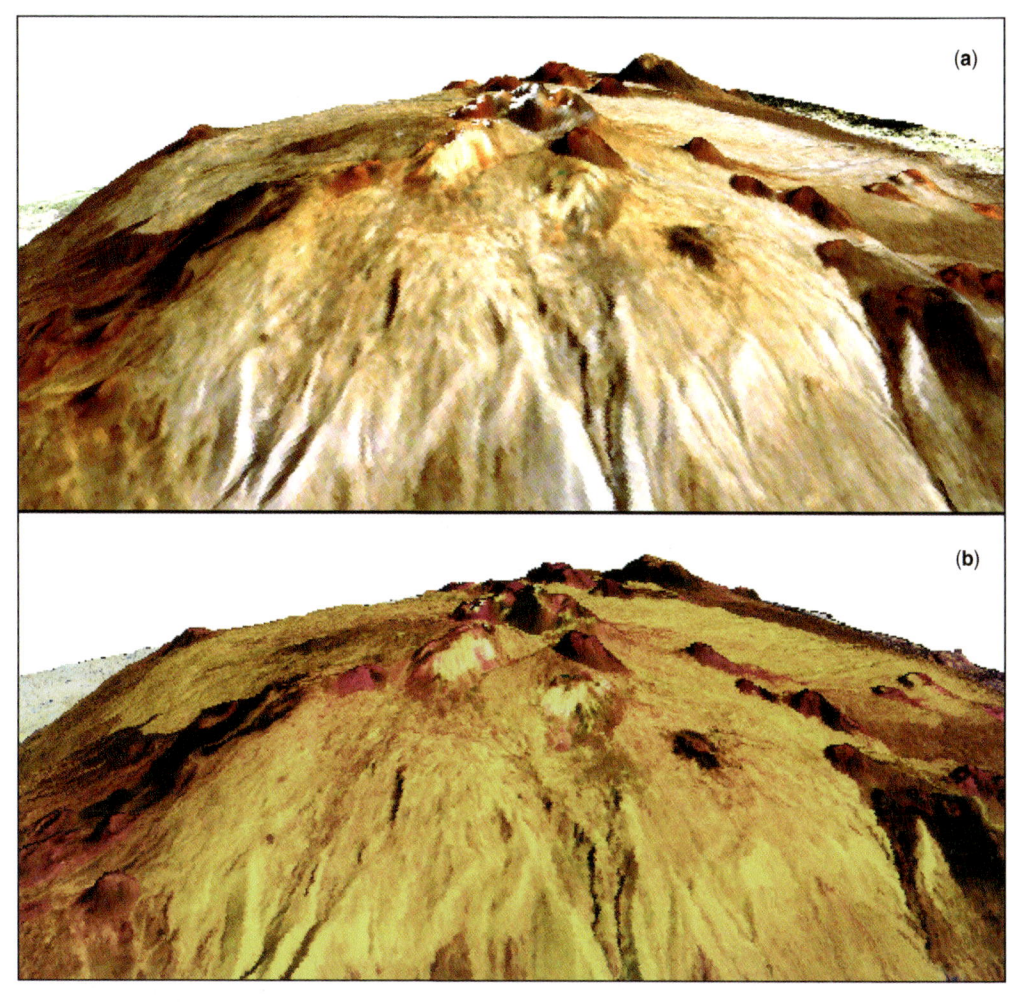

Fig. 6. (**a**) True colour composite of Landsat ETM+ bands draped over the 30 m SRTM DEM. (**b**) Colour composite of the three VNIR ASTER bands draped over the 30 m ASTER DEM.

in a geographical information system (GIS; ArcMap 8.1) for visual interpretation (Figs 4–6). The shaded relief view of the entire cone field (Fig. 4a) shows the high cone density and clustering in three main zones radiating away from the Mauna Kea summit, described as 'rift zones' by Porter (1972). The steep upper flanks are also clearly visible. The ASTER-derived DEMs likewise generated an accurate overall picture of the volcano morphology, although small-scale surface irregularities (i.e. noise in the data) limit identification of small topographic features, such as craters. Figure 5 compares oblique, SRTM- and 3D ASTER-derived views of the Mauna Kea summit cones. The 30 m spatial resolution SRTM DEM allows identification of individual cones and delineation of crater rims. At 90 m resolution, SRTM provides a much smoother representation of the topography. The smaller cones cannot be identified as specific features and craters can no longer be delineated or are no longer apparent. The ASTER shaded-relief DEM image, although at the same spatial resolution (30 m), is visually less attractive than that of the 30 m SRTM because of small-scale irregularities in the topography. From comparison with aerial photographs, these irregularities can be attributed to artefacts introduced through errors in the automated matching process. Nevertheless, small volcanic features and the largest craters can be recognized. Landsat colour composites (e.g. true colour, Fig. 4b) allow identification of volcanic features, contrasting vegetation cover, and volcanic and sedimentary surfaces such as glacial deposits in the Mauna Kea summit area.

Figure 6 illustrates the advantage of combining spectral data of a colour composite image and topographic data. The Landsat true colour composite draped over the SRTM DEM helps discriminate between cones, ash deposits and glacial moraines. To obtain a correctly draped image, georeferencing of the Landsat image had to be manually corrected using tie points, as it was offset by 200 m to the west. ASTER has the advantage that its spectral and topographic data are retrieved from the same dataset (and are therefore co-registered). The finer spatial resolution of the VNIR spectral bands (15 m) and the 30 m resolution ASTER DEM provide a combined high-quality volcano rendering. The comparison with an airphoto of the Mauna Kea summit with similar viewing conditions (Fig. 7) illustrates the level of detail that can be obtained through combination of multispectral (ASTER or Landsat) and topographic data (SRTM DEM or ASTER DEM).

Visual analysis of integrated optical and topographic data proved to be the best easily available RS product to map small-scale volcanic features. It also identified a major arcuate structural feature, currently a focus of study, relating to volcano-scale spreading. The arcuate feature is incised by deep valleys extending in the NNE and SSW upper flanks of Mauna Kea, and passing through the summit where the valleys join and are widest (Fig. 4). We interpret this as a superficial graben from volcano-scale spreading in a preferential WNW–ESE direction as a result of Mauna Kea's load over a mechanically weaker layer (Merle & Borgia 1996). Other, less well-defined graben structures, extending radially from the summit, are consistent with a volcano spreading interpretation. Directional spreading can be explained by the fact that the volcano is not buttressed by other large edifices in the WNW–ESE direction. This example illustrates how integration of topographic and multispectral RS data allows identification of major structural features that were not identified through field work or airphoto interpretation.

Morphometry of small-scale volcanic structures of Mauna Kea

Cone morphometry parameters were independently collected from the Landsat scene, ASTER-derived DEM draped with ASTER VNIR bands, SRTM DEM and topographic maps. First we assessed the capabilities of each of these datasets to identify

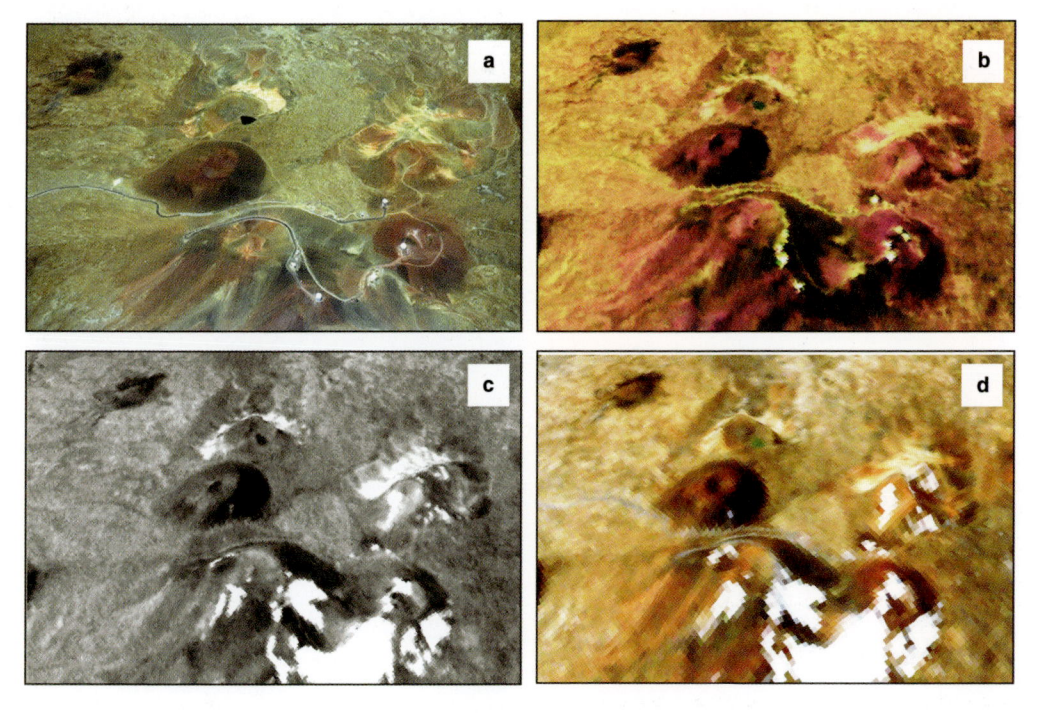

Fig. 7. Oblique view of Mauna Kea summit from north. (**a**) Colour airborne photography; (**b**) colour composite of the VNIR ASTER bands draped over the 30 m ASTER DEM; (**c**) 15 m resolution Landsat ETM+ panchromatic band and (**d**) colour composite of the three VIS Landsat ETM+ bands draped over the SRTM DEM at 30 m resolution. White areas in the bottom part of (c and d) are summit cones covered by snow.

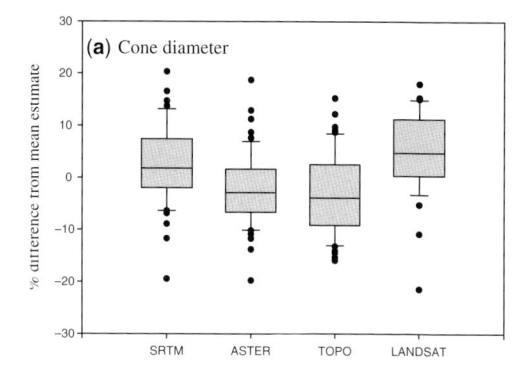

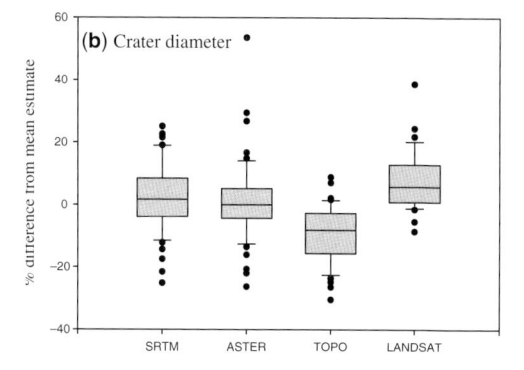

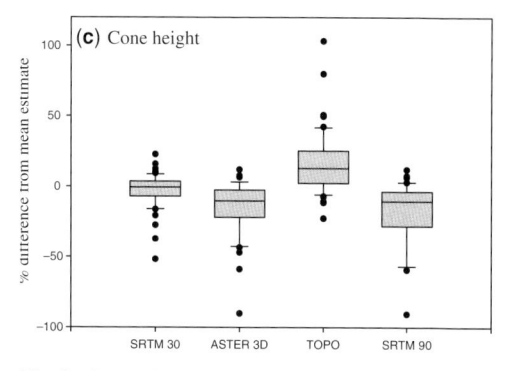

Fig. 8. Comparison of cone diameter (**a**), crater diameter (**b**), and cone height (**c**) estimations from four different data sources. The boxes illustrate the range of differences between estimates and the mean estimated value from all the datasets (SRTM 90 m not considered in (c) for average calculation). The vertical line in each box is the median of the distribution. The extent of each box includes 50% of the data around this median value. Horizontal lines above and below each box indicate the 90th and 10th percentiles. Outliers are plotted as separate dots. Datasets: SRTM, SRTM 30 m DEM; ASTER, combined DEM and multispectral imagery; TOPO, estimate from paper topographic maps; LANDSAT, estimate from Landsat ETM+ multispectral imagery; SRTM90, SRTM 90 m DEM.

pyroclastic constructs and vents, using the geological maps as a validation source. Pyroclastic constructs of the Mauna Kea cone field were identified by their common positive topographic relief. Importantly, specific spectral features of oxidized scoria also helped by allowing identification of constructs with little or no topographic relief and limited spatial extent (Davis *et al.* 1987). Quantitative parameters (e.g. namely the cone and crater diameters, the construct height, and the mean and maximum outer slope) were retrieved for all identified cones (*c.* 220) from the topographic maps and from the Landsat scene (i.e. elevation and slope estimates from topographic maps only). The same parameters were retrieved for a subset of 50 cones covering the entire range of cone sizes and shapes from the ASTER dataset and from the SRTM DEM. For the topographic maps, horizontal and vertical error ranges of *c.* 24 and 12 m, respectively, are expected as a result of small measurement errors and ambiguity in cone base location between two contour lines. For the RS datasets, an error range of 1 pixel (*c.* 15–30 m) is expected. The detailed volcanological interpretations will be the subject of another paper. Here we focus on methodological aspects and on comparison of the errors affecting the estimated parameters.

The geological map used as the validation dataset shows 310 vent constructs on Mauna Kea. Constructs that are not systematically detected as small volcanic features on the topographic and RS datasets are invariably those with a very limited spatial extent (<200 m), low height (<20 m) and no distinctive spectral expression. Most are interpreted as spatter mounds and ramparts. Several of these deposits are mapped as highly elongated features, the result of fissure eruptions. These constructs are usually less than 15 m high (e.g. Riedel *et al.* 2003). Several processes may conceal the original 'truncated cone' construct morphology, including intense weathering, compaction, mass wasting, and subsequent coverage of the lower slopes by lava flows or later pyroclastic products. These types of deposits, lacking a topographic expression, can only be identified by the combined use of high-resolution aerial photography and field work.

All the datasets allowed us to identify pyroclastic constructs with near-equal success. Ninety-five percent of the scoria cones (cone base >250 m) less than 70 ka old were identified and mapped for each dataset. Between 65 and 72% of all vents (i.e. including not only cone vents but also some small metre-scale vents along fissures) were unambiguously identified. Although the average vent construct size on Mauna Kea is small relative to other cone fields in the world (Wood 1980*a*), the above-mentioned results are similar to those obtained by previous RS mapping of volcanic cones (e.g. Davis *et al.* 1987; Chagarlamudi & Moufti 1991). The constructs that could not be mapped were small spatter constructs ($W_{co} < 250$ m) and a few larger cones (300–800 m diameter deposits on the geological map) whose morphology has been obliterated by erosion and by later lava flows. Topographic maps and the Landsat image allowed similar results to be derived: 230 and 220 pyroclastic constructs were identified, respectively, from these data sources. However, only 199 of these could be simultaneously identified and mapped on both topographic maps and Landsat imagery. Cloud and snow cover are important factors that prevent cone mapping on the Landsat (or ASTER) image. Another limitation is the moderate spatial resolution of Landsat data, which hinders recognition and/or delineation of constructs less than 150 m wide with no specific spectral expression. Several pyroclastic features could not be recognized on the topographic maps because of the lack of distinct positive topography, but could be mapped on the satellite imagery by taking advantage of the high MIR reflectance typical of near-vent oxidized volcanic products (i.e. red scoria or spatter).

The SRTM DEM allowed unambiguous identification of 218 of the 320 pyroclastic constructs. As with the Landsat data, the 30 m spatial resolution hindered recognition of the smallest constructs (those with width <200 m). The fact that the SRTM elevations correspond to the average height of the vegetation canopy also causes ambiguities in cone identification in places with discontinuous forest coverage. Small positive relief features can indeed be created by locally higher vegetation (e.g. a small forest in otherwise non-forested area). Ambiguity can be resolved by combining the SRTM DEM and multispectral data.

The ASTER DEM allowed recognition of 171 out of 260 mapped pyroclastic features within the area covered by the ASTER scene (a success level of 65%). This poorer result is explained by the fact that the ASTER scene includes only the summit region, which happens to contain most of the small-size spatter mounds. We conclude that the ASTER dataset (i.e. multispectral data and ASTER DEM) has a capability similar to the three other datasets for mapping Mauna Kea pyroclastic constructs. The availability of good quality SWIR bands combined with an ASTER-derived DEM constrained by high-quality GCPs would further improve the mapping capabilities of the ASTER data.

Quantitative morphological parameters

For all the Mauna Kea constructs identified on the topographic maps and Landsat scene, the basal diameter, W_{co}, was estimated (i.e. computed as the diameter of a circle having an area equivalent to that of the cone base). Results from the two datasets were similar over the entire cone field. On average, W_{co} estimates derived from the Landsat image are *c.* 40 m greater than those derived from topographic maps. However, differences range from −80 m (underestimation of Landsat-derived W_{co} relative to the topographic map-derived estimate) to 280 m. For 66% of the features, this difference ranges between −20 and +80 m. A key result is that the magnitude of the discrepancy is not dependent on construct size, but rather on the degree of irregularity of the cone outline. Using the geological map as a validation reference, a comparison of the spatial extent of each feature between the Landsat and topographic map datasets was undertaken for 75 features for which discrepancies in W_{co} values were high. For 40 cones the Landsat-derived W_{co} was found to be more accurate than the topographic map-derived value. Importantly, the results from the topographic map were usually ambiguous because of the lack of a well-defined basal break in slope for those cones. In only eight cases were the cones less accurately delineated with the Landsat data. This was due to misinterpretation of the cones' spatial extent, as a result of vegetation cover or absence of topographic information. The last 27 cones were very small features ($W_{co} < 200$ m) and the estimate from the geological map was halfway between those from the topographic map- and Landsat-derived estimates. The average 40 m discrepancy between estimates from the two data sources is to be expected. It falls within the estimated error ranges for the two estimates. Considering the 1:24 000 scale of the topographic map and the finite spatial resolution of the image, errors of *c.* 30 m are not abnormal, and a slightly larger error is likely to be related to limitations associated with defining the cone extent in those cases where the break in slope delimiting the cone base is not sharply defined.

Figure 8a illustrates the differences in cone base estimates from the four datasets for a subset of 50 cones. On average, all the datasets lead to similar cone size estimation, even if major discrepancies do sometimes occur as a result of misinterpretation in the spatial extent of the cones. The box-plots illustrate that for all the datasets, more than 50% of the estimates are within 10% of the mean estimate (i.e. average of the estimates from the four datasets). The same observation holds for the crater size estimation, although a larger relative error range is observed. Figure 8a and b suggests that topographic maps, with no spatially continuous data coverage, tend to lead to underestimated cone size values, whereas the opposite is true for the Landsat image, which lacks topographic information. Both the ASTER and SRTM datasets lead to estimates that are generally close to the average value. The discrepancy range is the smallest with ASTER, thanks to the combined use of multispectral and topographic data, although some large errors are obtained for a few cones for which the spatial extent of the cone and/or crater is ambiguous. These results suggest that reasonably accurate quantitative morphology description through W_{cr}/W_{co} ratios can be derived from any one of the datasets assessed here.

The height of volcanic cones is of interest because it is controlled both by the cone age (Wood 1980b) and by cone particle characteristics (Riedel et al. 2003). The height of 50 Mauna Kea scoria cones was estimated by subtracting the average cone base elevation from the average crater rim elevation. Results were independently extracted from the topographic maps, SRTM and ASTER DEMs. Results (Fig. 8c) illustrate that SRTM provides the smallest discrepancies (i.e. difference from the average of the three results). Elevations on topographic maps were recorded at only four locations each for the base and crater rims, and topographic map cone height estimates tend to be higher than the average. For 70% of the cones, the heights from the ASTER DEM are between 80 and 110% of the average estimated height. For constructs <50 m high (25% of the constructs), ASTER DEM height estimates are as much as 50% less than the average estimate. Although the ASTER-derived DEM has a lower vertical accuracy than the SRTM DEM, it still provides reasonable cone height estimates for large constructs ($H_{co} > 60$ m). Finally, it should be noted that the capabilities of the SRTM dataset would be greatly reduced if it were available only at 90 m spatial resolution (as is the case for non-USA locations; see Fig. 8c). The cone height estimates are significantly lower with the 90 m SRTM and the range of discrepancy compared with the average value is larger.

Rungwe Volcanic Province case study

Many volcanoes in developing countries remain poorly known and it is typically difficult to acquire good-quality maps and airphotos for them. Satellite RS allows us to rapidly explore volcano morphology and identify evidence of recent activity (e.g. Dubbi volcano, Ethiopia; Wiart et al. 2000). The Rungwe Volcanic Province illustrates the type of information that can be retrieved from different RS datasets, including SRTM DEM, multispectral datasets and SAR interferometry. The objective here is to illustrate what can be done with RS, including deriving hazard-relevant information that can be evaluated in the field, as well as to highlight some remaining challenges for RS.

The Rungwe Volcanic Province lies at the intersection of the west, east and south branches of the East African Rift in SW Tanzania (Fig. 9). The most recent eruption in the province was at Kiejo volcano 200 years ago. Little is known about other recent eruptions, either at Kiejo or at the other centres (especially at Rungwe or Ngosi volcanoes), although there are numerous Rungwe Volcanic Province ash layers in lakes within a few hundreds of kilometres, indicating frequent explosive eruptions in the province in the last 40 ka. In 2000–2001, a series of tectonic or volcanic earthquakes of moderate magnitude caused damage to villages. Tectonic activity is inferred along the Mbaka Fault (Fig. 9) but no surface rupture has been observed since the 2000–2001 event.

The only geological mapping of the Rungwe Volcanic Province was by Harkin (1960) in the 1950s at 1:250 000. A preliminary re-exploration of this poorly studied volcanic region was conducted using the 90 m SRTM DEM (Fig. 9) and a Landsat TM image. Combined use of these two datasets enhances our capability to understand the geology of the province. The three main eruptive centres are Ngozi volcano, directly south of Mbeya city, Rungwe volcano and the Kiejo volcano. The case of Rungwe volcano is here discussed for illustrative purposes. Rungwe culminates at 2962 m a.s.l. and, like Ngozi, is densely vegetated on its flanks. The Rungwe crater is bordered by a steep semi-circular wall on its NNW to SSE flanks, breached to the SW. Aligned pit craters and scarps, recognizable by their elevation and lack of vegetation, extend along the volcano flanks and are virtually uneroded. From the amphitheatre crater rim, Rungwe is much steeper and with a shorter slope to the NE compared with the SW. These features are interpreted as evidence for a large sector collapse (c. 2–3 km^3) that produced a debris avalanche. This is corroborated by Harkin's (1960) observations of a striking mound field including perfectly conical 10–20 m diameter hills made up of very poorly sorted, breccia-like material some c. 15-20 km SW from the crater. The RS data spatial resolution, especially that of the 90 m SRTM DEM, does not allow mapping of individual hummocks in itself, but our reinterpretation of Harkin's work using the RS data allowed the mapping of the debris avalanche extent. RS also allows us to establish that several large phonolite–trachyte domes were subsequently emplaced in the Rungwe amphitheatre crater with lava flowing down the breached flank. Lack of vegetation on the most recent flows points to a recent emplacement age, at most of the order of a few centuries. A field campaign is now necessary

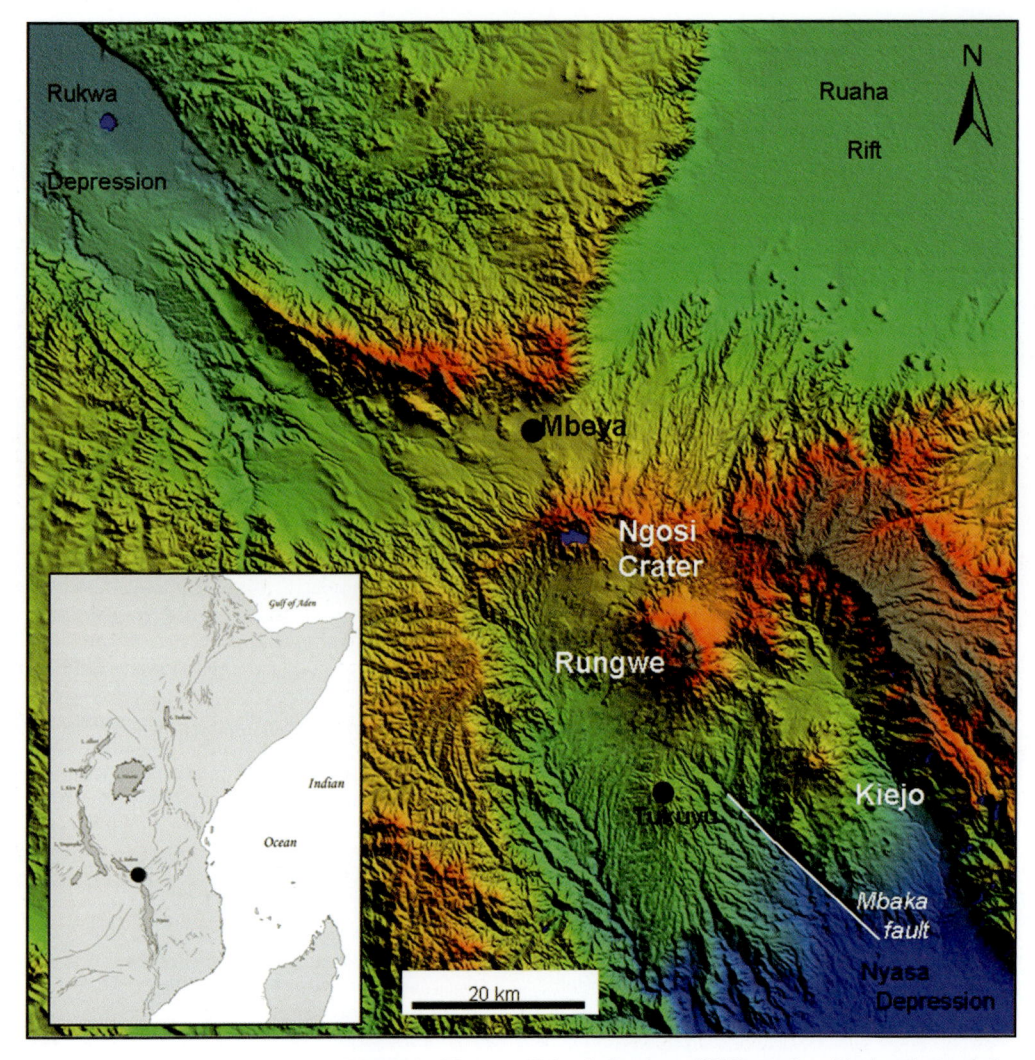

Fig. 9. Shaded relief of the 90 m SRTM DEM of Rungwe Volcanic Province, SW Tanzania, with elevation-scaled colour range. Three main phonolite volcanoes and Mbaka Fault are indicated on the image, together with the cities of Mbeya and Tukuyu. The Rift Valley trends NW–SE. Smaller-volume features include phonolite–trachyte domes located in the north and inside Rungwe crater. Inset shows location of the Rungwe Volcanic Province (●) with the East African Rift System.

to validate the hypothesis that this volcano went through a dramatic and hazardous flank collapse event. Sector collapse may have been triggered by the major NW–SE-trending Mbaka Fault extending SW of Kiejo volcano (Fig. 9). As suggested by van Wyk de Vries *et al.* (2003), offset strike-slip faults can trigger volcano-scale collapse. Examination of the Landsat scene and the DEM revealed many small phonolite domes and basaltic cones, which were targeted for 'ground truth' field visits. Using recent RS advances, poorly known terrains in the Rungwe Volcanic Province can be rapidly mapped, at both small and large scales, and hypotheses can be formulated about geohazards in the region.

Now let us consider the challenge of monitoring and studying changes in volcano topography over time, at poorly known and densely vegetated volcanoes, such as Rungwe. Eleven ERS SAR datasets acquired between December 1995 and February 2003 were processed to detect past movements including the 2001 seismic swarm. Dense vegetation causes a rapid loss of coherence in most part of the image and does not allow conventional

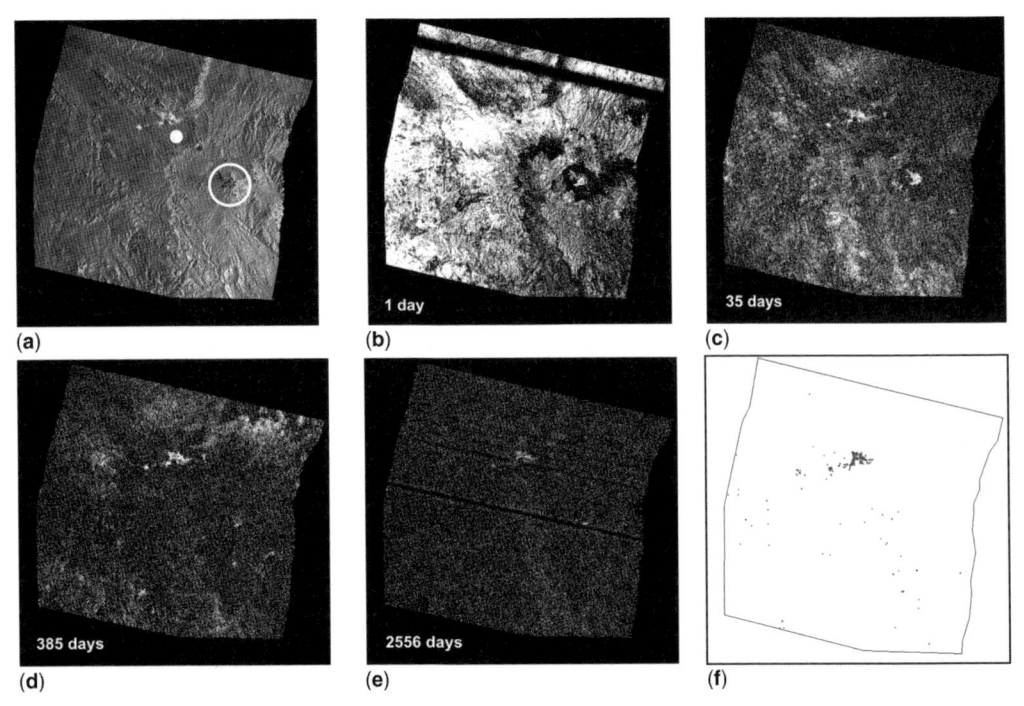

Fig. 10. (a) ERS average intensity image (11 looks); white circle and point show locations of Rungwe and of Mbeya, respectively. Bright area in top middle part corresponds to the city of Mbeya. This image was used in the field to locate the PS. (b–e) Maps of coherence above a threshold of 0.35 (white). (b) Coherence after 24 h obtained with the 26–27 September 1997 (dry season) tandem pair. Black patches correspond to densely vegetated areas. The stripe is related to missing lines in one of the two scenes. (c) Coherence after one orbital cycle. High-coherence level remains over urban areas and on bare surface at top of Rungwe (middle-right). (d) Coherence after 1 year. Coherence remains high over Mbeya urban area; top of Rungwe still visible. (e) Coherence after 7 years. Some coherence remains over Mbeya and the top of the Rungwe. (f) Persistent scatterers obtained from the series of coherence maps with a threshold at 0.55. Most of them are located in Mbeya; others correspond to isolated houses with a metal corrugated roof.

differential InSAR to be applied (Fig. 10). With the view of using the persistent scatterers (PS) technique, the data were filtered in such a way that c. 600 PS with coherence >0.5 were retained (Table 4). Although the area is densely populated, the only major urban centre is Mbeya, which is also one of the few places where coherence is preserved in the long term. The numerous villages scattered in the image do not provide enough phase stability. Hence, the selected PS distribution is poor and non-uniform (Fig. 10f). A field campaign was specifically dedicated to the PS identification and characterization.

The PS retrieved after the 11 SAR scenes were averaged to (1) produce a multi-look image with a drastic decrease of the speckle and (2) refine image quality. The enhanced geocoded image was then georeferenced with GCPs collected on 1:50 000 scale topographic maps with accuracy better than 2 pixels, i.e. 40 m (RMS). All the PS were then introduced within a hand-held GPS and tracked in the field. The PS with the highest coherence was found to correspond to a cement factory and nearby warehouses. Other PS correspond mainly to isolated houses or houses with corrugated metal roofs, grouped along two preferential directions (i.e. parallel or perpendicular to the incident radar beam).

For ground deformation monitoring, systematic measurements using the selected PS appear to be noisy and unreliable because of their intrinsic physical properties. The corrugated roofs are made of an assemblage of multiple metal panels fixed on a wooden structure, which itself rests on poor-quality brick walls. In addition to the poor PS density, dilatation–contraction of the metal panels with variations in diurnal insolation or the variable humidity-induced swelling of the wooden structures makes them unsuitable for accurate measurements of crustal deformation.

An alternative way to 'natural' PS could be to set up a network of specifically designed corner reflectors in well-selected sites. The advantage is

Table 4. *Number of pixels with coherence above given thresholds for all interferometric couples derived from 11 ERS SAR scenes*

	Coherence threshold					
Number of pixels	0.5 596	0.55 227	0.6 47	0.65 11	0.7 2	0.75 1

Very few PS points were found after dataset processing. PS number dramatically decreases with the coherence level. The highest coherence target corresponds to a cement factory SW of Mbeya.

the possibility of selecting the location of measurement points, as well as positions of reflectors that can be controlled to validate InSAR measurements. An array of such reflectors can be compared to a GPS array. Such an array of corner reflectors is, however, not easy or cheap to set up. The availability of only sparse reflectors in a large region also poses the problem of correcting for atmospheric water vapour gradients. The attempt to use InSAR to monitor or study ground deformation has helped identify that using the PS or corner reflector approaches does not offer practical solutions. Together with Stevens & Wadge (2004), we stress that the availability of an L-band radar sensor dedicated to ground deformation monitoring might be the only way to enhance the capability of differential InSAR in vegetated terrain. The Rungwe case study illustrates the capabilities of RS to provide valuable information on the distribution and morphology of eruptive centres, as well as insights into associated hazards, before embarking on field studies. Table 5 summarizes the capabilities and limitations of RS for the study of poorly known volcanoes in developing countries.

Concluding remarks and perspective

Eighty percent of historically active or dormant, potentially hazardous volcanoes (c. 450 and over 1200 volcanoes, respectively) are located in densely populated developing countries, where populations are also more at risk from natural hazards. The impact of natural disasters there is 20 times higher (as a percentage of GDP per head) than in industrialized countries (World Bank,

Table 5. *Summary of possibilities and limitations for RS studies of poorly known volcanoes in tropical developing countries*

Data Type	Multispectral	Digital Elevation Model		Synthetic Aperture Radar	Differential InSAR
Datasets	Landsat	ASTER – ASTER DEM SPOT – SPOT DEM	InSAR, e.g. SRTM DEM	ERS-1/2, JERS, RADARSAT, ENVISAT (single image)	ERS-1/2, JERS, RADARSAT (interferometric couple)
Derived products	Topomap at 1:25,000 or better				Ground Deformation Time Series & Monitoring
	Volcano-scale morphology				
		Elevation profiles			
		DEMs as baseline for risk models			
	Identification of most faults including large volcano-scale collapse scars				
	Contrasted terrain mapping (e.g. lava flows, pyroclastic flows)				
	Mapping small structures (e.g. domes, cones, vents, pit craters)				
	Collection of quantitative morphometric data for hazard modelling*				
Limitations Challenges	Cloud cover		Atmospheric effect†		Atmospheric effect
		Ground control point quality	Coherence loss in densely vegetated regions†		Coherence loss in densely vegetated regions
			Shadow effect due to angle of viewing – Lack of data in steep craters		Density of permanent coherent scatterers
	Spatial Resolution & Vegetation Cover: Disabling identification of small features (e.g. hummocks in distal zones of debris avalanches)				

*Lack of theory and experiments useful to invert morphometry for hazard assessment.
†For multiple-pass InSAR techniques.

Hazard Risk Management website 2005). For most of these volcanoes, the detailed geology has not been documented and there may be no or limited geological, geohazard or risk maps. Most of these volcanoes are not currently being monitored. In our experience, sets of aerial photographs may be impossible to obtain, particularly in sub-Saharan Africa. In this vacuum, low-cost satellite data offer the best means of rapidly mapping and monitoring hazardous terrain around volcanoes in developing countries.

We have illustrated that the combined approaches exemplified in this paper allow us to map the full diversity of volcanic terrains at 1:25 000 scale, such as volcanic cones, and pyroclastic or lava flows with contrasting surface textures, as well as structural features, indicating potential volcano-scale instability and geohazards. The combined approaches and individual datasets can be rigorously assessed against each other and are of high quality. Huge volumes of quantitative data can be generated quickly, providing new constraints on volcano shapes, sizes and vent distributions; thus opening up the possibility that these data could be used to mitigate volcanic geohazards.

Landsat and ASTER multispectral archive data, available at low cost, offer great capabilities for exploratory studies of poorly known volcanic terrains and quantitative morphological studies of various volcanic features. ASTER has the advantage over Landsat, in that high-resolution mapping (topography and multispectral) can be derived from the same imagery. However, the cost of near real-time data acquisition continues to limit the use of these data for rapid hazard assessment. ASTER and Landsat ETM+ sensors have also both already reached the end of their designed mission life, and follow-up data acquisition with similar sensors is not yet certain.

All the radar sensors penetrate cloud cover, a major advantage over optical sensors. The fact that only 90 m SRTM data are available worldwide is currently a limitation to mapping hazardous terrains and to mitigating geohazards. Multi-pass InSAR (compared with SRTM single-pass InSAR) looks set to become a powerful tool for deriving 3D topography, facilitating recognition of unstable volcanoes. To recognize flanks prone to instability would, however, require data at higher resolution than provided by current sensors (see Kerle et al. 2003). Multi-pass InSAR requires development before it can be effectively used in routine ground deformation monitoring. At present, the main limitations for multi-pass InSAR are changing atmospheric conditions and vegetation cover. As argued by Stevens & Wadge (2004), a dedicated L-band radar might be the way toward operational InSAR monitoring of densely vegetated active volcanoes. In the absence of ground monitoring at most

Table 6. *Characteristics of recently launched or planned multispectral and radar sensors*

Satellite or Sensor	Launch	Bands	Revisit time	Website
RapidEye constellation	2007	5 VNIR; 6.5 m	1 day	http://www.rapideye.de/
DMC	2002–2005	3 VNIR; 32 m	1 day	http://www.dmcii.com/
CartoSAT	2005	2 stereo PAN; 2.5 m	5 days	http://www.isro.org/Cartosat/
OrbView 5	2007	1 PAN; 0.41 m; 4 VNIR; 1.64 m	<3 days	http://www.orbimage.com/
ALOS (PRISM)	2006	1 PAN; 2.5 m	46 days	http://alos.jaxa.jp/index-e.html
ALOS (AVNIR-2)	2006	4 VNIR; 10 m	46 days	http://alos.jaxa.jp/index-e.html
ALOS (PALSAR)	2006	L-band; 10 m	46 days	http://alos.jaxa.jp/index-e.html
ENVISAT (ASAR)	2002	C-band; 25 m	35 days	http://envisat.esa.int/instruments/asar/
RADARSAT 2	2006	C-band; 3 m at best	24 days	http://www.radarsat2.info/
TerraSAR	2006	X-band; 1 m at best	11 days	http://www.terrasar.de/

The list is not exhaustive but presents sensors whose capabilities might be of greatest relevance for volcano hazard assessment. The RapidEye, CartoSat and ALOS (PRISM) sensors offer stereoscopic data. (For more information on these sensors, see the listed website.)

volcanoes in less developed countries, spaceborne monitoring of ground deformation is of crucial importance.

Table 6 summarizes new and planned sensors that should increase RS capabilities to assess hazards at volcanoes. Constellations of small satellites are currently launched to acquire daily multispectral images (e.g. RapidEye and the Disaster Monitoring Constellation, DMC). Sensors will provide data with higher spatial and temporal resolution, but with a limited number of spectral bands. RapidEye, CartoSat and ALOS (PRISM) will provide data with stereoscopic capabilities. High data costs from satellites launched by commercial companies, will, however, limit their usefulness for low-cost geohazard assessment. SAR sensors, with different wavelengths (X, C and L bands; Table 6) should also enhance the possibility of InSAR applications in volcanic terrains. Of special interest is the L-band sensor onboard ALOS, which will ensure enhanced coherence preservation over vegetated terrains.

High spatial resolution topographic data are crucial as a prerequisite to assist geological field work, for morphometric analyses and for volcanic hazard modelling. Satellite remote sensing provides one of the best ways to produce DEMs over extensive volcanic regions, especially in countries where aerial photographs are unavailable.

This paper is dedicated to our colleagues in developing countries who are facing geohazards, some of them in desperate need of acquiring RS monitoring capability. M.K and G.G.J.E are supported by the Belgian NSF (FWO-Vlaanderen) and the 'Fondation Belge de la Vocation'. This effort would not have been possible without the support of UGent colleagues, notably of P. Jacobs and J.-P. Henriet, and without the long-term support of S. Sparks and W. Rose. The Landsat, ASTER and SRTM imagery was obtained from the Global Land Cover Facility website, the Land Processes Distributed Active Archive Center (LPDAAC) website interface, and the Seamless Data Distribution System website.

References

ABRAMS, M. & HOOK, S. 2003. *ASTER user handbook version 2*. World Wide Web Address: http://asterweb.jpl.nasa.gov/documents/aster_user_guide_v2.pdf.

AirSAR Jet Propulsion Laboratory website 2005. World Wide Web Address: http://airsar.jpl.nasa.gov/.

ALTAIMER, A. & CHRISTOPH, K. 2002. Digital surface model generation from CORONA satellite image. *ISPRS Journal of Photogrammetry and Remote Sensing*, **56**, 221–235.

AMELUNG, F., JONSSON, S., ZEBKER, H. & SEGALL, P. 2000. Widespread uplift and 'trapdoor' faulting on Galapagos volcanoes observed with radar interferometry. *Nature*, **407**, 993–996.

ASTERGDS website 2005. World Wide Web Address: http://www.gds.aster.ersdac.or.jp/.

BALDI, P., BONVALOT, S., BRIOLE, P. ET AL. 2002. Validation and comparison of different techniques for the derivation of digital elevation models and volcanic monitoring (Vulcano Island, Italy). *International Journal of Remote Sensing*, **23**, 4783–4800.

BYRNES, J. M., RAMSEY, M. S. & CROWN, D. A. 2004. Surface unit characterization of the Mauna Ulu flow field, Kilauea Volcano, Hawaii, using integrated field and remote sensing analyses. *Journal of Volcanology and Geothermal Research*, **135**, 169–193.

CARN, S. A. 1999. Application of Synthetic Aperture Radar (SAR) imagery to volcano mapping in the humid tropics: a case study in East Java, Indonesia. *Bulletin of Volcanology*, **61**, 92–105.

CHAGARLAMUDI, P. & MOUFTI, M. R. 1991. The utility of Landsat images in delineating volcanic cones in Harrat Kishb, Kingdom of Saudi Arabia. *International Journal of Remote Sensing*, **12**, 1547–1557.

CONNOR, C. B. & CONWAY, F. M. 2000. Basaltic volcanic cones. *In*: SIGURDSSON, H., HOUGHTON, B. F., MCNUTT, S. R., RYMER, H. & STIX, J. (eds) *Encyclopedia of Volcanoes*. Academic Press, New York, 331–343.

COTTON, C. A. 1944. *Volcanoes as Landscape Forms*. Withcombe & Tombs, Christchurch, New Zealand.

CROWLEY, J. & ZIMBELMAN, D. 1997. Mapping hydrothermally altered rocks on Mount Rainier, Washington, with Airborne Visible Infrared Imaging Spectrometer (AVIRIS) data. *Geology*, **25**, 559–569.

DAVIS, P. A., BERLIN, G. L. & CHAVEZ, P. S. 1987. Discrimination of altered basaltic rocks in the southwestern United States by analysis of Landsat Thematic Mapper data. *Photogrammetric Engineering and Remote Sensing*, **53**, 51–73.

DREY, T., POETE, P., THAMM, H.-P., PANNENBECKER, A. & MENZ, G. 2004. Generation of a high resolution DEM with ASTER stereo data for the river Draa catchment in Morocco. IMPETUS Conference Abstracts, Ouarzazae.

Eurimage website 2005. World Wide Web Address: http://www.eurimage.com/.

FERRETTI, A., PRATI, C. & ROCCA, F. 2001. Permanent scatterers in SAR interferometry. *IEEE Transactions on Geoscience and Remote Sensing*, **39**, 8–20.

FERRETTI, A., NOVALI, F., BÜRGMANN, R., HILLEY, G. & PRATI, C. 2004. InSAR Permanent Scatterer analysis reveals ups and downs in San Francisco Bay area. *EOS Transactions, American Geophysical Union*, **85**, 317–324.

FRANCIS, P. W. & BAKER, M. C. W. 1978. Sources of two large volume ignimbrites in the Central Andes: some LANDSAT evidence. *Journal of Volcanology and Geothermal Research*, **4**, 81–87.

FRANCIS, P. W. & DE SILVA, S. L. 1989. Application of the Landsat Thematic Mapper to the identification of potentially active volcanoes in the Central Andes. *Remote Sensing of Environment*, **28**, 245–255.

FRANCIS, P. W. & WELLS, G. L. 1988. Landsat Thematic Mapper observations of debris avalanches deposits in the central Andes. *Bulletin of Volcanology*, **50**, 258–278.

FRANCIS, P. W., WADGE, G. & MOUGINIS MARK, P. J. 1996. Satellite monitoring of volcanoes. *In*: SCARPA, R. & TILLING, R. I. (eds) *Monitoring and Mitigation of Volcano Hazards*. Springer, Berlin, 257–298.

GADDIS, L. R. 1992. Lava-flows characterization at Pisgah volcanic field, California, with multiparameter imaging radar. *Geological Society of America Bulletin*, **104**, 695–703.

GLCF 2005. *Global Land Cover Facilities website.* World Wide Web Address: http://glcfapp.umiacs.umd.edu/index.shtml.

Global Volcanism Program—Smithsonian Institution 2005. World Wide Web Address: http://www.volcano.si.edu/.

GOOSSENS, R., SCHMIDT, M. & MENZ, G. 2003. High resolution DEM and ortho-photomap generation from TERRA-ASTER data—case study of Morocco. *In:* BENES, T. (ed.) *Geoinformation for European-wide Integration. Proceedings of the 22nd Symposium of the European Association of Remote Sensing Laboratories.* Millpress, prague,19–24.

HARKIN, D. A. 1960. *The Rungwe Volcanics at the Northern End of Lake Nyasa.* Geological Survey of Tanganyika Memoirs, **2**.

HARRIS, A. J. L., FLYNN, L. P., DEAN, K. ET AL. 2000. Real-time satellite monitoring of volcanic hot spots. *In:* MOUGINIS-MARK, P. J., CRISP, J. & FINK, J. F. (eds) *Remote Sensing of Active Volcanism.* Geophysical Monograph, American Geophysical Union, **116**, 139–159.

HARRIS, A. J. L., FLYNN, L. P., MATIOS, , O., ROSE,, W. I. & CORNEJO, J. 2004. The evolution of an active silicic lava flow field: an ETM+ perspective. *Journal of Volcanology and Geothermal Research*, **135**, 147–168.

HIRANO, A., WELCH, R. & LANG, H. 2003. Mapping from ASTER stereo image data: DEM validation and accuracy assessment. *ISPRS Journal of Photogrammetry and Remote Sensing*, **57**, 356–370.

HOOPER, A., ZEBKER, H., SEGALL, P. & KAMPES, B. 2004. A new method for measuring deformation on volcanoes and other natural terrains using InSAR persistent scatterers. *Geophysical Research Letters*, **31**, doi: 10.1029/2004GL021737.

IVERSON, R. M., SCHILLING, S. P. & VALLANCE, J. W. 1998. Objective delineation of lahar-inundation hazard zones. *Geological Society of America Bulletin*, **110**, 972–984.

KAHLE, A. B., GILLEPSIE, A. R., ABBOTT, E. A., ABRAMS, M. J., WALKER, R. E., HOOVER, G. & LOCKWOOD, J. P. 1988. Relative dating of Hawaiian lava flows using multispectral thermal infrared images: a new tool for geologic mapping of young volcanic terranes. *Journal of Geophysical Research*, **93**, 15239–15251.

KERLE, N., FROGER, J.-L., OPPENHEIMER, C. & VAN WYK DE VRIES, B. 2003. Remote sensing of the 1998 mudflow at Casita volcano, Nicaragua. *International Journal of Remote Sensing*, **24**, 4791–4816.

KERVYN, F. 2001. Modelling topography with SAR interferometry: illustrations of favourable and less favourable environment. *Computers & Geosciences*, **27**, 1039–1050.

KORMUS, W., ALAMUS, R., RUIZ, A. & TALAYA, J. 2004. Assessment of DEM accuracy derived from SPOT 5 High Resolution Stereoscopic imagery. International Archives of the Photogrammetry, Remote Sensing and Spatial Information Sciences, **35(B1)**, 445–452.

LEGELEY-PADOVANI, A., MERING, C., GUILLANDE, R. & HUAMAN, D. 1997. Mapping of lava flows through SPOT images—an example of the Sabancaya volcano (Peru). *International Journal of Remote Sensing*, **18**, 3111–3133.

LPDAAC 2005. *Land Processes Distributed Active Archive Center, Datapool @ LPDAAC.* World Wide Web Address: http://edcdaac.usgs.gov/datapool/datapool.asp.

LU, Z., FIELDING, E., PATRICK, M. & TRAUTWEIN, C. 2003. Estimating lava volume by precision combination of multiple baseline spaceborne and airborne interferometric synthetic aperture radar: the 1997 eruption of Okmok volcano. *IEEE Transactions on Geoscience and Remote Sensing*, **41**, 1428–1436.

MASSONNET, D. & FEIGL, K. L. 1995. Discrimination of geophysical phenomena in satellite radar interferograms. *Geophysical Research Letters*, **22**, 1537–1540.

MCDONALD, R. 1995. CORONA: success for space reconnaissance, a look into the Cold War, and a revolution for intelligence. *Photogrammetric Engineering and Remote Sensing*, **61**, 689–720.

MCKAY, M. E. & MOUGINIS-MARK, P. J. 1997. The effect of varying acquisition parameters on the interpretation of SIR-C radar data: the Virunga volcanic chain. *Remote Sensing of Environment*, **59**, 321–336.

MERLE, O. & BORGIA, A. 1996. Scaled experiments of volcanic spreading. *Journal of Geophysical Research*, **101**, 13805–13817.

MOUGINIS-MARK, P. J., CRISP, J. & FINK, J. (eds) 2000. *Remote Sensing of Active Volcanism.* Geophysical Monograph, American Geophysical Union, **116**.

NEWHALL, C. G. & PUNONGBAYAN, R. S. (eds) 1996. *Fire and Mud: Eruptions and Lahars of Mount Pinatubo, Philippines.* Philippines Institute of Volcanology and Seismology, Quezon City; University of Washington Press, Seattle.

OPPENHEIMER, C. 1998. Volcanological application of meteorological satellite data. *International Journal of Remote Sensing*, **19**, 2829–2864.

PATRICK, M. R., DEHN, J., PAPP, K. R. ET AL. 2003. The 1997 eruption of Okmok Volcano, Alaska: a synthesis of remotely sensed imagery. *Journal of Volcanology and Geothermal Research*, **127**, 87–105.

PIERI, D. & ABRAMS, M. J. 2004. ASTER watches the world's volcanoes: a new paradigm for volcanological observations from orbit. *Journal of Volcanology and Geothermal Research*, **135**, 13–28.

PORTER, S. C. 1972. Distribution, morphology and size frequency of cinder cones on Mauna Kea volcano, Hawaii. *Geological Society of America Bulletin*, **83**, 3607–3612.

RABUS, B., EINEDER, M., ROTH, A. & BAMLER, R. 2003. The Shuttle Radar Topography Mission—A new class of digital elevation models acquired by spaceborne radar. *ISPRS Journal of Photogrammetry and Remote Sensing*, **57**, 241–262.

RAMSEY, M. & DEAN, J. 2004. Spaceborne observations of the 2000 Bezymianny, Kamchatka eruption: the integration of high-resolution ASTER data into near real-time monitoring using AVHRR. *Journal of Volcanology and Geothermal Research*, **135**, 127–146.

RAMSEY, M. & FINK, J. H. 1999. Estimating silicic lava vesicularity with thermal remote sensing: a new technique for volcanic mapping and monitoring. *Bulletin of Volcanology*, **61**, 32–39.

RAMSEY, M. S. & FLYNN, L. P. 2004. Strategies, insights and the recent advances in volcanic monitoring and mapping with data from NASA's Earth Observing System. *Journal of Volcanology and Geothermal Research*, **135**, 1–11.

RIEDEL, C., ERNST, G. G. J. & RILEY, M. 2003. Controls on the growth and geometry of pyroclastic constructs. *Journal of Volcanology and Geothermal Research*, **127**, 121–152.

ROSE, W. I., BLUTH, G. J. S. & ERNST, G. G. J. 2000. Integrating retrievals of volcanic cloud characteristics from satellite remote sensors: a summary. *Philosophical Transactions of the Royal Society of London, Series A*, **358**, 1585–1606.

ROWLAND, S. K. 1996. Slope, lava flow volumes, and vent distribution on Volcano Fernandina, Galapagos Islands. *Journal of Geophysical Research*, **101**, 27657–27672.

ROWLAND, S. K. & GARBEIL, H. 2000. The slopes of oceanic basalt volcanoes. *In*: MOUGINIS-MARK, P. J., CRISP, J. & FINK, J. (eds) *Remote Sensing of Active Volcanism*. Geophysical Monograph, American Geophysical Union, **116**, 223–247.

ROWLAND, S. K., SMITH, G. A. & MOUGINIS-MARK, P. J. 1994. Preliminary ERS-1 observations of Alaska and Aleutian volcanoes. *Remote Sensing of Environment*, **48**, 358–369.

ROWLAND, S. K., MCKAY, M. E., GARBEIL, H. & MOUGINIS-MARK, P. J. 1999. Topographic analysis of Kilauea Volcano, Hawaii, from interferometric airborne radar. *Bulletin of Volcanology*, **61**, 1–14.

ROWLAND, S. K., HARRIS, A. J. L., WOOSTER, M. J., GARBEIL, H., MOUGINIS-MARK, P. J., AMELUNG, F. & WILSON, L. 2003. Volumetric characteristics of lava flows from interferometric radar and multi-spectral satellite data. *Bulletin of Volcanology*, **65**, 311–330.

SCHMIDT, M., MENZ, G. & GOOSSENS, R. 2002. Processing techniques for CORONA satellite images in order to generate high-resolution digital elevation models (DEM). *In*: BÉGNI, G. (ed.) *Observing our Environment from Space: New Solutions for a New Millennium*. Proceeding of the 21st EARSeL Symposium. Balkema, Rotterdam, 191–196.

Seamless Data Distribution System. 2005. World Wide Web Address: http://seamless.usgs.gov/.

SHERIDAN, M. F., HUBBARD, B., CARRASCO-NUNEZ, G. & SIEBE, C. 2004. Pyroclastic flow hazard at Volcan Citlaltépetl. *Natural Hazards*, **33**, 209–221.

SIMKIN, T. & SIEBERT, L. 1994. *Volcanoes of the World*, 2nd edn. Geosciences Press, Tucson, AZ.

SRTM mission, JPL website, 2005. World Wide Web Address: http://www2.jpl.nasa.gov/srtm/.

STEVENS, N. F. & WADGE, G. 2004. Towards operational repeat-pass SAR interferometry at active volcanoes. *Natural Hazards*, **33**, 47–76.

STEVENS, N. F., MANVILLE, V. & HERON, D. W. 2002. The sensitivity of a volcanic flow model to digital elevation model accuracy: experiments with digitised map contours and interferometric SAR at Ruapehu and Taranaki volcanoes, New Zealand. *Journal of Volcanology and Geothermal Research*, **119**, 89–105.

STEVENS, N. F., GARBEIL, H. & MOUGINIS-MARK, P. J. 2004. NASA EOS Terra ASTER: volcanic topographic mapping and capability. *Remote Sensing of Environment*, **90**, 405–414.

USGS Earth Explorer website 2005. World Wide Web Address: http://edcsns17.cr.usgs.gov/EarthExplorer/.

VAN WYK DE VRIES, B., WOOLLER, L., CECCHI, E. & MURRAY, J. 2003. Spreading volcanoes: The importance of strike-slip faults, EGU General Assembly Conference Abstracts. Nice, Abstract 2480.

VASSILOPOULOU, S., HURNI, L., DIETRICH, V., BALTSAVIAS, E., PATERAKI, M., LAGIOS, E. & PARCHARIDIS, I. 2002. Orthophoto generation using IKONOS imagery and high-resolution DEM: a case study on volcanic hazard monitoring of Nisyros Island (Greece). *ISPRS Journal of Photogrammetry and Remote Sensing*, **57**, 24–38.

WADGE, G., FRANCIS, P. W. & RAMIREZ, C. F. 1995. The Socompa collapse and avalanche event. *Journal of Volcanology and Geothermal Research*, **66**, 309–336.

WIART, P. A. M., OPPENHEIMER, C. & FRANCIS, P. 2000. Eruptive history of Dubbi volcano, northeast Afar (Eritrea), revealed by optical and SAR image interpretation. *International Journal of Remote Sensing*, **21**, 911–936.

WILLIAMS, D. 2003. *Landsat 7, Science data user handbook*. Landsat Project Science Office. World Wide Web Address: http://ltpwww.gsfc.nasa.gov/IAS/handbook/handbook_toc.html.

WOLFE, E. W., WISE, W. S. & DALRYMPLE, G. B. 1997. The Geology and Petrology of Mauna Kea, a Study of Postshield Volcano. *US Geological Survey, Professional Papers*, **1557**.

WOOD, C. A. 1980*a*. Morphometric evolution of cinder cones. *Journal of Volcanology and Geothermal Research*, **7**, 387–413.

WOOD, C. A. 1980*b*. Morphometric analysis of cinder cone degradation. *Journal of Volcanology and Geothermal Research*, **8**, 137–160.

World Bank, Hazard Risk Management website 2005. World Wide Web Address: www.worldbank.org/hazards/.

WRIGHT, R., FLYNN, L. P., GARBEIL, H., HARRIS, A. J. L. & PILGER, E. 2004. MODVOLC: near-real time thermal monitoring of global volcanism. *Journal of Volcanology and Geothermal Research*, **135**, 29–49.

X-SAR SRTM website 2005. World Wide Web Address: http://www.dlr.de/SRTM/SRTM_en.html.

ZEBKER, H. & VILLASENOR, J. 1992. Decorrelation in interferometric radar echoes. *IEEE Transactions on Geoscience and Remote Sensing*, **30**, 950–959.

ZEBKER, H. A., WERNER, C. L., ROSEN, P. A. & HENSLEY, S. 1994. Accuracy of topographic maps derived from ERS-1 interferometric radar. *IEEE Transactions on Geoscience and Remote Sensing*, **32**, 823–836.

ZEBKER, H. A., AMELUNG, F. & JONSSON, S. 2000. Remote sensing of volcano surface and internal processes using radar interferometry. *In*: MOUGINIS-MARK, P. J., CRISP, J. & FINK, J. (eds) *Remote Sensing of Active Volcanism*. Geophysical Monograph, American Geophysical Union, **116**, 179–205.

Comparison and validation of Airborne Thematic Mapper thermal imagery using ground-based temperature data for Grímsvötn caldera, Vatnajökull, Iceland

S. F. STEWART[1], H. PINKERTON[2], G. A. BLACKBURN[1] & M. T. GUÐMUNDSSON[3]

[1]*Department of Geography, Lancaster University, Lancaster LA1 4YB, UK*

[2]*Department of Environmental Science, Lancaster University, Lancaster LA1 4YB, UK*
(e-mail: s.stewart@Lancaster.ac.uk)

[3]*Institute of Earth Sciences, University of Iceland, Sturlugata 7, IS-101 Reykjavik, Iceland*

Abstract: Grímsvötn, Iceland's most active volcano, is also one of the most powerful geothermal areas in Iceland. This subglacial volcano is located in the centre of Vatnajökull, Europe's largest temperate ice cap, and it erupted most recently in 1998 and 2004. As part of continuing research on heat flux, morphological changes and volcanic processes at Grímsvötn, thermal anomalies were mapped using remote sensing Natural Environmental Research Council (NERC) Airborne Research and Survey Facility (ARSF) data. The 2001 Airborne Thematic Mapper (ATM) thermal images of the Grímsvötn subglacial caldera reveal distinct areas of geothermal activity and provide an overview of the thermal anomalies associated with water and rock exposures. A crater lake located on the 1998 eruption site is shown to have a surface temperature of 30–35 °C. There is a good correlation between the ARSF data and ground-based temperature measurements. The thermal images also revealed previously undetected areas of high heat flow. Factors that complicate the interpretation and comparison of different datasets from an ice-covered area include recent cornice collapses and variations in atmospheric humidity. To reduce uncertainty in future missions, temperature measurements should be made at points whose position is well constrained using differential global positioning system. In addition, humidity and temperature measurements should be made at the time of flight.

Subglacial volcanoes are found both beneath cold-based ice sheets in the polar regions and beneath temperate glaciers, such as in Iceland. The type of eruptive activity on a subglacial volcano depends on a number of factors including glacier type, ice thickness and magma chemistry (Guðmundsson *et al.* 1997, 2002; Höskuldsson & Sparks 1997; Bourgeois *et al.* 1998; Einarsson 1999; Smellie 2000, 2002; Tuffen *et al.* 2002; Björnsson 2003). When magma reaches the base of the glacier through a fissure or circular vent underneath glacial ice, it starts to melt the ice. On the surface a depression known as an ice cauldron (Björnsson 1975) is formed as a result of subsidence, with large crevasses forming in a circular pattern in response to ice strain caused by ice melting and the subsequent release of water. Active subglacial volcanic eruptions cannot be studied using traditional field-based methods, because of poor accessibility. Thus, remote sensing (RS) and observations from the air have been the main tools to study such events (Rothery *et al.* 1988; Guðmundsson *et al.* 2004). Moreover, in some cases most of the volcanic edifice may be concealed under hundreds of metres of ice. Many stratovolcanoes with small summit ice caps are often near populated regions, and although not completely subglacial, have similar accessibility problems. However, even remote ice-covered volcanoes are hazardous because of the rapid formation of melt water, which can travel large distances from the volcano. Temperate glaciers are at their pressure melting temperature throughout their thickness, and water can migrate underneath the glacier allowing melt water to flow from the geothermal heat source of an eruption to the edges of the glacier, causing devastating floods known by the Icelandic term, jökulhlaups (Smellie 2000). RS offers a low-risk, global coverage of volcanoes that allows the study of a range of phenomena such as thermal, temporal and topographic variations (Oppenheimer & Rothery 1991). This has the potential to improve our understanding of subvolcanic processes, which are difficult to study using conventional methods, and could lead to better forecasting of eruptions and hazardous events such as jökulhlaups. The objective of this research is to build on present ground-based calorimetric work on thermal anomalies, specifically for inaccessible and hazardous terrain.

Geological setting

The Neovolcanic Zone in Iceland has been subdivided into three rift systems (Fig. 1a), the western, eastern and northern volcanic zones (Einarsson 1999). In southern Iceland, accretion is controlled by NE–SW-trending fissures and faults, whereas mainly north–south-trending lineations are dominant to the north of Vatnajökull. The volcanic fissures of the eastern volcanic zone (EVZ) lie beneath the western side of the Vatnajökull ice sheet (Einarsson et al. 1997), Europe's largest temperate ice sheet (Fig. 1b) covering an area of 8300 km^2 in SE Iceland.

Two of the most active volcanic systems in the EVZ are Bárðarbunga and Grímsvötn (Guðmundsson & Björnsson 1991). Both systems are composed of a central volcano and one or two fissure swarms (Saemundsson 1979). Although Bárðarbunga has erupted more magma, Grímsvötn is the more active of the two, with about 60 small eruptions that have deposited tephra over Vatnajökull in the last 800 years (Guðmundsson & Björnsson 1991; Grönvold et al. 1995; Larsen et al. 1998; Konstantinou et al. 2000). Mapping of subglacial topography using radio echo sounding has shown that Grímsvötn is a caldera-type volcano (Björnsson & Einarsson 1990).

Within the Grímsvötn caldera, intense geothermal activity continuously melts the surrounding ice at a rate of 0.2–0.5 km^3 year^{-1} (Björnsson 2003), creating permanent depressions or ice calderas. The meltwater accumulates in the subglacial caldera lake, lifting the water level by 50–100 m over a period of a few years. It is subsequently released in jökulhlaups. When this occurs, the lake level is lowered by 50–100 m in 1–2 weeks, typically releasing 1–2 km^3 of meltwater (Björnsson 2003). Björnsson (1998) described the caldera as being 6–10 km diameter, bordered by the mountain ridge Grímsfjall to the south and subglacial mountains to the north and east (Fig. 1c). The Grímsvötn caldera is divided into three smaller calderas: the main (or south), north and east calderas (Guðmundsson & Milsom 1997). The subglacial lake covers the main caldera and extends into the northern depression at high water levels. The lake covers the areas of highest geothermal activity, and the eruptions in 1922, 1934, 1983, 1998 and 2004 all took place at its southern margin, where geothermal activity is also most intense (Guðmundsson & Björnsson 1991).

The ice-covered Grímsvötn caldera lake has been used as a calorimeter for measuring the heat transfer from magma to ice and melt water, and the rate of accumulation of melt water (Guðmundsson 2003). Studies of geothermal power using calorimetry to convert ice melting rates into heat transfer rates have shown that Grímsvötn is one of the most powerful geothermal areas in the world. The heat output has mostly been in the range of 2000–4000 MW over recent decades (Björnsson & Guðmundsson 1993). However, this work was carried out on the Grímsvötn caldera lake prior to the 1998 eruption, when the lake was believed to be an enclosed system with no leakage. Since the 1998 eruption, increased melting at the ice dam that used to seal Grímsvötn caldera lake has led to leakage from the lake (Guðmundsson 2003). Consequently, calorimetry can no longer be effectively used. Moreover, at the 1998 eruption site, heat has been lost directly to the atmosphere via bodies of open water. Thermal RS has the potential to provide an alternative method of calculating the thermal budget for this volcano. This paper provides the preliminary analysis carried out to extract temperatures and validate the airborne thermal images that will later be used, along with meteorological estimates, to calculate heat flux.

ATM data

The UK Natural Environmental Research Council (NERC) Airborne Remote Sensing Facility's (ARSF) Airborne Thematic Mapper (ATM) sensor is a Daedalus 1268 AZ-16, a passive remote sensor designed to collect radiation reflected and emitted from the Earth's surface from an airborne platform. The scan head optics and detector layout separates the incoming radiation into 11 spectral bands (Table 1) ranging from blue in the visible (VIS) parts of the spectrum to thermal infrared (IR) (Azimuth System 2001). Bands 1–8 cover the VIS and near-IR parts of the spectrum. Bands 9 and 10 are shortwave infrared (SWIR) and band 11 is thermal IR. The thermal band is calibrated in 'real time' using two on-board blackbodies (NERC ARSF 2002), which are imaged by the sensor during each scan before and after the scene pixels (5 m × 5 m) on a scan line. Aerial survey data along three flight lines were collected on 10 and 14 June 2001. The central line was chosen from each flight as these covered the entire Grímsfjall ridge area, an area of 16 km^2. The first stage in processing the ATM data was geometric correction, carried out using AZGCORR software on a UNIX workstation (NERC ARSF 2002). AZGCORR combines the scanned image data with pre- or post-navigation records and then interpolates a map projection, referenced to the output image that has been corrected for aircraft position and altitude (NERC ARSF 2002). The navigation is converted from geographical co-ordinates on global positioning system (GPS) satellite datum, to a suitable survey map projection. The current GPS datum is the WGS84 (World Geodetic System agreed in

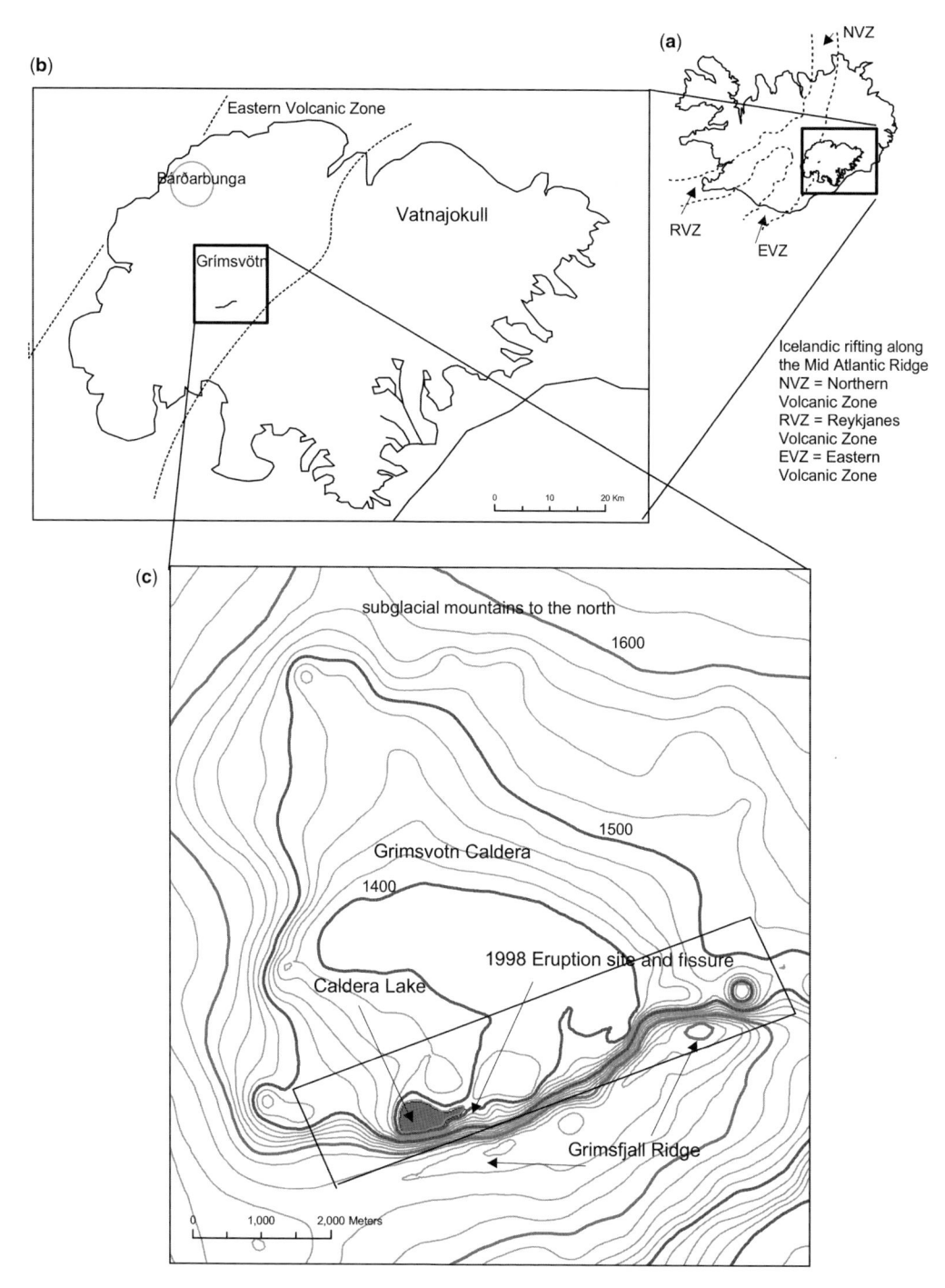

Fig. 1. (a) Map of Iceland showing the location of the Vatnajökull ice sheet in relation to the three volcanic zones. (b) Map of Vatnajökull showing the location of Grímsvötn caldera together with the 1996 Gjálp fissure and the subglacial volcano, Bárðarbunga. (c) Contour map of the Grímsvötn caldera created in 2001 by Icelandic collaborators, using differential GPS, with the Grímsfjall ridge, crater lake, and 1998 eruption site labelled. The rectangular box over the Grímsfjall ridge shows the land area covered by the aerial survey.

Table 1. *Airborne Thematic Mapper (ATM) bands and relevant parts of the electromagnetic spectrum*

Band	Wavelength (μm)	
1	0.42–0.45	Visible
2	0.45–0.52	
3	0.52–0.60	
4	0.60–0.62	
5	0.63–0.69	
6	0.69–0.75	Near-infrared
7	0.76–0.90	
8	0.91–1.05	
9	1.55–1.75	Shortwave infrared
10	2.08–2.35	
11	8.5–13.0	Thermal infrared

1984). For geocorrection of the ARSF imagery of Grímsvötn, the WGS 1984 geodetic datum was chosen, along with UTM zone 28, as this is the most widely used projection used in this region. A pixel resolution of 5 m was chosen for each image.

Converting radiance to temperature

The pre-processing carried out by the ARSF converts sensor radiances to surface radiances. In this conversion, it is assumed that the ground surface acts as a blackbody. Consequently, an emissivity of one is used in Planck's law to estimate temperature. As this is an unrealistic approximation for most materials, correct temperature data require the inclusion of more realistic emissivities. The 8–14 μm range includes an atmospheric window and is the area of peak energy emissions for most surfaces at normal Earth surface temperatures (Kahle & Alley 1992; Lillesand *et al.* 2004). Therefore land surface materials are often treated as greybodies in this wavelength range. Close examination of Earth surface materials shows that emissivity can also vary with other conditions, such as whether the material is wet or dry (Liang 2001; Lillesand *et al.* 2004). Early work showed that, to determine accurate temperatures and emissivities, corrected radiance measurements were necessary (Kahle & Alley 1992; Kealy & Hook 1993), and this required atmospheric corrections across the image. No atmospheric correction was carried out on the images as there was a lack of atmospheric data collected at the time of the survey. For the purposes of this research an average figure for emissivity was used because of the lack of ground truth observations for land cover type and atmospheric data.

In the current analysis, a map of emissivity was derived from land cover based on the visible bands, where a specific emissivity value was allocated to a land cover class using a supervised classification technique (parallelepiped and maximum likelihood rules) based on 27 training clusters in nine land cover classes. Figure 2 shows the supervised classification image used to allocate each emissivity. Each emissivity is based on published emissivities (Table 2) taken from Lillesand *et al.* (2004). Using ENVI (version 3.5), an array was generated allowing each pixel to be allocated an appropriate emissivity, dependent on its land cover class. A second array was produced, to derive a new blackbody radiance based on radiance emissivity principles:

$$M_g = M_b \epsilon \qquad (1)$$

where M_g is greybody radiance, M_b is blackbody radiance and ϵ is emissivity.

According to Planck's law, the temperature is equivalent to the area beneath a spectral radiance curve. Using the program of A. K. Wilson (pers. comm.) the spectral radiance curve was separated into 141 wavelength divisions based on the sensor response. These wavelength divisions are then used as the wavelengths for temperature conversion. Using Planck's law, a temperature of 273 K is assumed initially, from which the program calculates a radiance for the given wavelength band, sensor response and temperature. Equation (2) is a variation of Planck's formula, with the added sensor response from the program of A. K. Wilson (pers. comm.):

$$M_\lambda = \frac{C_1 \lambda^{-5} \text{ resp}}{\exp(C_2/\lambda T) - 1} \qquad (2)$$

where M_λ is the radiant exitance, the total energy radiated in all directions by a unit area in a unit time (Curran 1992); spectral radiance constants $C_1 = 119095879.96$ Wm^2 and $C_2 = 14387.75225$ Wm^2; resp. is the ATM sensor response, divided into 141 wavelengths; λ is the wavelength (μm); T is temperature in Kelvin. The program continues for every pixel in the image until all pixels have a temperature assigned within a new array.

The images were exported from ENVI as Geo-Tiffs, for subsequent analysis in ERDAS Imagine and ArcGIS. The images were imported into ArcMap where, to aid further interpretation, they were classified into 14 temperature bands, each with a specific colour attributed to a 5 °C range. Figure 3 shows a colour composite ATM image, overlain by a temperature image with colour palette of 5 °C intervals. All temperatures below 0 °C are shown as colourless, to reflect the snow and ice cover. The higher-temperature areas are

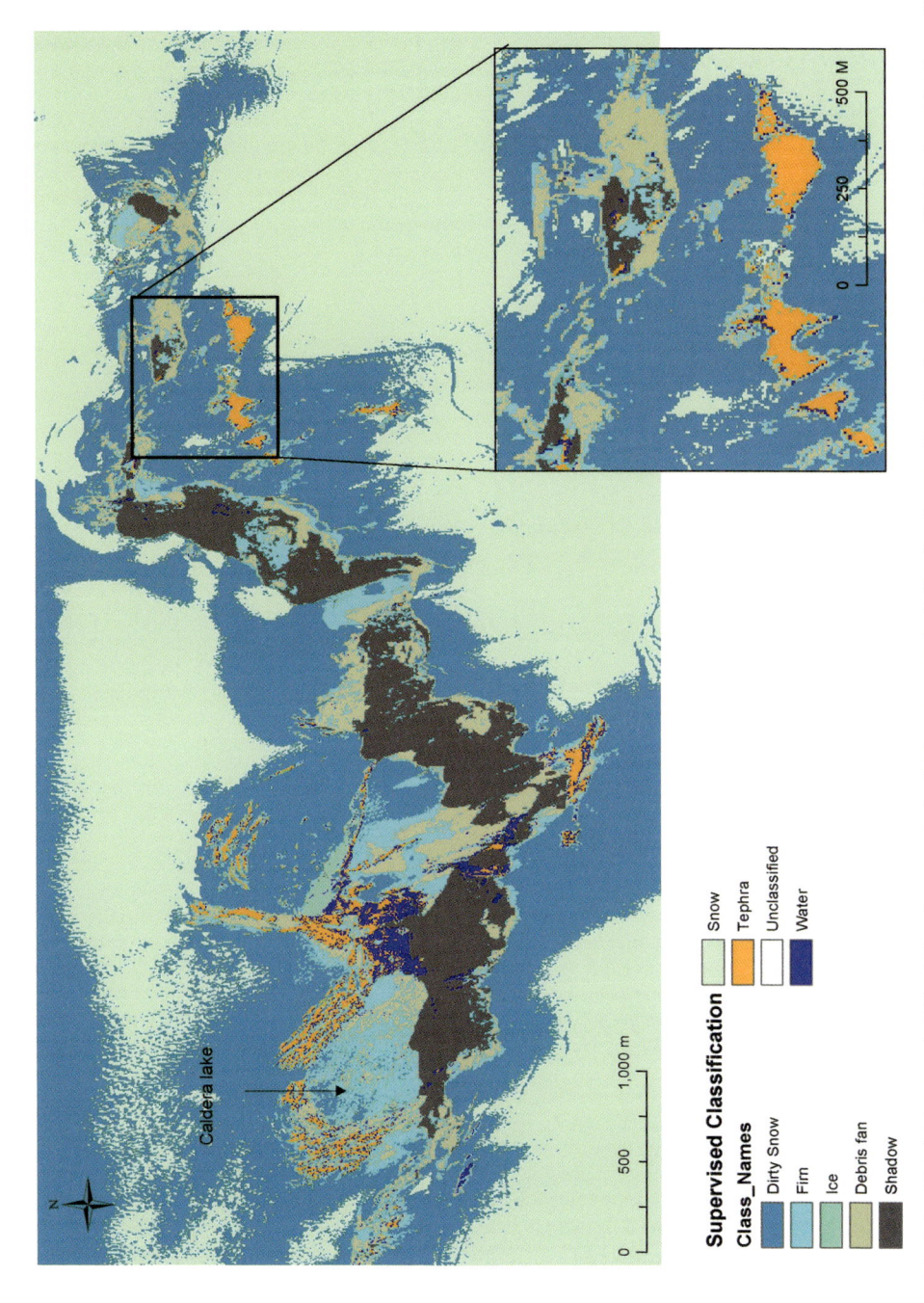

Fig. 2. Supervised classification image using 27 training classes over nine land cover types, including firn (intermediate class between snow and ice) and screefan (a combination of scree slopes and debris fan deposits). The inset shows an example of the increased pixel detail along the Grimsfjall ridge.

Table 2. *Emissivity values allocated to land cover classes (Lillesand* et al. *2004)*

Land cover class	Emissivity
Ice	0.975
Dirty snow	0.975
Basalt	0.96
Firn	0.975
Water	0.985
Screefan	0.96
Tephra	0.96
Snow	0.98
Shadow	1

either in the crater lakes, or are areas of rock or ice covered by tephra, such as the 1998 eruption site, shown in the inset of Figure 3. The dark area is the topographic shadow of the Grímsfjall ridge for which no thermal data can be collected.

Thermal analysis

Ground-based temperature data collected from unconsolidated tephra within the 1998 eruption crater on 4 June 2001 by Icelandic collaborators were overlain on the thermal images. The co-ordinates of the sample points were collected using a Trimble Pathfinder differential global positioning system (DGPS) instrument with an accuracy of 1–2 m. The point source temperatures were collected from the top 10 cm of tephra, using an Ebro platinum thermometer probe, with a precision of 0.1 °C and accuracy of 0.3 °C, along seven NW–SE lines, with approximately three points on each line, depending on accessibility. Unfortunately, only a small number of temperature measurements were collected, because of the hazardous terrain.

An attempt was made to compare the airborne and ground-based temperature data by identifying specific pixels (5 m × 5 m in size) within which each of the ground sample points were located. However, this revealed that the geometric correction of the ATM imagery was insufficiently accurate to allow this point within specific pixel comparison. The inadequate geometric correction of the imagery was attributable to the inability to incorporate a digital elevation map (DEM) of adequate spatial resolution into the geocorrection procedure. This problem could not be resolved using ground control points to perform further geocorrection of the imagery, as no ground survey points could be found that were also identifiable in the imagery. A further difficulty in comparing airborne and ground targets was that each ground measurement recorded the temperature of the small local area in contact with the thermometer, whereas the temperature recorded in the ATM data was an integrated value derived from a 5 m × 5 m area of ground. Within each 5 m × 5 m area considerable variability in surface temperatures may exist, as a result of spatial variations in surface properties, in which case a single point temperature measurement is unlikely to be representative of the area as a whole. Figure 3 shows how variable the temperatures across a specific area of the ground can be, especially in areas of high geothermal activity, where there are numerous fumaroles and steam vents. In an attempt to overcome this problem and the difficulties associated with geocorrection of the ATM imagery, the following method was adopted for comparing ground and airborne temperature data.

An image composed of pixels of the same size as the ARSF data (5 m × 5 m), was constructed by interpolating the ground-based point temperature data. A number of interpolation techniques are currently used in geographical information systems (GISs) (Oliver & Webster 1990). Some of the more statistically simple systems lose detail because of smoothing (Oliver & Webster 1990). Kriging is a technique based on the statistical approach of regionalized variable theory (Mather 1999). It was selected because it allows for optimization of unbiased results along with an indication of the confidence limits in the analysis of individual datasets shown by a variogram created during this type of analysis. A variogram is a measure of the variance between data as a function of distance (Mather 1999). The ordinary linear kriging method available within ArcMap was used, with the z field being temperature, a search radius of 10 and output cell size of 5 m, to produce a pixelated rectangular grid.

The kriging method assumes that each pixel measurement is representative of a specific point within that pixel. However, for most measurements this will be invalid. The ATM image in Figure 3 shows that adjacent pixels can have a temperature difference of up to 25 °C. Consequently, temperature variations greater than 25 °C may not be unusual within 1 pixel. There is also a natural tendency, when making measurements in the field, to be drawn to the highest temperatures even when attempting random sampling. There is less confidence in regions away from the sampling sites, as those areas may have the highest variance. Hence a linear 200 m profile diagonally through the temperature points was chosen (i.e. the area within which we can have most confidence in the interpolation results as indicated by minimal variance). Figure 4 shows the interpolated temperature points and the location of the profile. Blocks of 3 pixels astride the profile line were averaged to produce the temperature value for each location

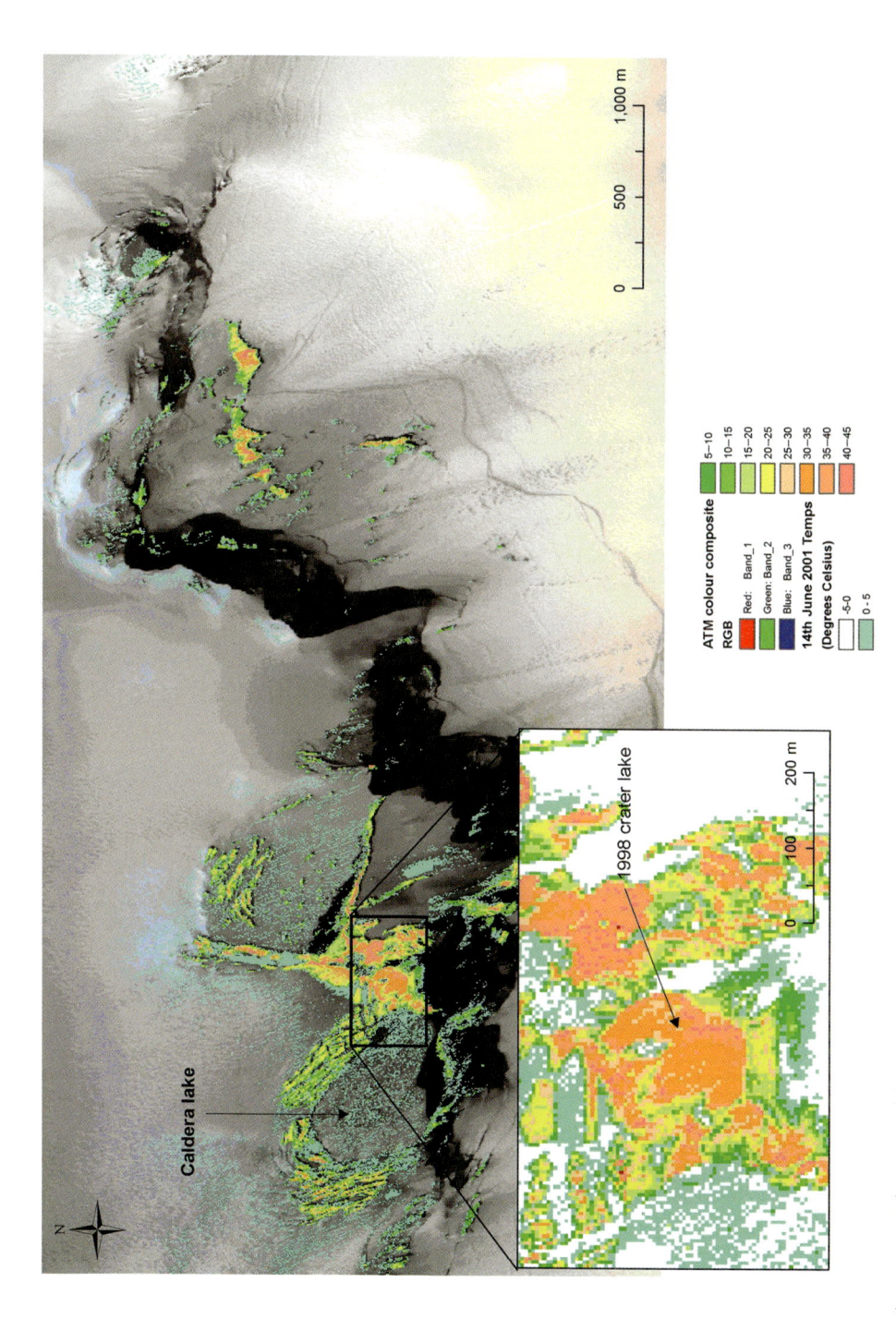

Fig. 3. A temperature image, overlying a colour composite ATM image (Bands 1, 2 and 3 in red, green and blue, respectively). All the temperatures above 0 °C are colour coded in 5 °C intervals. The enlarged area is the 1998 eruption site.

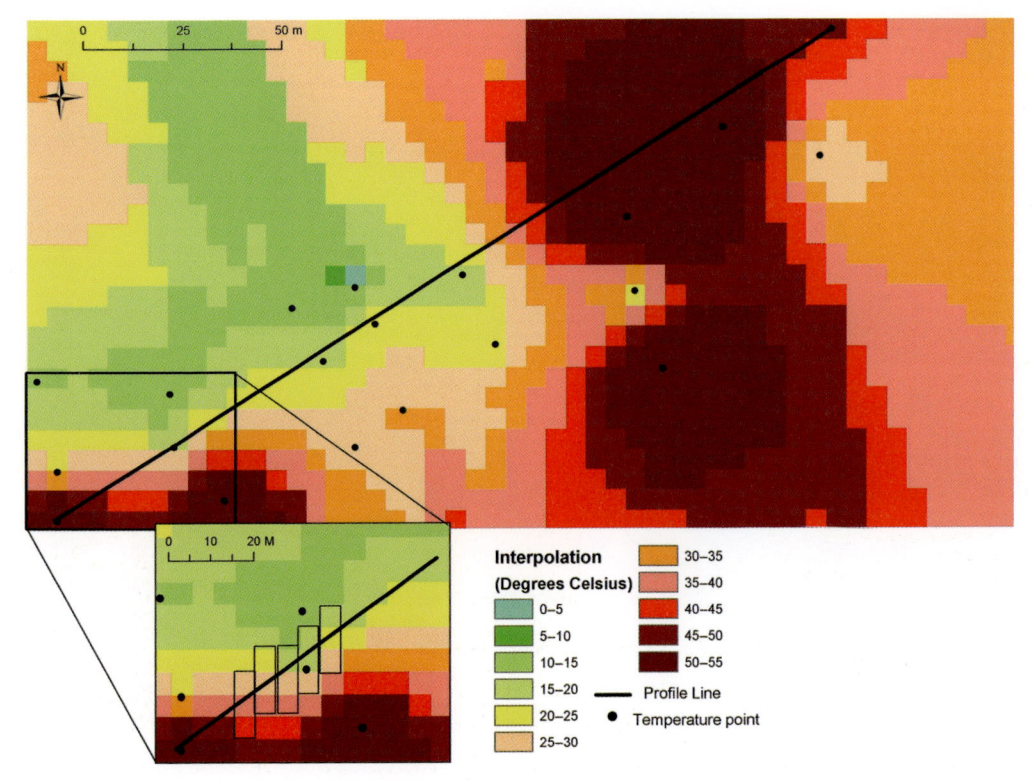

Fig. 4. Interpolation of ground-based temperature points displayed as 5 m pixels using the RGB palette in 5 °C intervals. The locations of the temperature points and line of profile are shown. In the inset, boxes of 3 pixels length show how the average temperatures were taken along the profile.

along the profile (Fig. 4 inset). In this way errors associated with the registration of the temperature points with the thermal image and errors associated with uncertainty within pixels were minimized. Figure 5 shows the 1998 eruption site, the location of the 2001 ground-based temperature points, and the overlaid interpolation. Comparison of the two images using the coloured pixels shows that the two approaches, in some cases, produce similar temperatures. The difference in colour depicted in the two images is explained by the fact that the ground-based images have overall higher temperatures because of the high-temperature sampling points.

Figure 6a shows a comparison between the ground-based and remote sensing temperatures. The graph shows that, along the central part of the profile from 40 to 120 m, the temperatures are similar, with the ATM profiles showing slightly elevated temperatures compared with the ground-based ones. The temperatures correlate reasonably well, considering the two datasets were acquired almost a week apart. The maximum variation is of 10 °C. The high correlation is supported by the visual spatial pattern in Figure 5. However, there are anomalous high-temperature areas in the first 20 m and the last 80 m of the ground-based temperature profile compared with the ATM data. Within the first 20 m of the profile, there are a number of anomalous points (Fig. 6a) that have good coverage by ground measurements. The thermal images show lower temperatures than the ground-based interpolation. This could be explained by pixel-integrated temperatures within the imagery that appear to lower the overall temperature, especially in the last 80 m of the profile, where ground temperatures are substantially higher than the aerial image temperatures. An alternative explanation for this could be related to the low atmospheric pressure measured at the Grímsvötn meteorological station at 160 m above sea level (m a.s.l.), as this can cause an increase in steam emissions along the Grímsfjall ridge. The air pressure was measured as 808 hPa on 4 June at the time of the ground survey, as opposed to 823 hPa on 10 June 2001.

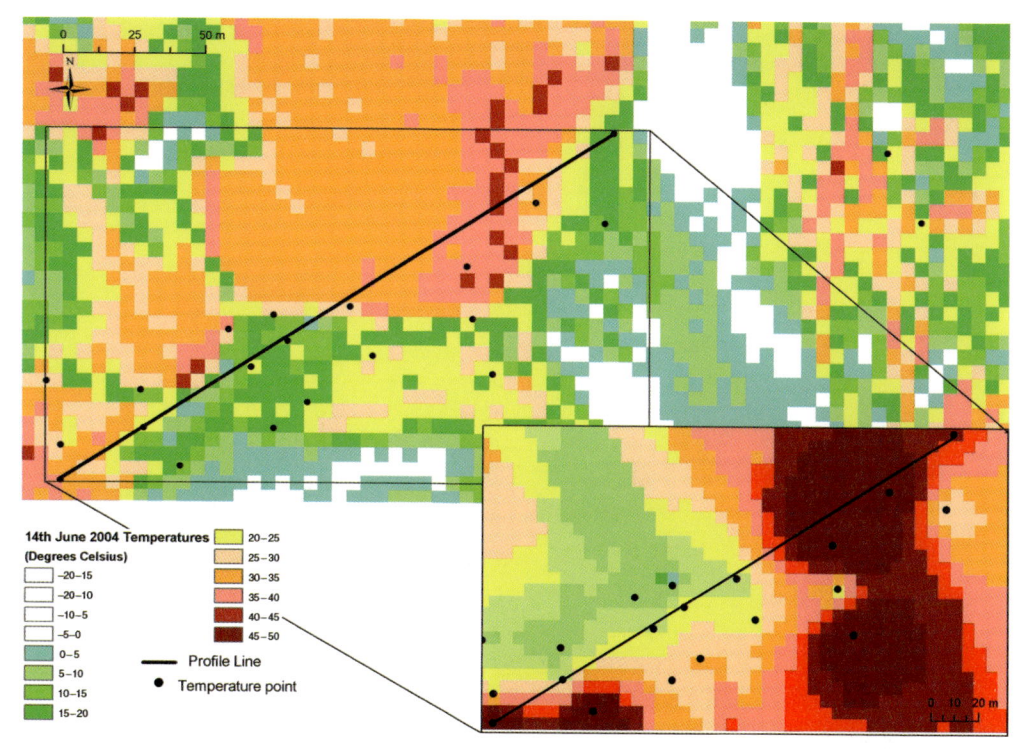

Fig. 5. Locations of the temperature points collected on 4 June 2001 within the 1998 eruption crater, shown on a temperature image for 14 June 2001, together with the profile line used for the analysis. The smaller inset shows the interpolated 2001 temperatures using the same colour scheme.

Figure 8b is a graph comparing the temperatures of the two ATM images (10 and 14 June 2001). Overall the temperatures are in agreement, except in the first 50 m of the profile. An explanation for the difference between the ATM temperatures as well as the ground-based temperatures could be related to differences in the conditions in which the images were collected; that is, the height difference of aircraft (214 m); the look angle because of the position of the flight line (the ATM sensor was almost directly above the lake on 14 June (at an angle of 6.46°), but was at an angle of 37.87° from normal on 10 June 2001; time of day (10:00 h on 14 June and 16:00 h on 10 June) causing changes in the angle of the Sun and solar reflectance; rather than a change in the ground conditions. Atmospheric attenuation is a major contributor to these differences, as the two surveys were flown from different altitudes. Although no atmospheric correction was made on the images, an atmospheric attenuation model derived from a FLIR Systems ThermoCAM was used to make an atmospheric transmission correction following analysis. For a relative humidity range of 30–70%, a maximum error of up to 17 °C is possible for an altitude of 1469.03 m (for 10 June 2001), and 15.3 °C at an altitude of 924.37 m (for 14 June 2001). However, based on this information the error could be as much as 17 °C if there was an extreme change in humidity over the 4 days between flights. Additional errors will arise as a consequence of instrument drift, geometric correction, and the co-registration of temperature points to the ATM images. Figure 7 shows the error applied to the 200 m profile shown in Figure 6a and b. Given the size of the error bars, the temperatures are very similar, save for a few high ground-based temperatures at the beginning and end of the profile.

Comparisons of ground-based temperature data with ATM temperature imagery reveal similarities in results. However, there are problems combining the different techniques, as it is difficult to compare temperatures at specific locations and times because of pixel integrated temperatures and topographic variations. It is possible to derive spatial variations and change over time, but there ideally needs to be a longer time frame between the acquisitions of temperature imagery. However, there is still a need for additional ground observations, including

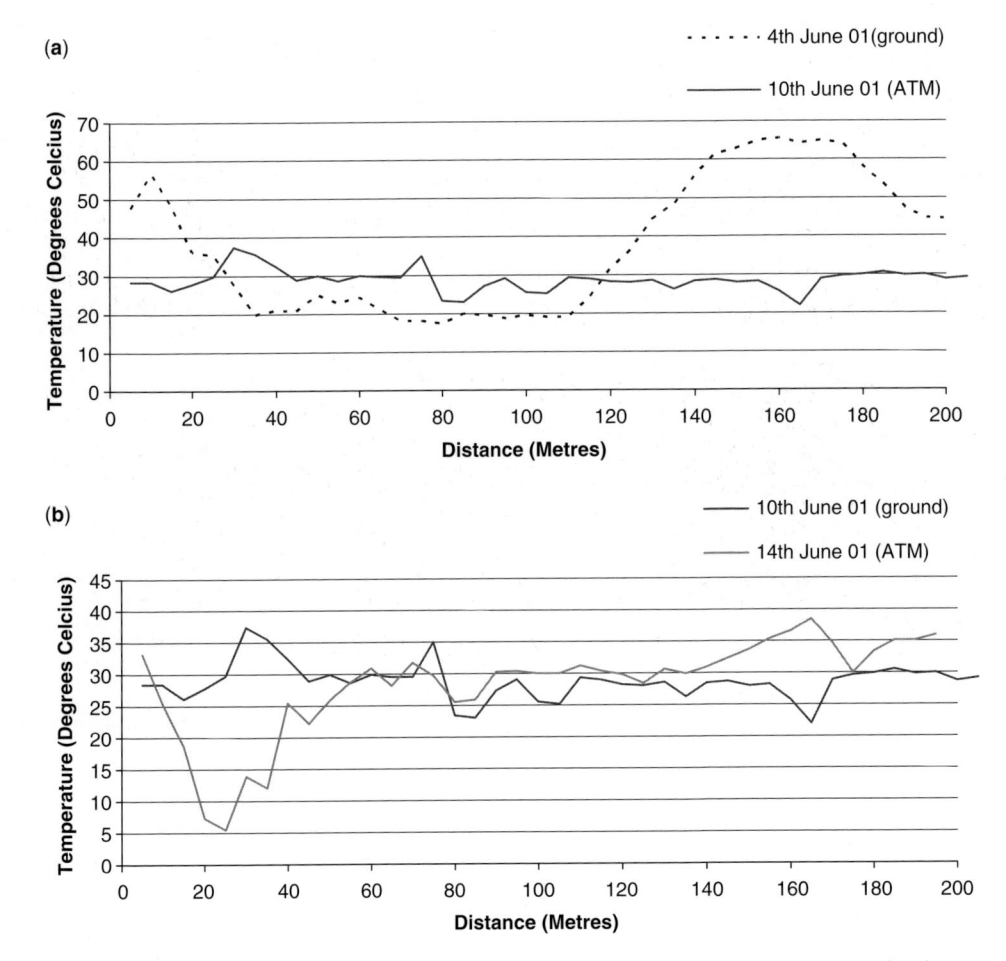

Fig. 6. (**a**) The 200 m long least variance temperature profiles, within the 1998 eruption crater, comparing the ground-based (4 June 2001) and airborne (10 June 2001) temperature data. (**b**) Comparison of ATM temperature profiles for 10 and 14 June 2001.

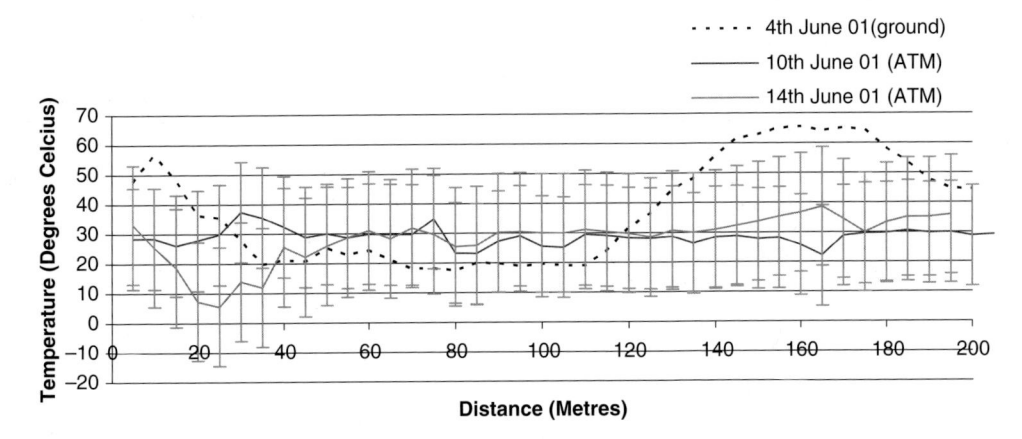

Fig. 7. Graph to show the maximum error (of 17 °C) that could be caused by atmospheric attenuation, along the 200 m profile.

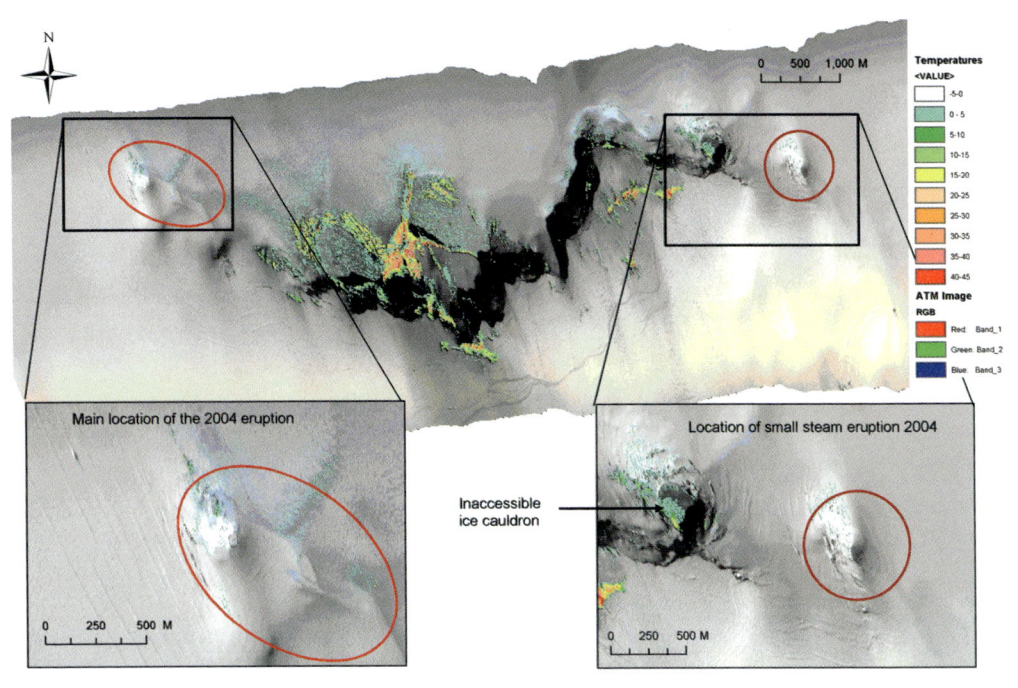

Fig. 8. A temperature image, overlying a colour composite ATM image (Bands 1, 2 and 3 in red, green and blue, respectively). All the temperatures above 0 °C are colour-coded in 5 °C intervals. Also shown are the main locations where the 2004 eruption broke through the ice (red circles).

meteorological and atmospheric measurements, and a larger number of temperatures.

Evaluating usefulness

A primary project aim was to compare 'modelled' temperature data from the ARSF images with ground truthing measurements and to use the ARSF imagery to observe and map thermal anomalies at Grímsvötn. Once the analysis is complete, it is hoped that the data will lead to an analysis of heat flux, as this will lead to improved understanding of the processes at work within such a dynamic region.

The NERC ARSF visible and thermal IR images reveal the presence of a number of ice cauldrons around the Grímsvötn caldera and in areas that cannot be measured in the field. This includes inaccessible parts of the Grímsfjall ridge, such as the small water-filled ice cauldron on the left-hand side on the eastern inset (Fig. 8). As the ice cauldron has steep-sided ice walls there is no way, other than by using RS, to obtain water temperature data. From the thermal images pixel integrated temperatures range from -0.96 to 25.09 °C on both days. A high-temperature trend along the southern section of both the ice cauldron lake and the 1998 crater lake, following the trend of the 1998 eruption fissure, is visible. When compared with ground-based photographs there is steam visible across these sections of the lakes.

Guðmundsson (2003) showed the role of heat transfer from magma to ice and melt water using calorimetry. Precursory evidence of an eruption would be in the melt water rather than on the rock or ice surface. Therefore temperatures recorded from the crater lake or ice cauldrons are fundamental to monitoring Grímsvötn. Figure 9 is a frequency distribution curve of the 1998 eruption site crater lake temperatures from the two thermal images. This shows that the overall distribution of temperatures is similar for the two flights. As noted above, the sensor was almost directly above the lake on 14 June (at an angle of $6.46°$), but was at an angle of $37.87°$ from normal on 10 June. The oblique nature of the view affects the area imaged. The lake water data show that there is similarity between the two flight days. However, there are still discrepancies shown within the ground-based profile data. Atmospheric attenuation is a major contributor to these differences, as the two surveys were flown from different altitudes (i.e. 1469.03 m a.s.l. and 924.37 m a.s.l., respectively). Additional errors will arise as a consequence of instrument drift and related to geometric correction.

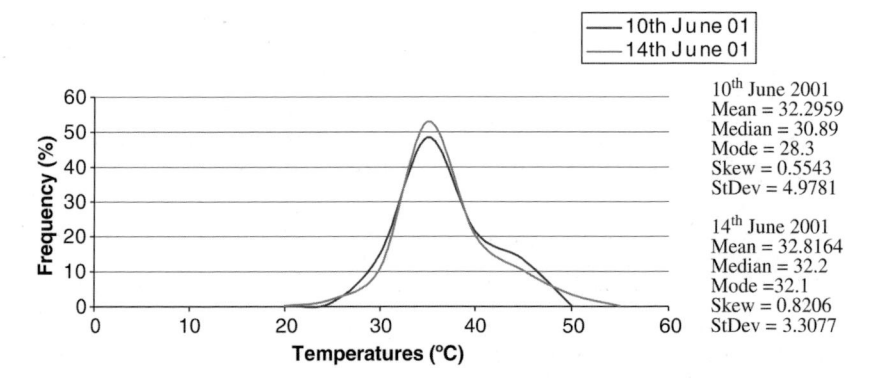

Fig. 9. Frequency distribution curve of the 1998 eruption site crater lake temperatures from the two thermal images. This shows that the overall distribution of temperatures is similar for the two flights. Percentage frequency was used to standardize the histograms.

10th June 2001
Mean = 32.2959
Median = 30.89
Mode = 28.3
Skew = 0.5543
StDev = 4.9781

14th June 2001
Mean = 32.8164
Median = 32.2
Mode = 32.1
Skew = 0.8206
StDev = 3.3077

Following the November 2004 eruption, it has been possible to assess whether the 2001 imagery contains any information that could be useful in the identification of the 2004 eruption site. Figure 8 shows the thermal image overlying the ATM colour composite image, with the locations of the main and smaller steam eruptions marked. The main eruption site (red oval) in the southwestern corner of the cauldera shows the presence of an ice cauldron but no thermal anomaly. The same is true for the smaller steam eruption (red circle) in the southeastern corner. This is not unexpected, as the 2004 eruption was accompanied by the rapid opening of a fissure system, possibly as a consequence of dyke injections; thus it is unlikely that a precursory thermal region would be detectable 3 years before the 2004 eruption.

Conclusion

Thermal imagery collected using instruments deployed on the NERC ARSF aircraft in June 2001 can be used to give an estimate of temperature distribution along the Grímsfjall ridge. The thermal images give an overall view of thermal anomalies along the area of rock and water exposures at Grímsvötn and reveal elevated temperature regions that were not detected during ground-based measurements. These images allow inaccessible areas to be monitored. As two thermal images collected 4 days apart have shown, there are sometimes large apparent temperature differences on this time scale. The reasons for these are varied, including atmospheric attenuation and image acquisition differences. This will have to be taken into account during any future surveys, along with the need to ensure adequate ground control, good on-board GPS, day and night imagery, and measurement of humidity and atmospheric conditions at the time of the flight. This would allow an appropriate atmospheric correction to be included as part of the data processing, and this would reduce uncertainties within the data. In conclusion, in spite of the errors, the ground-based and ARSF temperatures are in reasonable agreement where ground control temperatures were measured, allowing us to have confidence in the temperatures collected from this volcano. The temperature data collected during the 2001 survey will be used to constrain the total heat budget from Grímsvötn, using meteorological data to estimate heat flux from areas of open water.

We wish to thank the NERC ARSF aircrew and ground staff for acquiring the data in 2001, and W. Mockridge and A. K. Wilson for all their assistance in pre-processing the data. We thank M. Ball for his assistance in converting radiance, and G. Davies for technical support. We thank volunteers of the Iceland Glaciological Society, led by K. Langley, who collected the ground truth data on 4 June 2001. Finally, we thank the reviewers for their useful and constructive comments on this paper.

References

AZIMUTH SYSTEM 2001. AZ-16 airborne remote sensing control display and data acquisition system. *AZGCORR User Guide 2001.* Nerc, Swindon.
BJÖRNSSON, H. 1975. Subglacial water reservoirs, jökulhlaups and volcanic eruptions. *Jökull,* **25**, 1–15.
BJÖRNSSON, H. 1998. Hydrological characteristics of the drainage system beneath a surging glacier. *Nature,* **395**(6704), 771–774.
BJÖRNSSON, H. 2003. Subglacial lakes and jökulhlaups in Iceland. *Global and Planetary Change,* **35**, 255–271.
BJÖRNSSON, H. & EINARSSON, P. 1990. Volcanoes beneath Vatnajökull, Iceland: evidence from radio echo-sounding, earthquakes and jökulhlaups. *Jökull,* **40**, 147–168.

BJÖRNSSON, H. & GUÐMUNDSSON, M. T. 1993. Variations in the thermal output of the subglacial Grímsvötn Caldera, Iceland. *Geophysical Research Letters*, **20**, 2127–2130.

BOURGEOIS, O., DAUTEUIL, O. & VAN VLEIT-LANÖE, B. 1998. Pleistocene subglacial volcanism in Iceland: tectonic implications. *Earth and Planetary Science Letters*, **164**, 165–178.

CURRAN, P. J. 1992. *Principles of Remote Sensing*. Longman Scientific and Technical, Harlow, England.

EINARSSON, P. 1999. *Geology of Iceland: Rocks and Landscape*. Mál og Menning, Reykjavik, Iceland.

EINARSSON, P., BRANDSDÓTTIR, B., GUÐMUNDSSON, M. T., BJÖRNSSON, H., GRÖNVOLD, K. & SIGMUNDSSON, F. 1997. Centre for the Icelandic hotspot experiences volcanic unrest. *EOS Transactions, American Geophysical Union*, **78**(35), 369 and 374–375.

GRÖNVOLD, K., OSKARSSON, N., JOHNSEN, S., CLAUSEN, H., HAMMER, C. U., BOND, G. & BARD, E. 1995. Ash layers from Iceland in the Greenland GRIP ice core correlated with oceanic and land sediments. *Earth and Planetary Science Letters*, **135**, 149–155.

GUÐMUNDSSON, M. T. 2003. Melting of ice by magma–ice–water interactions during subglacial eruptions as an indicator of heat transfer in subaqueous eruptions. *In*: ÍWHITE, J. D. L., SMELLIE, J. L. & CLAGUE, D. (eds) *Explosive Subaqueous Volcanism*, Geophysical Monograph, American Geophysical Union, **140**, 61–72.

GUÐMUNDSSON, M. T. & BJÖRNSSON, H. 1991. Eruptions in Grímsvötn 1934–1991. *Jökull*, **41**, 21–46.

GUÐMUNDSSON, M. T. & MILSOM, J. 1997. Gravity and magnetic studies of the subglacial Grímsvötn volcano, Iceland. Implications for crustal and thermal structure. *Journal of Geophysical Research*, **102**, 7691–7704.

GUÐMUNDSSON, M. T., SIGMUNDSSON, F. & BJÖRNSSON, H. 1997. Ice–volcano interaction of the 1996 Gjálp subglacial eruption, Vatnajökull, Iceland. *Nature*, **389**, 954–957.

GUÐMUNDSSON, M. T., PÁLSSON, F., BJÖRNSSON, H. & HÖGNADÓTTIR, TH. 2002. The hyaloclastite ridge formed in the subglacial 1996 eruption in Gjálp, Vatnajökull, Iceland: present-day shape and future preservation. *In*: SMELLIE, J. L. & CHAPMAN, M. G. (eds) *Volcano–Ice Interaction on Earth and Mars*. Geological Society, London, Special Publications, **202**, 319–335.

GUÐMUNDSSON, M. T., SIGMUNDSSON, F., BJÖRNSSON, H. & HÖGNADÓTTIR, TH. 2004. The 1996 eruption at Gjálp, Vatnajökull ice cap, Iceland: efficiency of heat transfer, ice deformation and subglacial water pressure. *Bulletin of Volcanology*, **66**, 46–65.

HÖSKULDSSON, A. & SPARKS, R. S. J. 1997. Thermodynamics and fluid dynamics of effusive subglacial eruptions. *Bulletin of Volcanology*, **59**, 219–230.

KAHLE, A. B. & ALLEY, R. E. 1992. Separation of temperature and emittance in remotely sensed radiance measurements. *Remote Sensing of Environment*, **42**, 107–111.

KEALY, P. S. & HOOK, S. J. 1993. Separating temperature and emissivity in thermal infared multispectral scanner data: Implications for recovering land surface temperatures. *IEEE Transactions on Geoscience and Remote Sensing*, **13**, 1155–1164.

KONSTANTINOU, K. I., NOLET, G., MORGAN, W. J., ALLEN, R. M. & PRITCHARD, M. J. 2000. Seismic phenomena associated with the 1996 Vatnajökull eruption, central Iceland. *Journal of Volcanology and Geothermal Research*, **102**, 169–187.

LARSEN, G., GUÐMUNDSSON, M. T. & BJÖRNSSON, H. 1998. Eight centuries of periodic volcanism at the center of the Iceland hot spot revealed by glacier tephrastratigraphy. *Geology*, **26**, 943–946.

LIANG, S. 2001. An optimisation algorithm for separating land surface temperature and emissivity from multispectral thermal infrared imagery. *IEEE Transactions on Geoscience and Remote Sensing*, **39**(2), 264–274.

LILLESAND, T. M., KIEFER, R. W. & CHIPMAN, J. W. 2004. *Remote Sensing and Image Interpretation*, 5th edn. Wiley, New York.

MATHER, P. M. 1999. *Computer Processing of Remotely Sensed Images. An Introduction*, 2nd edn. Wiley, Chichester.

NERC ARSF 2002. *User Manual. Airborne Thematic Mapper (ATM) and Integrated Data System (IDS)*. Version 2, Nerc, Swindon.

OLIVER, M. A. & WEBSTER, R. 1990. Kriging: a method of interpolation for geographical information systems. *International Journal of Geographical Information Systems*, **4**(3), 313–332.

OPPENHEIMER, C. & ROTHERY, D. A. 1991. Infrared monitoring of volcanoes by satellite. *Journal of the Geological Society, London*, **148**, 563–569.

ROTHERY, D. A., FRANCIS, P. W. & WOOD, C. A. 1988. Volcano monitoring using short wavelength infrared data from satellites. *Journal of Geophysical Research*, **93**(B7), 7993–8008.

SAEMUNDSSON, K. 1979. Outline of the geology of Iceland. *Jökull*, **29**, 7–28.

SMELLIE, J. L. 2000. Subglacial eruptions. *In*: SIGURDSSON, H. (ed.) *Encyclopaedia of Volcanoes*. Academic Press, New York, 403–418.

SMELLIE, J. L. 2002. The 1969 subglacial eruption on Deception Island (Antarctica): events and processes during an eruption beneath a thin glacier and implications for volcanic hazards. *In*: SMELLIE, J. L. & CHAPMAN, M. G. (eds). *Volcano–Ice Interaction on Earth and Mars*. Geological Society, London, Special Publications, **202**, 59–79.

TUFFEN, H., PINKERTON, H., MCGARVIE, D. W. & GILBERT, J. S. 2002. Melting of the glacier base during a small-volume subglacial rhyolite eruption: evidence from Bláhnúkur, Iceland. *Sedimentary Geology*, **149**, 183–198.

Developments in synthetic aperture radar interferometry for monitoring geohazards

M. RIEDMANN & M. HAYNES

NPA Group, Crockham Park, Edenbridge TN8 6SR, UK
(e-mail: michael.riedmann@npagroup.com)

Abstract: In 1993 synthetic aperture radar (SAR) interferometry (InSAR) was introduced to the wider remote sensing community with the publication of the interferogram depicting the ground deformation caused by the Landers earthquake. Although the power of interferometry was demonstrated, the conventional technique has not always been applicable in all operational scenarios. Over the last few years, however, a number of technical developments have emerged that provide a higher precision of motion rates, the extraction of specific motion histories, and precise targeting. This paper examines uses of differential SAR interferometry (DifSAR) for monitoring geohazards. Limitations of DifSAR will be discussed: lack of coherence, atmospheric refraction and targeting. It will be shown how some of these limitations can be overcome with persistent scatterer interferometry (PSI), which detects slow ground motion with annual rates of as little as a few millimetres, reconstructing a motion history based on the European Space Agency's SAR image archive. The technique permits the estimation and removal of the atmospheric phase, achieving higher accuracies than DifSAR. PSI relies on the availability of pre-existing ground features that strongly and persistently reflect back the signal from the satellite. However, in highly vegetated regions, PSI may not be applicable because of the lack of natural scatterers. To ensure motion measurement of the ground or structures at targeted locations, the NPA Group is developing InSAR using artificial radar reflectors, such as Corner Reflectors (CRs) or Compact Active Transponders (CATs). Both reflector types are still undergoing validation tests, but results show a high phase stability in both cases.

Geohazards such as landslides, rockslides, earthquakes and sinkholes can pose a significant danger to humans and built infrastructure. Areas of extensive subsidence, such as that associated with underground coal mining, or the extraction of petroleum, brine or groundwater, can also cause costly damage to buildings and infrastructure. Large-scale measurement of ground deformation in endangered areas is therefore in the interest of the safety of the public and built environment.

Precision ground surveys can be carried out over sites to measure the stability of the terrain; however, such surveys are inherently expensive and in some cases can be dangerous to human life. Furthermore, some unstable areas can remain undetected by geoscientists, as a result of unfavourable survey conditions (e.g. thick vegetation, unsuitable weather conditions, absence of clear line of sight) or because ground movements are so slow that they are difficult to detect on the ground.

Monitoring ground movement with radar satellites has evolved in the last decade from conventional imaging InSAR to improved techniques such as persistent scatterer interferometry (PSI). The following two sections will give a short introduction to InSAR and illustrate applications for monitoring geohazards.

InSAR principles

Satellite synthetic aperture radar (SAR) systems transmit electromagnetic radiation signals at microwave and radio frequencies and measure the intensity backscatter and the time delay (phase) of the signals that are reflected back from objects in the signal path. The resulting SAR image has a spatial resolution of 10–20 m. Its brightness (i.e. the intensity of the measured backscatter) depends on the surface roughness, dielectric constant, moisture content and the slope of the local topography. The advantage of radar is that it is generally unaffected by atmospheric conditions, such as rain, dust and cloud cover, and can be used day or night. For more information the reader is referred to Hanssen (2001).

SAR interferometry (InSAR) is a technique in which the phase component of the returning radar signals of two or more radar scenes of the same location (see Fig. 1) are compared to allow the detection of ground movements to sub-centimetric precision (Gabriel *et al.* 1989). Although satellites' orbits are precisely controlled to allow for repeat-track missions, there will be slight differences in the position of the satellites when two images of the same ground location are taken from two different satellite passes in the same nominal orbital

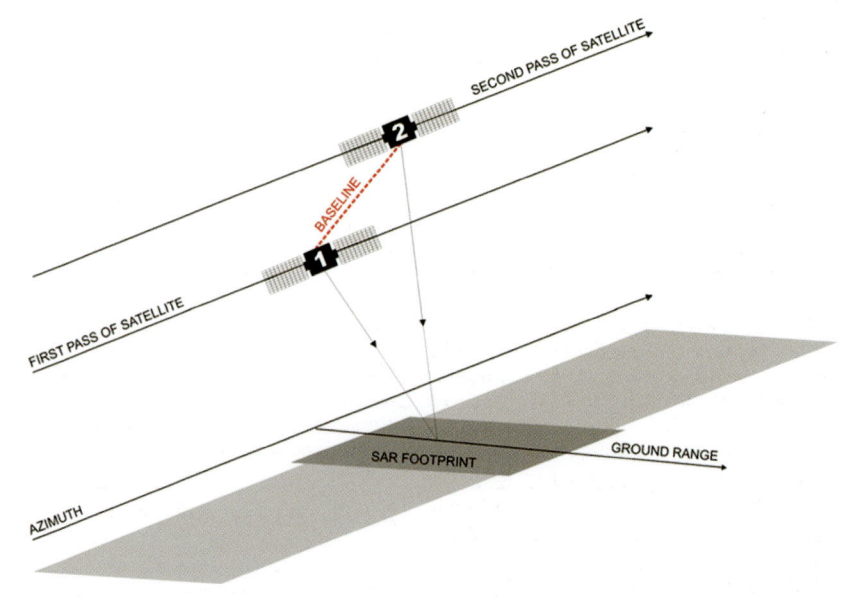

Fig. 1. Geometry of a satellite interferometric SAR system. The orbit separation is called the 'interferometer baseline' and its projection perpendicular to the satellite radar viewing direction is one of the key parameters to allow SAR interferometry analysis. The baseline is much smaller than the satellites' altitude, typically by about three orders of magnitude.

position. These differences allow for angular measurement similar to the principle used in optical photogrammetry, here with the angles not being measured directly, but inferred from distance measurements using trigonometry (Hanssen 2001, p. 17). For InSAR, the phase rather than the amplitude information is used from the returning signal to measure any change in ground height.

Data sources and issues

Currently three C-band SAR satellites are in operation: ERS-2, Radarsat-1 and Envisat. Table 1 presents some characteristics of these satellites and planned missions. ERS-2 and its precursor ERS-1 have built up a regularly updated archive of more than 1.5 million images worldwide. Envisat ensures continuity of SAR image acquisitions worldwide, and builds up a regular archive over some important and critical regions of the world. The Canadian Radarsat-1 instrument works on image request only.

InSAR measurements are generally limited by the characteristics of the sensor used to acquire the data. For example, measurements are possible only in the line-of-sight (LOS; i.e. viewing direction) of the sensor and scene updates depend on the repeat cycle frequency of the satellite. For long-term historical measurements over a given area, the data archive of the sensor needs to be checked for availability of sufficient and appropriate SAR data.

Also shown in Table 1 are details of the planned Radarsat-2 and TerraSar-X commercial missions. In addition, a C-band SAR satellite mission (Sentinel 1) is currently under discussion for the

Table 1. *Current and planned SAR imaging satellites*

Platform	Sensor	Country	Launch	Wavelength	Repeat pass (days)	Sensor incidence angle (deg)	Resolution (m)
ERS-2	AMI	Europe	1995	C-band	35	23	20
Radarsat-1	SAR	Canada	1995	C-band	24	20–50	≤ 8
Envisat	ASAR	Europe	2002	C-band	35	15–45	20
ALOS	PALSAR	Japan	2006	L-band	46	10–51	≤ 10
Radarsat-2	SAR	Canada	2006	C-band	24	20–50	3
TerraSar-X	SAR	Europe	2006	X-band	11	20–55	≤ 1

Table 2. *Three interferometric techniques used for ground motion measurements: DifSAR (differential SAR interferometry), PSI (persistent scatterer interferometry) and CR (Corner Reflector) or CAT (Compact Active Transponder) interferometry*

Method	Measurement periods	Need for archive data	Extents	Precision	Cost
DifSAR	Historical/present	Low	Map	Sub-centimetric	Low
PSI	Historical/present	High	Map of points	Millimetric	High
CRInSAR	Present	None	Specific locations	Sub-centimetric	Low to medium

European Space Agency's (ESA's) Global Monitoring for Environment and Security (GMES) Earth Observation (EO) component to provide continuity of InSAR applications beyond the lifespan of Envisat.

One of the GMES applications is the Terrafirma project (www.terrafirma.eu.com): this aims to establish a pan-European ground motion information service to detect millimetric ground displacements using PSI. Initially the service focuses on urban subsidence but it will eventually include earthquake zones, landslides, coastlines and flood plains. Terrafirma is one of a number of Service Element projects being run under ESA's GMES initiative, distributed throughout Europe via the national geological surveys.

InSAR techniques

Three methods for ground motion measurements by InSAR have evolved over the years. They are employed according to particular operational applications and are summarized in Table 2.

Differential InSAR (DifSAR)

DifSAR maps wide-area relative ground deformation and can cover an area of 100 km by 100 km in a single process. The output is a map of ground deformation showing sub-centimetric displacements in the LOS of the satellite. A key requirement is that the response characteristics of the ground cover in the area of interest have not changed significantly

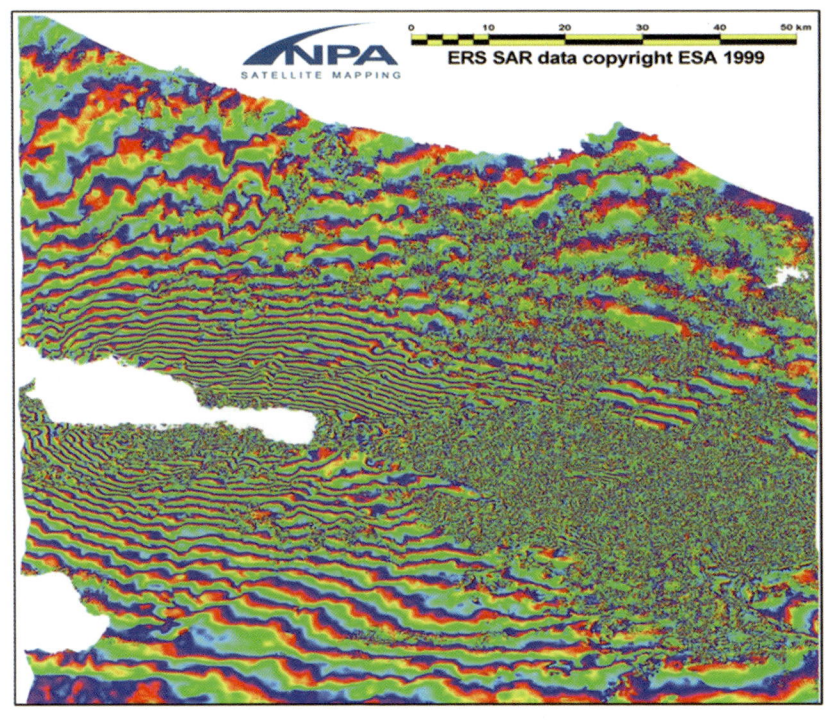

Fig. 2. Displacement map for the Izmit earthquake on 17 August 1999. The colour-coded contour cycles correspond to displacement of 2.8 cm in the line-of-sight of the satellite. Actual relative ground movement across the fault was 4 m horizontally. Image copyright NPA Group 1999; SAR data copyright ESA 1999.

between two image acquisitions. Depending on the ground cover, measurement periods from 24 days (rural environment) to 5 years or more (urban or arid areas) can be analysed. Key DifSAR studies include those by Gens & Van Genderen (1996), Bamler & Hartl (1998), Madsen & Zebker (1998), Massonnet & Feigl (1998) and Bürgmann et al. (2000). DifSAR has been applied successfully to map ground displacements resulting from:

(1) earthquakes: measurement of the build-up of elastic strain between earthquakes, as well as the actual deformation caused by an earthquake (Massonnet et al. 1993; Zebker et al. 1994; Peltzer & Rosen 1995; Wright 2002);

(2) volcanic deformation: inflation and deflation of volcanoes before and during eruptions, respectively (Amelung et al. 1999; Massonnet & Sigmundsson 2000);

Fig. 3. Map of persistent scatterer points, with their calculated average annual motion rates (mm year^{-1}, colour-coded) for St. Petersburg, Russia. PSI data copyright NPA Group 2005; ERS data copyright ESA 1992–2004; background image Landsat ETM + Band 8. (Data processed by NPA for ESA's GMES Terrafirma service.)

(3) water, brine, oil, or gas extraction and underground coal mining: ground subsidence measurements during extraction (Usai 1997);

(4) ice motion: mapping the motion of glaciers, ice streams, ice sheets (Goldstein et al. 1993).

An example for earthquakes is given in Figure 2, which shows an 'interferogram' or deformation map, generated after the Izmit (Turkey) magnitude of 7.4 earthquake on 17 August 1999. ERS SAR images from 13 August 1999 and 17 September 1999 were used to generate this map. Each fringe cycle corresponds to a specific amount of relative motion in the LOS of the satellite. This amount is a function of the radar wavelength; in this case each cycle represents 28 mm of motion. A total of 4 m displacement in the satellite's LOS was measured. Further information on the use of DifSAR with the Izmit earthquake is available from Wright et al. (2001).

Persistent scatterer interferometry (PSI)

The PSI technique was first introduced by Ferretti et al. (1999) and different algorithms have been developed since then (e.g. Werner et al. 2003). PSI uses about 30–100 co-registered SAR images to identify time-persistent radar scatterer points and to derive an atmospheric phase screen for each scene. Correction for atmospheric effects produces much finer measurements than the DifSAR technique. For each one of these persistent scatterers, a motion history is available for the time span of the available data, which could stretch back to 1992 using combined ERS-1, ERS-2 and Envisat data.

PSI maps wide-area relative ground movements with sub-centimetric precision along the satellite's LOS and its vertical precision is beyond that achievable with the Global Positioning System (GPS). Using ERS data, the absolute spatial accuracy is about 15 m, and the relative spatial accuracy is about ±5 m in east–west and ±2 m in north–south direction. 'Spatial accuracy' refers in this context to the accuracy of locating the persistent scatterer on the ground. PSI represents a rapid and cost-effective measure of ground motion: over large areas with built infrastructure; in areas undergoing slow and steady subsidence (smaller than 10 cm year^{-1}); for long measurement periods (>5 years).

Urban areas are best suited as PSI application areas, and Figure 3 shows an example of PSI output for St. Petersburg, giving the average annual motion rate of ground points for the period between 1992 and 2004. For every persistent scatterer shown in Figure 3, an individual time series is available (see Fig. 4).

It should be noted that it is the movement of the persistent scatterer (e.g. a building) that is measured and not that of the ground (although in many instances these will be interrelated). Furthermore, for ground motion to be resolved unambiguously in the resulting PSI maps, ground movement between two SAR acquisitions (in the range of 24–35 days for the current missions Radarsat-1 and ERS-2–ENVISAT, respectively) should not exceed a quarter of the wavelength of the sensor. For example, for ERS with a wavelength of 5.6 cm, subsidence rates should not be larger than 1.4 cm per shortest consecutive repeat image acquisition (35 days).

PSI requires a large number of ERS SAR scenes (minimum 30 but ideally as many as are available). A feature of the technique is that the number and location of persistent scatterers cannot be predicted before processing, and measurement success can be guaranteed only over built-up urban areas or over dry and rocky regions. To complement the

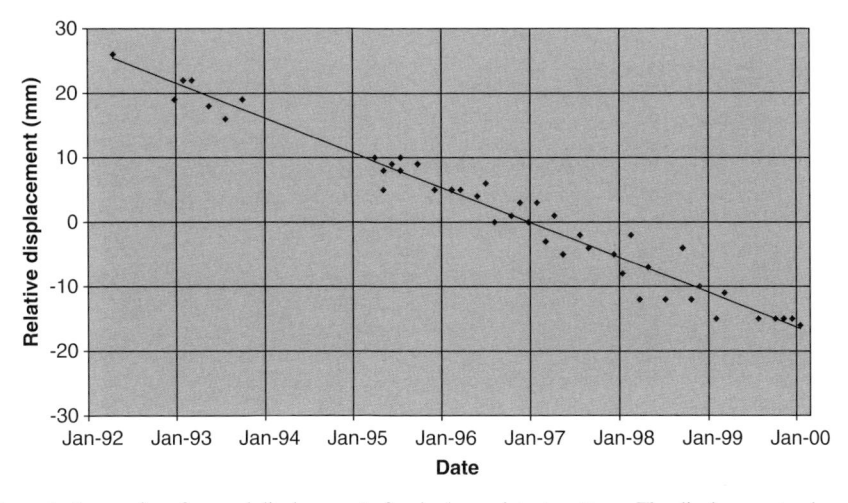

Fig. 4. Example time series of ground displacement of a single persistent scatterer. The displacement values are relative to a chosen reference point.

Fig. 5. A metallic Corner Reflector (CR), with a Compact Active Transponder (CAT) in the foreground.

distribution of persistent scatterer points, artificial radar reflectors can be installed at locations of interest.

Corner Reflectors and Compact Active Transponders

Corner Reflectors (CRs) are purpose-built triangular reflecting metal plates angled upwards towards the satellite and installed at specific locations of interest (see Fig. 5). The size of the CR is less than 1.2 m in all three dimensions and attaches to a flat base-plate, which is anchored into the ground, by concreting and/or ground spikes. Sub-centimetric ground movements are detectable at each CR location. The absolute spatial accuracy is about 20 m for the current Radarsat-1 and Envisat missions, but can be precisely ascertained at the time of installation by GPS surveying. To receive a clear CR response,

Fig. 6. Intensity responses from a network of NPA's Compact Active Transponders, deployed for monitoring subsidence with InSAR. Image copyright NPA Group 2005.

CRs need to be sited away from other potential scatterers such as buildings or metallic structures, or overhead obstructions. CRs may be used to map slow landslip or structural instability (e.g. dams, bridges) with sub-centimetric precision in height. Operability is best in remote areas, where the CRs are not subjected to vandalism.

An alternative to CRs are Compact Active Transponders (CATs; see Fig. 5), which are more compact than CRs and do not suffer as much from environmental impact such as strong winds and the accumulation of debris or snow. Whereas CRs can only be oriented to suit either the ascending or descending viewing modes of the satellite (i.e. when orbiting south to north or north to south, respectively), CATs can be used for the two modes in one setup, and are responsive to all line-of-sight modes of radar satellites.

Figure 6 shows the radar responses from part of a network of NPA transponders (CATs) deployed in a region of subsidence. The transponders are c. 150 m apart and their intensity responses in the radar imagery are overlaid on optical data. Through InSAR analysis of their SAR phase component, motion at these locations can be measured and monitored over time.

Summary

Within a decade, imaging radar interferometry has matured into a widely used geodetic technique for measuring the topography and deformation of the Earth. There are three relative ground motion measurement techniques that complement each other for monitoring geohazards: (1) differential interferometry to map wide-area movements at low cost; (2) persistent scatterer interferometry to provide a time series (dating back to 1992) of ground movement for each persistent radar reflector found in the scene; (3) Corner Reflector and Compact Active Transponder interferometry to measure ground motion at specific locations.

References

AMELUNG, F., GALLOWAY, D. L., BELL, J. W., ZEBKER, H. A. & LACZNIAK, R. J. 1999. Sensing the ups and downs of Las Vegas: InSAR reveals structural control of land subsidence and aquifer-system deformation. *Geology*, 27, 483–486.

BAMLER, R. & HARTL, P. 1998. Synthetic aperture radar interferometry. *Inverse Problems*, 14, R1–R54.

BÜRGMANN, R., ROSEN, P. & FIELDING, E. 2000. Synthetic aperture radar interferometry to measure Earth's surface topography and its deformation. *Annual Review of Earth and Planetary Sciences*, 28, 169–209.

FERRETTI, A., ROCCA, F. & PRATI, C. 1999. Non-uniform motion monitoring using the permanent scatterers technique. *In: FRINGE '99: Second ESA International Workshop on ERS SAR Interferometry, 10–12 November 1999, Liège*. ESA, 1–6.

GABRIEL, A. K., GOLDSTEIN, R. M. & ZEBKER, H. A. 1989. Mapping small elevation changes over large areas: differential radar interferometry. *Journal of Geophysical Research*, 94(B7), 9183–9191.

GENS, R. & VAN GENDEREN, J. L. 1996. SAR Interferometry—issues, techniques, applications. *International Journal of Remote Sensing*, 17(10), 1803–1835.

GOLDSTEIN, R. M., ENGELHARDT, H., KAMB, B. & FROLICH, R. M. 1993. Satellite radar interferometry for monitoring ice-sheet motion: application to an Antarctic ice stream. *Science*, 262, 1525–1530.

HANSSEN, R. F. 2001. *Radar Interferometry—Data Interpretation and Error Analysis, Remote Sensing and Digital Image Processing, Vol. 2*. Kluwer, Dordrecht.

MADSEN, S. N. & ZEBKER, H. A. 1998. Imaging radar interferometry. *In*: HENDERSON, F. M. & LEWIS, A. J. (eds) *Principles and Applications of Imaging Radar*, Wiley, New York, 359–380.

MASSONNET, D., ROSSI, M., CARMONA, C., ADRAGNA, F., PELTZER, G., FEIGL, K. & RABAUTE, T. 1993. The displacement field of the Landers earthquake mapped by radar interferometry. *Nature*, 364, 138–142.

MASSONNET, D. & FEIGL, K. L. 1998. Radar Interferometry and its application to changes in the earth's surface. *Reviews of Geophysics*, 36(4), 441–500.

MASSONNET, D. & SIGMUNDSSON, F. 2000. Remote sensing of volcano deformation by radar interferometry from various satellites. *In*: MOUGINIS-MARK, P. J., CRISP, J. & FINK, J. (eds) *Remote Sensing of Active Volcanism*. Geophysical Monograph, American Geophysical Union, 116, 207–221.

PELTZER, G. & ROSEN, P. 1995. Surface displacement of the 17 May 1993 Eureka Valley, California earthquake observed by SAR interferometry. *Science*, 268, 1333–1336.

USAI, S. 1997. The use of man-made features for long time scale INSAR. *In: Proceedings of IGARSS 1997, Vol. IV*. IEEE Operations Center, Piscataway, NJ, 1542–1544.

WERNER, C., WEGMÜLLER, U., STROZZI, T. & WIESMANN, A. 2003. Interferometric point target analysis for deformation mapping. *In: Proceedings of IGARSS 2003, Toulouse, France, 21–25 July 2003*. IEEE, New York, 4359–4361.

WRIGHT, T. 2002. Remote monitoring of the earthquake cycle with Satellite Radar Interferometry. *Philosophical Transactions of the Royal Society of London, Series A*, 360, 2873–2888.

WRIGHT, T., FIELDING, E. & PARSONS, B. 2001. Triggered slip: observations of the 17 August 1999 Izmit (Turkey) earthquake using radar interferometry. *Geophysical Research Letters*, 28, 1079–1082.

ZEBKER, H. A., ROSEN, P. A., GOLDSTEIN, R. M., GABRIEL, A. & WERNER, C. L. 1994. On the derivation of coseismic displacement fields using differential radar interferometry: the Landers earthquake. *Journal of Geophysical Research*, 99(B10), 19617–19634.

Aerial photography and digital photogrammetry for landslide monitoring

J. WALSTRA, J. H. CHANDLER, N. DIXON & T. A. DIJKSTRA

Department of Civil & Building Engineering, Loughborough University, Loughborough LE11 3TU, UK (e-mail: J.Walstra@lboro.ac.uk)

Abstract: A review is given of the techniques that are available to extract relevant information from multi-temporal aerial photographs for use in the monitoring stage of landslide assessments. It is shown that aerial photograph interpretation reveals qualitative information on surface characteristics, which is helpful in detecting landslide features and inferring the mechanisms involved. Photogrammetrically derived products can be used to quantify these processes, providing distinctive advantages. Comparison of digital elevation models (DEMs) from different times provides detailed information on changes in surface topography, whereas orthophotos can be used to measure horizontal displacements. The various factors influencing the quality of the products are also identified. Examples from a case study on the Mam Tor landslide are used to illustrate the benefits of the different approaches.

Aerial photographs are a generally accepted tool used in landslide studies. They not only provide a metric model from which quantitative measurements can be obtained, but also give a qualitative description of the Earth surface. These two capabilities are irrefutably related to each other, as 'one must know what one is measuring' (Lo 1976).

The application of aerial photographs to landslide investigation provides a number of distinct advantages. Reconnaissance of the study area can greatly benefit from the 3D representation that is provided by stereoscopic viewing, thereby showing relationships between the various landscape elements more obviously than from a ground perspective. Furthermore, photographically based derivatives provide a suitable base on which boundaries can be delineated accurately. In addition, photographs support the efficient planning of field investigations and sampling schemes, without the need for visiting the site physically, which is especially useful in remote and inaccessible areas (Crozier 1984, Van Zuidam 1985). A final and important advantage is the quantitative topographic information contained, which can be unlocked by appropriate photogrammetric techniques. However, quantitative use of aerial photographs create some difficulties, such as the requirement of experienced analysts and appropriate equipment, combined with sufficient knowledge of the site under investigation (Lo 1976).

Aerial photographs can be used in various stages of landslide investigations (Mantovani *et al.* 1996), and have been extensively used in the detection and classification of landslides. When properly interpreted they allow the identification of diagnostic surface features, such as morphology, vegetation cover, soil moisture and drainage pattern. Furthermore, recent photographs can be compared with historical imagery to assess landslide conditions over different periods of time and allow the progressive development to be examined. Characteristics of mass movements that can be monitored by sequential photographs are, for example, the areal extent of the landslide body, regression rate of the head scar, displacement velocity, surface topography, succession of vegetation and soil moisture conditions. Accurate quantification of change requires the application of rigorous photogrammetric techniques (Chandler 1989). Finally, aerial photography can be helpful in hazard mapping. The purpose of landslide hazard mapping is to analyse the susceptibility of the terrain to slope movements. Aerial photographs can be used to delimit terrain units and map the controlling factors affecting slope stability.

The aim of this paper is to give an overview of the ways in which aerial photographs and digital photogrammetric techniques can be used in the monitoring stage of landslide assessments. Particular attention will be paid to the quality of data derived from aerial photographs of differing type, when using the different techniques available. The various approaches can be roughly divided into three categories: those based on simple aerial photograph interpretations (APIs), those involving the extraction of digital elevation models (DEMs), and those based on the creation of orthophotos. The underlying techniques will be described and illustrated with some results that were obtained from a case study focusing on the Mam Tor landslide (Derbyshire, UK).

The study area: Mam Tor

The landslide of Mam Tor is situated on the eastern flank of this 517 m high hill, at the head of the Hope Valley, Derbyshire, UK [SK135835]. The former main road between Sheffield and Manchester (A625) was constructed across the slide, but abandoned in 1979 as a consequence of continuous damage caused by the moving ground mass (Fig. 1). The slope consists of predominantly sandstone sequences (Mam Tor Beds) overlying predominantly shale units (Edale Shales). The layers dip slightly inwards of the slope. From scarp to toe, the landslide measures c. 1000 m, and elevation varies from 510 to 230 m. The mean slope of the slipped mass is 12° and the maximum thickness 30–40 m (Skempton et al. 1989).

The initial rotational failure has been dated back to 3600 BP (Skempton et al. 1989). While advancing downslope the mass broke into a complex of blocks and slices. Disintegration of the front slices created a debris mass, which slid further down.

The unstable transition zone, overlying the steepest part of the basal shear, is the most active part, moving on average 0.35 m a^{-1} over the last century (Rutter et al. 2003). There is evidence that the movements are not continuous but accelerate during wet winters, when rainfall exceeds certain limits; that is, more than 250 mm rain in a single month and over 750 mm in the preceding 6 months (Waltham & Dixon 2000).

There are several information sources available that quantify displacements that have taken place over the last century. Notes about regular disturbance and repairs of the road, from 1907 until the final closure in 1979, are kept by Derbyshire and stability analysis was carried out (Skempton et al. 1989). Since closure of the road, temporary monitoring schemes were set up by Sheffield University (1981–1983; Al-Dabbagh & Cripps 1987), Nottingham Trent University (1990–1998; Waltham & Dixon 2000) and Manchester University (since 1996; Rutter et al. 2003).

Aerial photographs

A conventional photo search for aerial photography of Mam Tor revealed that there are numerous image epochs available, both oblique and vertical, from 1948 until the present. Vertical imagery from eight epochs was acquired and processed (Table 1). The images are of varying quality and scales, and can be used in an assessment of the potential of the various techniques applied to a range of commonly available material.

Photogrammetric data processing was achieved using the IMAGINE OrthoBASE Pro 8.6 software package (ERDAS LLC 1991–2002). During photogrammetric processing the relationship between photo co-ordinates and the Ordnance Survey national grid co-ordinate system was

Fig. 1. Damaged road section at Mam Tor.

Table 1. *Characteristics of the acquired image epochs of Mam Tor*

Date	Source	Scale	Scan resolution (μm)	Ground resolution (m)	Media
1953	NMR*	1/10 700	42	0.45	Scanned contact prints
1971	NMR	1/6 400	42	0.27	Scanned contact prints
1973	CUCAP†	1/4 300	15	0.065	Scanned diapositives
1973	CUCAP	Oblique	15	–	Scanned diapositives
1984	ADAS‡	1/27 200	15	0.41	Scanned diapositives
1990	CUCAP	1/12 000	15	0.18	Scanned diapositives
1995	CUCAP	1/16 400	15	0.25	Scanned colour negatives
1999	Infoterra	1/12 200	21	0.26	Scanned colour negatives

*National Monuments Record.
†Cambridge University Collection of Air Photos.
‡Agricultural Development and Advisory Service.

established. An independent module performing a self-calibrating bundle adjustment was used for estimating the camera's interior parameters (camera constants), if the original calibration certificate was unavailable (as described by Chandler & Clark 1992). Ground control was collected by means of a differential global positioning system (GPS) survey.

Aerial photograph interpretation (API)

Photo-interpretation involves the systematic examination of photographic images for the purpose of identifying objects and judging their significance (Colwell 1960). Although aerial photographs can be interpreted with a specific theme in mind, interpretation relies on using the same basic characteristics of the surface: tone, texture, pattern, shape, context and scale, which were created by reflection of natural electromagnetic light energy from the objects that make up the scene and their arrangement. The use of these qualitative attributes is very much a matter of experience and personal bias (Drury 1987).

The quality of an API is affected by several factors, which can be separated into four main categories: photographic parameters, natural factors, equipment and analysis techniques, and the qualification of the interpreter (Rib & Liang 1978).

Photographic parameters

The effects of the different photographic parameters on landslip detection have been described by Norman et al. (1975) and Soeters & Van Westen 1996). Natural colour and panchromatic (black-and-white) films are the most widely available film types. Colour film is especially valuable for outlining differences in soil conditions, drainage and vegetation. Colour IR films are most suitable for detecting landslides, mainly because of the capability of identifying the presence of water and thus showing the vigour of vegetation cover. Panchromatic films, on the other hand, provide a better image resolution (Lo 1976) and are generally less expensive. Most historical imagery is of this form, although resolution tends to degrade with photo age, as a result of developments in photographic emulsion that have subsequently occurred.

Aerial photographs for mapping purposes are typically vertical, with 60% overlap between successive frames to provide stereoscopic coverage. Stereoscopic viewing is important, as landslide features are most frequently recognized by their morphology. Vertical exaggeration, when viewing stereoscopically, can be enhanced if a super wide angle lens is used during photo acquisition. The lower flying height increases the base/distance ratio. However, this may create problems of 'dead ground' on far side of hills and in narrow valleys. Oblique photographs sometimes can provide a more unobstructed view of steep slopes and cliffs (Rib & Liang 1978), and give a more familiar perspective for the less experienced interpreter (Chandler 1989).

The most suitable scale is inevitably a compromise. Large-scale photographs provide a high level of detail, but may require many frames to cover the study area. Small-scale photography provides less detail, but allows a better interpretation of the overall context. Quoted optimum scales for site studies are in the range between 1/5000 and 1/15 000 (Norman et al. 1975; Mantovani et al. 1996; Soeters & Van Westen 1996).

The time of the day when photographs are taken determines the length of shadows. In general, photographs taken when the sun is high and shadows on the hillsides and slopes are minimal are best for interpretation. However, in areas of low topography, the relief will be enhanced by long shadows. The time of the year influences the effects of soil moisture and vegetation (Norman et al. 1975; Soeters & Van Westen 1996). The quality of photographs depends on the various processes the images go through. Norman et al. (1975) used the following criteria for assessing photo quality: sharpness, over- or under-exposure, cloud cover, shadow and print quality.

In spite of recent developments in the field of airborne digital sensors (Eckhardt et al. 2000), the most common way of obtaining digital imagery is by scanning the original film. Photogrammetric scanners have a high geometric resolution, but radiometric performance may be rather poor (Baltsavias 1999). Modern software packages allow digital images to be easily adjusted to the needs of the user, for example, zooming in on particular areas or enhancing the contrast.

Natural factors

Photo-interpretation is also influenced by natural factors. Steep slopes, forest canopy and shadows may hide certain surface features. Optimal conditions for detecting anomalies in vegetation may be expected in either the very early or very late stages of the growing season. Differences in drainage conditions are most pronounced shortly after the start of the wet season or shortly after the snow-melt period in spring (Soeters & Van Westen 1996). Weather conditions have an important influence on photo quality: clouds and snow cover may obscure the ground surface, haze decreases contrast, and solar angle influences shadowing (Rib & Liang 1978).

Qualification of interpreter

The quality of API is also influenced by the capability of the human interpreter, particularly experience in photo-interpretation, and knowledge of the phenomena and processes being studied. Various researchers have shown the large subjective element in photo-interpretation by comparing maps of the same landslide area created by different interpreters (Van Westen 1993; Carrara *et al.* 1995). Identification of the exact positions of morphological features can be difficult, especially delineation of the boundaries. Moreover, different classes may be assigned to a specific feature as a result of different interpretation (Chandler 1989). Obviously, different mapping legends will lead to very different maps (Van Westen *et al.* 1999).

Diagnostic features

The interpretation of landslides from aerial photographs is mainly based on features indirectly related to slope movements, such as characteristic morphology, anomalous vegetation and drainage conditions, or disturbed infrastructure. Soeters & Van Westen (1996) have provided an extensive overview of terrain features associated with landslides and their characterization on aerial photographs. Based on these diagnostic features, statements can be made on the type of movement, degree of activity and depth of movement (Mantovani *et al.* 1996).

Geomorphological mapping

A useful tool in presenting photo-interpreted information is a geomorphological map. Geomorphological maps are transmitters of information about the form, origin, age and distribution of landforms together with their formative processes, rock type and surface materials (Brunsden *et al.* 1975). They are not only a way of presenting data, but also the result of a method of research, revealing associations of landforms, which is essential for understanding of both individual landforms and landscapes (De Graaff *et al.* 1987). Parise (2003) pointed out the importance of large-scale geomorphological mapping, and especially its repetition in time, in the study of active mass movements. A combination of detailed multi-temporal mapping of surface features, indirect indicators of deformation and displacements may result in better understanding of the landslide and its zonation in different elements, characterized by different styles of deformation.

Geomorphological maps can emphasize different aspects of landforms: form, origin, age and relations. Dependent on the purpose of the map, these aspects can be depicted by coloured area symbols, patterns and line symbols (Van Zuidam 1985). A great variety in geomorphological mapping systems have been designed in the course of time. Savigear 1965) developed a purely morphological legend, aiming to describe the form of slopes, without reference to the origin. The ITC system (Van Zuidam 1985), designed for multipurpose use at all scales, distinguishes the highest level on the basis of morphogenesis. The importance of geomorphological mapping has also been recognized by engineering geologists, judging from the proposed legend for engineering geological maps by the Geological Society Engineering Group Working Party, which contains a large number of symbols for geomorphological features (Anonymous 1972).

API is a valuable tool in geomorphological mapping, although field work remains necessary for checking the photo-interpretation and mapping of small features (Hayden 1986). Accurate definition and coding of geomorphological boundaries by rigorous photogrammetric techniques combines the benefits of geomorphological interpretation with positional relevance (Chandler & Brunsden 1995). Photogrammetric measurements allow quantitative comparison between photo-interpreted maps from different periods (Chandler & Cooper 1989). Multi-temporal geomorphological maps from sequential aerial photographs have been used to document the evolution of the Black Ven landslide, UK (Chandler & Brunsden 1995), and the Tessina landslide, Italy (Van Westen & Getahun 2003). Kalaugher *et al.* (1987) used oblique aerial photographs to identify geomorphological processes on sea cliffs in East Devon, UK.

Figure 2 displays a simple geomorphological map of the Mam Tor landslide, created by photo-interpretation of the 1990 images. The mapped features were placed in the exact spatial context by using photogrammetry. Contour lines were obtained from a DEM, extracted from the same photographs.

Digital elevation models (DEMs)

Digital photogrammetric techniques have the capability of automatically deriving very high-resolution DEMs from stereo photographs, providing a detailed representation of the surface topography (Chandler 1999). Photogrammetry is based on the concept of collinearity, whereby a point on the object, centre of lens and resultant image point lie on a single line in 3D space. Based on this principle, 3D co-ordinates representing the object can be extracted from a stereopair of

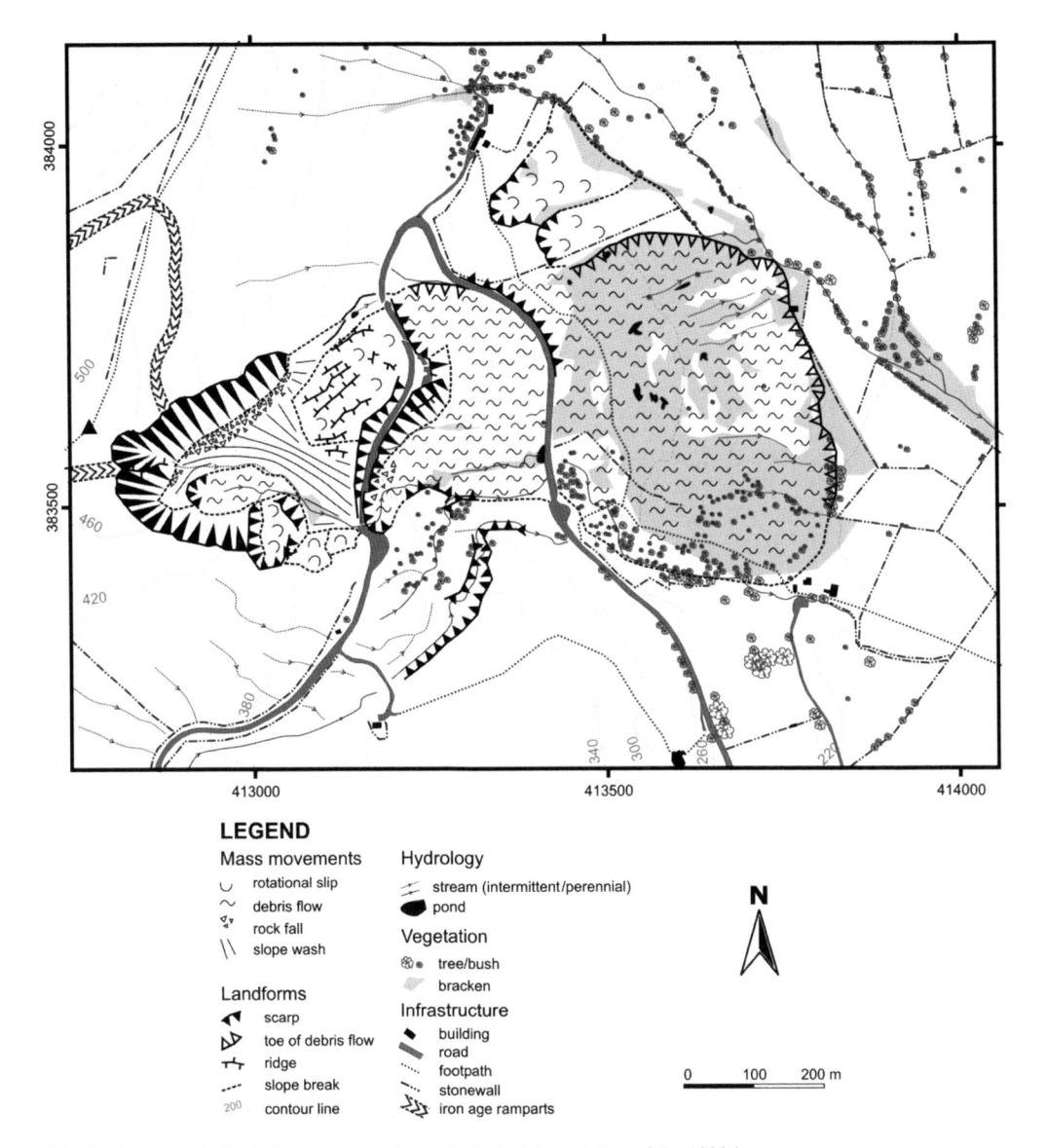

Fig. 2. A geomorphological map, created through photo-interpretation of the 1990 images.

photographs, provided that the inner geometry (interior orientation) and the position and orientation of the camera at the moment of exposure (exterior orientation) are known. The exterior orientation parameters of all frames in a block can be simultaneously estimated in a bundle block adjustment, with the help of ground control points of which both ground and image co-ordinates are known (Wolf & Dewitt 2000).

Once the relationship between the photographs and ground surface has been established, co-ordinates can be extracted from anywhere on the site, and used to create a DEM. A significant recent development is the automation of this process. Automatic generation of DEMs from a stereomodel comprises three tasks: image matching, surface fitting and quality control (Schenk 1996). The process of image matching involves the identification of conjugate points in the overlap portion of the images. A commonly applied matching strategy is area-based cross-correlation, in which small image patches are compared according to their grey-level distribution. Perfect matches will never occur in reality because of noise, small differences in illumination, and small geometric distortions. Because a regular gridded DEM is often required, surface fitting needs to be performed. This procedure comprises the interpolation of intermediate points, as the points obtained by image matching do not represent the entire surface.

The quality of a DEM is a function of the accuracy, reliability and precision of the photogrammetric measurements and the block bundle adjustment itself (Butler et al. 1998). As defined by Cooper & Cross 1988), precision is related to random errors, inherent in the measurement process. The bundle adjustment procedure is capable of propagating stochastic properties through the model, providing an estimation of precision. Reliability can be related to gross errors. Fortunately, gross errors are normally easy to detect and eradicate because of their size. Accuracy is related to the presence of undetected systematic errors, which are more difficult to isolate and generally provide a limiting constraint on the quality of the derived data. The mean and standard deviation of the discrepancies with independent check points provide a measure of DEM accuracy (Butler et al. 1998).

Controls on DEM quality

As pointed out by Fryer et al. (1994) and Lane et al. (2000), the ease with which terrain data may be generated using digital photogrammetric techniques has focused attention more on analysis and interpretation of the acquired results than on issues of data quality. In addition to the conventional controls on photogrammetry, the automated algorithms in digital processing have important influences on the quality of results.

The precision that can be achieved by photogrammetric measurements is mainly dependent on the quality of the source data (i.e. the aerial photographs). Photographic resolution is a function of the optical quality of an image, and influenced by the resolving power of the film and camera lens, image motion during exposure, atmospheric conditions and the conditions of film processing. The effects of scale and resolution can be combined in terms of ground resolution distance, which determines the level of detail that is visible on the photographs (Lillesand & Kiefer 1994). When using digital images, scan resolution and the quality of the scanner (geometric and radiometric) are important controls. To preserve an original film resolution of 30–60 lines per mm, a scanned pixel size of 6–12 μm would be needed. For many practical applications, such as DEM generation, good results can be achieved with 25–30 μm resolution (Baltsavias 1999). The height precision of photogrammetric measurements is also dependent on the geometry provided by stereo-photographs. A strong convergence (high base/distance ratio), and large relief displacement, gives rise to highly precise object coordinates (Wolf & Dewitt 2000). According to Fryer et al. (1994), the best vertical precision that can be expected using standard mapping configurations is about 1–3 parts per 10 000 of flying height.

Systematic errors are always inherent in the stereo-model, arising from a variety of sources including lens distortion, atmospheric effects, film deformation, scan distortions, and inaccurate or poorly distributed control points, or result from errors during the image matching procedure (Chandler 1989; Buckley 2003). If camera calibration parameters are not available, which is sometimes the case when using archival imagery, these can be estimated in a self-calibrating bundle-adjustment. However, accounting for all systematic effects is difficult, because many systematic errors cannot be modelled explicitly, and there is usually high correlation between the modelling parameters (Granshaw 1980).

Control points should be evenly distributed over the images to gain a strong geometry. Ideal locations tie frames together and surround the volume of interest. A minimum of two planimetric and three height points is needed to define a datum, but more control points are desirable as redundancy provides appropriate checks (Wolf & Dewitt 2000). Automated image matching is affected by surface texture and geometric distortion caused by different viewing angle. These controls upon automated generation of elevation data are of special relevance to complex terrain surfaces (Lane et al. 2000). If there is insufficient texture, the software is unable to match two points successfully and an interpolated estimate may be created. Surface roughness has a positive effect on texture, and consequently on matching. However, this effect may be countered by the increasing differences in the viewing of areas, which thus reduce the level of correlation between the images. In addition, interpolation will be least effective in areas of great roughness. DEM collection parameters can be optimized, but these control individual matches rather than affecting the resulting surface accuracy (Lane et al. 2000). Surface quality is also affected by its point density. An increase in grid spacing will smooth the topography; the minimum grid spacing is, however, bound by the object space pixel dimension (Lane et al. 2000).

Gross errors are genuine mistakes or blunders that arise during photogrammetric measurement (Cooper & Cross 1988). They can be detected by increasing the redundancy of measurements

Fig. 3. A 3D view of Mam Tor, created by draping an orthophoto over a DEM, obtained from the 1990 images.

(Hottier 1976), which gives rise to datasets that are 'internally reliable' (Cooper & Cross 1988). Gross error sources that commonly affect the determination of exterior orientation include misidentified or mistyped control points. Fortunately, these errors give rise to large residuals at the block bundle adjustment stage and, if data redundancy is high, are normally readily identifiable. In the context of DEM generation a measure of internal reliability can be derived by comparing two DEMs of the same area but extracted from different stereopairs (Butler *et al.* 1998). DEMs can provide a perspective view of the area from any specified position. In combination with an orthophoto realistic views can be created, which are useful in analyses (Fig. 3).

Multi-temporal DEMs

DEMs not only provide a useful tool to enhance data analysis by perspective viewing, but also contain quantitative topographical data. Subtracting a DEM of one epoch from an earlier DEM creates a grid surface representing the change of form over the period. This surface of change, or 'DEM-of-difference', quantifies the effects of geomorphological processes. Areas experiencing removal of material will be indicated by depressions, whereas those receiving material are indicated by peaks. Caution should be taken, as areas exhibiting no change are not necessarily inactive regions; they can represent areas where input of material has equalled output during the time interval (Chandler & Brunsden 1995).

A photogrammetric method was developed in the late 1980s that was able to derive quantitative spatial information from historical aerial photographs (Chandler & Cooper 1989; Chandler & Clark 1992). This offered potential to unlock the photographic archive for obtaining quantitative terrain data, covering a time span of more than 50 years. Developments in digital photogrammetry allowed the acquisition of much denser DEMs (Brunsden & Chandler 1996). Following the efforts by Brunsden & Chandler in their studies on the Black Ven landslide (Chandler & Brunsden 1995; Brunsden & Chandler 1996), the use of multi-temporal DEMs has become more widely adopted in landslide research (e.g. Cheng 2000; Adams & Chandler 2002; Gentili *et al.* 2002; Van Westen & Getahun 2003; Ager *et al.* 2004; Bitelli *et al.* 2004).

Figure 4 shows a 'DEM-of-difference' of the central part of the Mam Tor landslide, created by subtracting DEMs of 1990 and 1973. The change in elevation is draped over a standard DEM for a better interpretation. Dark areas represent a lowering in elevation, whereas bright areas depict an increase in height.

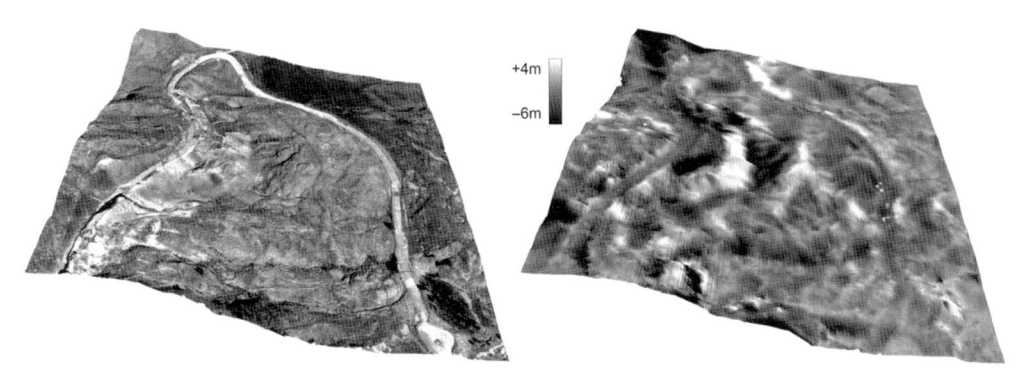

Fig. 4. On the right a 'DEM-of-difference' of the central part of the Mam Tor landslide, created by subtracting DEMs of 1990 and 1973; the elevation change is draped over a standard DEM for better interpretation. Left image is an orthophoto of the same area.

Orthophotos

Orthophotos combine the image characteristics of a photograph with the geometric qualities of a map. Unlike normal aerial photographs, relief displacement is removed so that all ground features are displayed in their true ground position. This allows the direct measurement of distances, areas, angles and positions. Orthophotos are created through differential rectification, which eliminates image displacements caused by photographic tilt and terrain relief. The rectification procedure requires a photograph with known orientation parameters and a DEM. The collinearity concept can be used to determine the corresponding photo co-ordinates of all DEM points (Wolf & Dewitt 2000). The geometric quality of orthophotos is dependent on its source data; that is, the original photographs, the functional model that relates photo to ground co-ordinates, and the quality of the DEM (Krupnik 2003). Hence, the quality controls are similar to those for DEMs. The minimum grid spacing is bounded by the resolution of the original photographs, as a higher resolution would imply over-sampling.

The combination of interpretative capabilities of the original photographs with the positional relevance of a map makes orthophotos particularly valuable for Earth scientists (Chandler 2001). Several researchers have shown the use of multi-temporal orthophotos to map horizontal surface displacements. Powers *et al.* (1996) measured movements of the Slumgullion landslide by determining the displacement of surface features, such as trees and rocks, between two orthophotos acquired at different times. Gentili *et al.* (2002) measured the displacements of building corners on the Corniglio landslide, Italy, from orthophotos.

Automatic extraction of displacement vectors

Digital techniques allow the potential of automatic measurement of objects on images. Kääb & Vollmer (2000) used an area-based cross-correlation algorithm to automatically map the velocity field of a rock glacier in the Swiss Alps from multi-temporal, orthorectified photographs. The high density and accuracy of the velocity data provided by the technique make it possible to extract meaningful strain-rate information. In related studies their approach was successfully applied to other types of superficial movements,

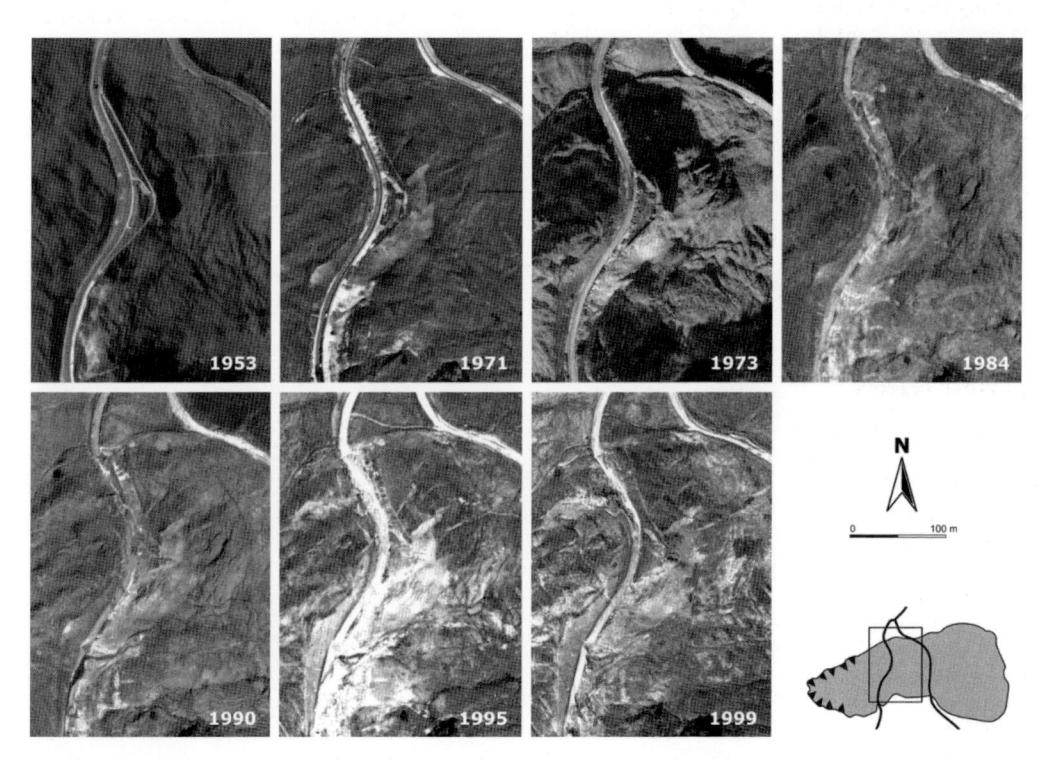

Fig. 5. A sequence of orthophotos obtained from different epochs, showing the progressively changing terrain surface in the central part of the Mam Tor landslide.

such as those of glaciers and rockslides (Kääb 2002).

Nevertheless, some researchers have indicated that the accuracy of displacement vectors is limited by the relatively poor DEMs used in the orthorectification process. Casson *et al.* (2003) developed an alternative approach, allowing the creation of better DEMs than by most commercial software packages. Their approach corrects for topographic distortion by estimating local slopes, to improve the image matching performance. Kaufmann & Ladstädter (2002, 2004) used the concept of pseudo-orthophotos, which, in combination with a rough DEM, still contain the same stereo-information as the original photos, thus allowing strict 3D reconstruction. Pseudo-orthophotos are better suited for matching than original photo scans, as perspective distortions have been removed to a great extent. An additional advantage of this approach is that the obtained vectors are 3D, rather than horizontal.

Orthophotos were created from all photographic epochs of Mam Tor (Fig. 5). Some clear features can be identified throughout the entire series, and can be manually measured to obtain displacement vectors. The mean displacement of the landslide over the period from 1953 to 1990 is 0.32 m a^{-1}, varying from 0.11 m a^{-1} at the toe to 0.81 m a^{-1} in the central, most active part. These values are of comparable size to movement rates found by Rutter *et al.* (2003), $0.04-0.35$ m a^{-1} over the last century and up to 0.50 m a^{-1} in recent years. Automatic measurements can provide much denser displacement vector fields, allowing more detailed comparison with the alternative surveys. However, in the case of the Mam Tor images, differences in image quality, illumination conditions and vegetation cover hampered this procedure.

Further research

The applied techniques have the ability to provide both qualitative and quantitative information on landslide movements from aerial photographs. Continuing research aims at a thorough analysis of their extra value to landslide investigations in terms of photographs available, techniques available and the usefulness of the products (type and quality) in the analysis of landslide mechanisms. Further research will focus on how the different datasets can be integrated to allow a comprehensive analysis of the evolution of a landslide. Eventually, the techniques that have been developed at Mam Tor will be validated in a case study in South Wales, which should confirm their practical applicability.

Conclusion

This review has demonstrated the valuable information, qualitative as well as quantitative, that is captured by aerial photographs. It has also been shown that these data can be extracted relatively easily, using established techniques, and are well suited for use in the monitoring stage of landslide assessments. API reveals qualitative information on surface characteristics, which is very helpful in detecting landslide features and in formulating statements about the mechanisms involved. Photogrammetrically derived products can provide quantification of the inferred processes: DEMs provide detailed data on surface topography, whereas orthophotos can be used to measure surface displacements. Automated methods allow very dense and accurate data to be collected. In this way, the photographic archive can provide invaluable data on landslide evolution, thus leading to a better understanding of landslide mechanisms.

References

ADAMS, J. C. & CHANDLER, J. H. 2002. Evaluation of Lidar and medium scale photogrammetry for detecting soft-cliff coastal change. *Photogrammetric Record*, **17** 405–418.

AGER, G., MARSH, S., HOBBS, P., CHILES, R., HAYNES, M., THURSTON, N. & PRIDE, R. 2004. Towards automated monitoring of ground instability along pipelines. *In*: *Terrain and Geohazard Challenges Facing Onshore Oil and Gas Pipelines*. Thomas Telford Ltd., London.

AL-DABBAGH, T. H. & CRIPPS, J. C. 1987. Data sources for planning: geomorphological mapping of landslides in north-east Derbyshire. *In*: CULSHAW, M. G., BELL, F. G., CRIPPS, I. C. & O'HARA, M. (eds) *Planning and Engineering Geology*. Engineering Geology Special Publication 4, Geological Society, London, 101–114.

Anonymous 1972. The preparation of maps and plans in terms of engineering geology. Report by the Geological Society Engineering Group Working Party. *Quarterly Journal of Engineering Geology*, **5**, 297–367.

BALTSAVIAS, E. P. 1999. On the performance of photogrammetric scanners. *In*: FRITSCH, D. & SPILLER, R. (eds) *Photogrammetric Week '99*, Wichmann Verlag, Heidelberg, Stuttgart, 155–173.

BITELLI, G., DUBBINI, M. & ZANUTTA, A. 2004. Terrestrial laser scanning and digital photogrammetry techniques to monitor landslide bodies. *International Archives of Photogrammetry and Remote Sensing*, **XXXV**(B5), 246–251.

BRUNSDEN, D. & CHANDLER, J. H. 1996. Development of an episodic landform change model based upon the Black Ven mudslide, 1946–1995. *In*: ANDERSON, M. G. & BROOKS, S. M. (eds) *Advances in Hillslope Processes*. Wiley, Chichester, 869–896.

BRUNSDEN, D., DOORNKAMP, J. C., FOOKES, P. G., JONES, D. K. C. & KELLY, J. M. H. 1975. Large scale geomorphological mapping and highway engineering design. *Quarterly Journal of Engineering Geology*, **8**, 227–530.

BUCKLEY, S. J. 2003. A geomatics data fusion technique for change monitoring PhD thesis, University of Newcastle upon Tyne.

BUTLER, J. B., LANE, S. N. & CHANDLER, J. H. 1998. Assessment of DEM quality for characterizing surface roughness using close range digital photogrammetry. *Photogrammetric Record*, **16**, 271–291.

CARRARA, A., CARDINALI, M., GUZZETTI, F. & REICHENBACH, P. 1995. GIS-based techniques for mapping landslide hazard. *In*: CARRARA, A. & GUZZETTI, F. (eds) *Geographical Information Systems in Assessing Natural Hazards*. Kluwer, Dordrecht, 135–176.

CASSON, B., DELACOURT, C., BARATOUX, D. & ALLEMAND, P. 2003. Seventeen years of the 'La Clapière' landslide evolution analysed from orthorectified aerial photographs. *Engineering Geology*, **68**, 123–139.

CHANDLER, J. H. 1989. The acquisition of spatial data from archival photographs and their application to geomorphology. PhD thesis, The City University, London.

CHANDLER, J. H. 1999. Effective application of automated digital photogrammetry for geomorphological research. *Earth Surface Processes and Landforms*, **24**, 51–63.

CHANDLER, J. H. 2001. Terrain measurement using automated digital photogrammetry. *In*: GRIFFITHS, J. S. (ed.) *Land Surface Evaluation for Engineering Practice*. Engineering Geology Special Publication 18, Geological Society, London, 13–18.

CHANDLER, J. H. & BRUNSDEN, D. 1995. Steady state behaviour of the Black Ven mudslide: the application of archival analytical photogrammetry to studies of landform change. *Earth Surface Processes and Landforms*, **20**, 255–275.

CHANDLER, J. H. & CLARK, J. S. 1992. The archival photogrammetric technique: further application and development. *Photogrammetric Record*, **14**, 241–247.

CHANDLER, J. H. & COOPER, M. A. R. 1989. The extraction of positional data from historical photographs and their application to geomorphology. *Photogrammetric Record*, **13**, 69–78.

CHENG, H.-H. 2000. Photogrammetric digital data processing of Tsau-Lin big landslide. *In*: *Proceedings 21st Asian Conference on Remote Sensing*, Taipei, Taiwan, http://www.gisdevelopment.net/aars/acrs/2000/ps1/ps102.asp.

COLWELL, R. N. 1960. *Manual of Photographic Interpretation*. American Society of Photogrammetry, Washington, DC.

COOPER, M. A. R. & CROSS, P. A. 1988. Statistical concepts and their application in photogrammetry and surveying. *Photogrammetric Record*, **12**, 637–663.

CROZIER, M. J. 1984. Field assessment of slope instability. *In*: BRUNSDEN, D. & PRIOR, D. B. (eds) *Slope Instability*. Wiley, Chichester, 103–142.

DE GRAAFF, L. W. S., DE JONG, M. G. G., RUPKE, J. & VERHOFSTAD, J. 1987. A geomorphological mapping system at scale 1:10 000 for mountainous areas. *Zeitschrift für Geomorphologie*, **31**, 229–242.

DRURY, S. A. 1987. *Image Interpretation in Geology*. Allen & Unwin, London.

ECKHARDT, A., BRAUNECKER, B. & SANDAU, R. 2000. Performance of the imaging system in the LH Systems ADS40 airborne digital sensor. *International Archives of Photogrammetry and Remote Sensing*, **XXXIII**(B), 104–109.

FRYER, J. G., CHANDLER, J. H. & COOPER, M. A. R. 1994. On the accuracy of heighting from aerial photographs and maps: implications to process modellers. *Earth Surface Processes and Landforms*, **19**, 577–583.

GENTILI, G., GIUSTI, E. & PIZZAFERRI, G. 2002. Photogrammetric techniques for the investigation of the Corniglio landslide. *In*: ALLISON, R. J. (ed.) *Applied Geomorphology*. Wiley, Chichester, 39–48.

GRANSHAW, S. I. 1980. Bundle adjustment methods in engineering photogrammetry. *Photogrammetric Record*, **10**, 181–207.

HAYDEN, R. S. 1986. Geomorphological mapping. *In*: SHORT, S. & BLAIR, R. W. (eds) *Geomorphology from Space*. NASA. World Wide Web Address: http://daac.gsfc.nasa.gov/geomorphology/GEO_11/GEO_CHAPTER_11.HTML (accessed 18 April 2006).

HOTTIER, P. 1976. Accuracy of close range analytical solutions. *Photogrammetric Engineering & Remote Sensing*, **42**, 345–375.

KÄÄB, A. 2002. Monitoring high-mountain terrain deformation from repeated air- and spaceborne optical data: examples using digital aerial imagery and aster data. *ISPRS Journal of Photogrammetry & Remote Sensing*, **57**, 39–52.

KÄÄB, A. & VOLLMER, M. 2000. Surface geometry, thickness changes and flow fields on creeping mountain permafrost: automatic extraction by digital image analysis. *Permafrost and Periglacial Processes*, **11**, 315–326.

KALAUGHER, P. G., GRAINGER, P. & HODGSON, R. L. P. 1987. Cliff stability evaluation using geomorphological maps based upon oblique aerial photographs. *In*: CULSHAW, M. G., BELL, F. G., CRIPPS, I. C. & O'HARA, M. (eds) *Planning and Engineering Geology*. Engineering Geology Special Publication 4, Geological Society, London, 163–170.

KAUFMANN, V. & LADSTÄDTER, R. 2002. Monitoring of active rock glaciers by means of digital photogrammetry. *International Archives of Photogrammetry and Remote Sensing*, **XXXIV**(3B), 108–111.

KAUFMANN, V. & LADSTÄDTER, R. 2004. Documentation of the movement of the hinteres langtalkar rock glacier. *International Archives of Photogrammetry and Remote Sensing*, **XXXV**(B7), 893–898.

KRUPNIK, A. 2003. Accuracy prediction for ortho-image generation. *Photogrammetric Record*, **18**, 41–58.

LANE, S. N., JAMES, T. D. & CROWELL, M. D. 2000. Application of digital photogrammetry to complex topography for geomorphological research. *Photogrammetric Record*, **16**, 793–821.

LILLESAND, T. M. & KIEFER, R. W. 1994. *Remote Sensing and Image Interpretation*. Wiley, New York.

LO, C. P. 1976. *Geographical Applications of Aerial Photography*. David & Charles, Newton Abbot.

MANTOVANI, F., SOETERS, R. & VAN WESTEN, C. J. 1996. Remote sensing techniques for landslide studies and hazard zonation in Europe. *Geomorphology*, **15**, 213–225.

NORMAN, J. W., LEIBOWITZ, T. H. & FOOKES, P. G. 1975. Factors affecting the detection of slope instability with air photographs in an area near Sevenoaks, Kent. *Quarterly Journal of Engineering Geology*, **8**, 159–176.

PARISE, M. 2003. Observation of surface features on an active landslide, and implications for understanding its history of movement. *Natural Hazards and Earth System Sciences*, **3**, 569–580.

POWERS, P. S., CHIARLE, M. & SAVAGE, W. Z. 1996. A digital photogrammetric method for measuring horizontal surficial movements on the Slumgullion earthflow, Hinsdale Co., Colorado. *Computers & Geosciences*, **22**, 651–663.

RIB, H. T. & LIANG, T. 1978. Recognition and identification. *In*: SCHUSTER, R. L. & KRIZEK, R. J. (eds) *Landslides Analysis and Control*. National Academy of Sciences, Washington, DC, 34–80.

RUTTER, E. H., ARKWRIGHT, J. C., HOLLOWAY, R. F. & WAGHORN, D. 2003. Strains and displacements in the Mam Tor landslip, Derbyshire, England. *Journal of the Geological Society, London*, **160**, 735–744.

SAVIGEAR, R. A. G. 1965. A technique of morphological mapping. *Annals of the Association of American Geographers*, **55**, 514–538.

SCHENK, A. F. 1996. Automatic generation of DEMs. *In*: CARY, T., JENSEN, J. & NYQUIST, M. (eds) *Digital Photogrammetry. An Addendum to the Manual of Photogrammetry*. American Society for Photogrammetry and Remote Sensing, Bethesda, MD, 247–250.

SKEMPTON, A. W., LEADBEATER, A. D. & CHANDLER, R. J. 1989. The Mam Tor landslide, North Derbyshire. *Philosophical Transactions of the Royal Society of London, Series A*, **329**, 503–547.

SOETERS, R. & VAN WESTEN, C. J. 1996. Slope instability recognition, analysis, and zonation. *In*: TURNER, A. K. & SCHUSTER, R. L. (eds) *Landslides Investigation and Mitigation*. National Academy Press, Washington, DC, 129–177.

VAN WESTEN, C. J. 1993. *GISSIZ Training Package for Geographic Information Systems in Slope Instability Zonation. Part 1: Theory*. ITC, Enschede.

VAN WESTEN, C. J. & GETAHUN, F. L. 2003. Analyzing the evolution of the Tessina landslide using aerial photographs and digital elevation models. *Geomorphology*, **54**, 77–89.

VAN WESTEN, C. J., SEIJMONSBERGEN, A. C. & MANTOVANI, F. 1999. Comparing landslide hazard maps. *Natural Hazards*, **20**, 137–158.

VAN ZUIDAM, R.A. 1985. *Aerial Photo-Interpretation in Terrain Analysis and Geomorphological Mapping*. Smits, The Hague.

WALTHAM, A. C. & DIXON, N. 2000. Movement of the Mam Tor landslide, Derbyshire, UK. *Quarterly Journal of Engineering Geology and Hydrogeology*, **33**, 105–123.

WOLF, P. R. & DEWITT, B. A. 2000. *Elements of Photogrammetry, with Applications in GIS*. McGraw–Hill, Boston, MA.

Geomorphology and urban geology of Bukavu (R.D. Congo): interaction between slope instability and human settlement

P. TREFOIS[1], J. MOEYERSONS[1], J. LAVREAU[1], D. ALIMASI[2], I. BADRYIO[2], B. MITIMA[2], M. MUNDALA[2], D. O. MUNGANGA[2] & L. NAHIMANA[3]

[1]*Musée Royal de l'Afrique Centrale, Leuvensesteenweg 13, B-3080 Tervuren, Belgium*
(e-mail: philippe.trefois@africamuseum.be)

[2]*Centre Universitaire de Bukavu, B.P. 570, Bukavu, Democratic Republic of Congo*

[3]*Université du Burundi, B.P. 2700, Bujumbura, Burundi*

Abstract: The city of Bukavu, on the south coast of Lake Kivu in the Democratic Republic of Congo, suffers from frequent landsliding, which leads to continual damage and destruction of buildings, roads, waterworks and sewerage infrastructure. Thirty-one landslides identified on aerial photographs of 1954 and 1973 are located outside the Bukavu micro-rift and are related to actively incising rivers. Their origin is thought to be due to increases in hydrostatic pressure. Six landslides occur within the Bukavu micro-rift, four of which are much larger and wider than other slope failures in the study region. These large landslides are adjacent to, or crossed by, active faults. They fall far below the topographic threshold envelope, a relationship of slopes at the head of the slide and the surface area drained into it, indicating seismic or anthropomorphic triggering. The Bukavu landslides still reactivate occasionally. Deforestation, followed by a large increase in the population, have been indirect causes of the reactivated mass wasting processes. On the steep slopes in the south of the city the high density of newly built houses has led to reduced water infiltration and enhanced runoff, causing landslides and mudflows. Very high spatial resolution IKONOS satellite images have recently been used as base maps for geohazard assessments of Bukavu. A geographical information system (GIS) has been developed for Bukavu's planners: this shows the locations of buildings, roads and tracks, the river network, the water distribution system and the sewerage infrastructure, as well as areas of slope instability.

The city of Bukavu on the west bank of the Ruzizi outlet of Lake Kivu (Fig. 1) has always been affected by slow ground movements. Accelerated landsliding and sudden gully development also occur. In many districts, houses have to be rebuilt frequently and roads are generally in bad condition. The town is in a continual state of repair and rebuilding, with disruption to both the water supply infrastructure and the sewerage system. However, this has not discouraged people from settling in Bukavu (many seeking refuge from regional military conflicts), with the population increasing from 147 647 in 1977 to some 450 000 in 2002 (UNESCO 2002).

Slope instability and soil erosion in and around Bukavu have long been of concern to the local government. As early as 1945, the Belgian authority installed the so-called 'Mission anti-érosive' in Kivu. However, the subsequent deployment of soil conservative measures was not very successful in halting soil movements in Bukavu. First, the main action taken was the installation of intersection trenches for water and eroded soil, positioned at distances of tens of metres parallel to slope contours. By 1959 most hills in Bukavu were trenched in this way. Such trenches can stop soil wash, but have been proven in neighbouring Rwanda to actively contribute to mass wasting (Moeyersons 1989, 2003). By the 1980s it was realized that the main problem in Bukavu was mass wasting, rather than soil erosion (Lambert 1981). Furthermore, it took time to realize that mass wasting at Bukavu was driven not only hydrogeologically, but also tectonically–seismically. Ground investigations led to the realization that Bukavu is crossed by a north–south-trending micro-graben; moreover, that it lies at the intersection of tectonic lineaments trending SSE–NNW (Tanganyika trend) and SSW–NNE (Albertine trend) (Cahen 1954; Ilunga 1989; UNESCO 2002).

The debate concerning the origin of landsliding in Bukavu continues today (Ilunga 1978, 1991; Moeyerson *et al.* 2004). It was first thought that low-quality construction of house roofs and leakages in waterworks contributed to high soil water contents, leading to landsliding. To lower the impact of humans on the environment, the 1958 Mission d'Etalement de la Population redistributed the inhabitants from high population density areas

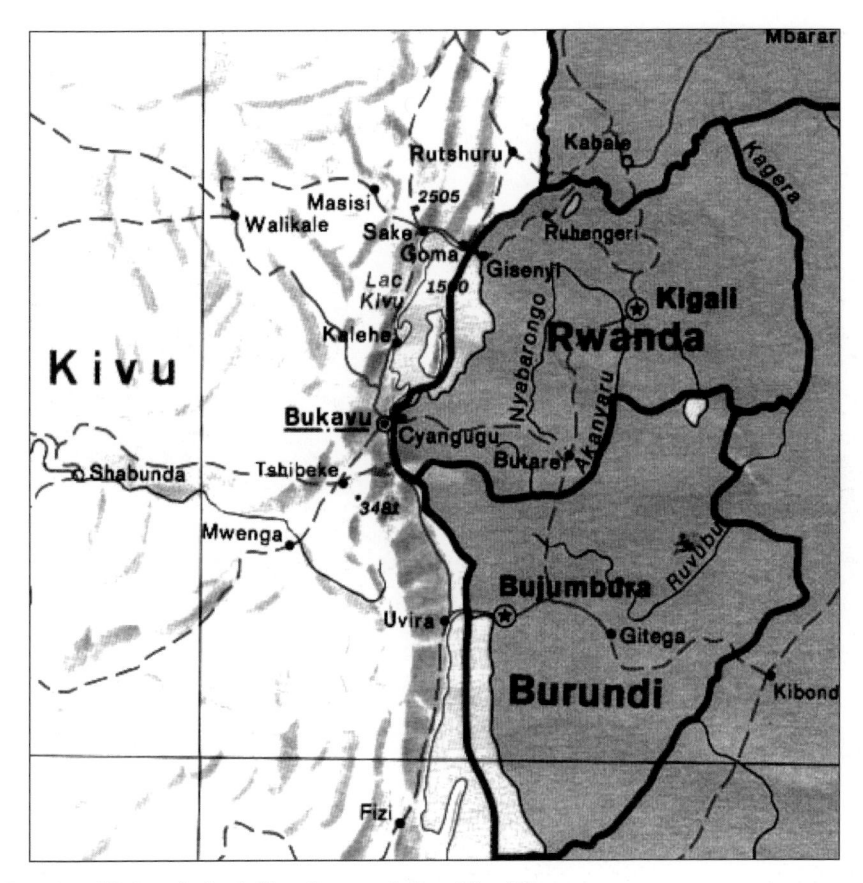

Fig. 1. Location of Bukavu in South Kivu, Democratic Republic of Congo.

to low-density districts. In the proceedings of the 'Erosion at Bukavu' conference (Ischebeck *et al.* 1984), there was still a tendency to consider even the largest mass movements to be the result of human mismanagement. However, geomorphological surveys of Bukavu have revealed an apparently active double fault step. More recently, a seismic origin for the larger landslides at Bukavu has been suggested by Munyololo *et al.* (1999).

Detailed geological mapping and geotechnical investigations have only recently started and synoptic geomorphological descriptions, as well as movement rate measurements, are still lacking. To redress this situation, the 2002 UNESCO project 'Géologie urbaine de Bukavu: interaction entre la stabilité du sol et la pression démographique', was initiated, based on the following.

(1) A geomorphological map of Bukavu, based on a stereoscopic study of panchromatic aerial photographs from 1954 and 1973. This included the mapping of active tectonic structures and landslides; the determination of spatial relationships between tectonic structures, such as fault steps and landslides; and extensive use of the existing geomorphological data to unravel the origin of the landslides. The first practical outcome of this study has been the identification of landslide mechanisms and triggering factors for dangerous sectors of the town.

(2) An updating of Bukavu city maps in geographical information system (GIS) format, based on recent very high spatial resolution remote sensing data (IKONOS) and a digital elevation model, as well as field surveys focused on geology, weathering facies, geomorphology, hydrology, slope instability, mud flows, degradation of infrastructure and pollution.

Geology and tectonics

In the Bukavu area, the folded and faulted Precambrian substrate is covered by thick Tertiary and Quaternary basalt traps, resulting mainly from fissural outflow. The oldest series, not present at Bukavu itself, predates the local rifting and has

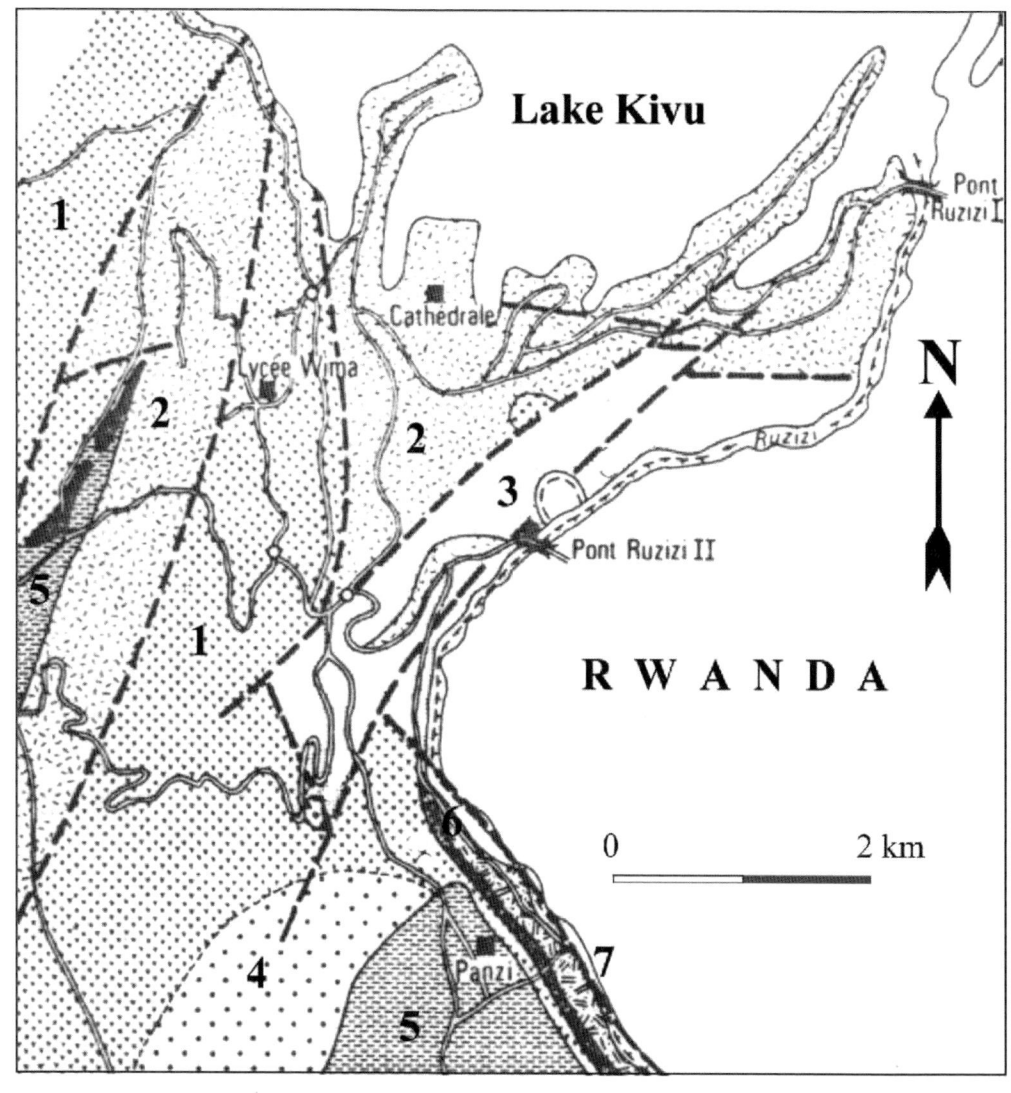

Fig. 2. Geological map of Bukavu, from Kampunzu *et al.* (1983). 1, Basalts of the upper series; 2–7, Mio-Pliocene alkaline lavas (2, basanite; 3, hawaiites and mugearites; 4, ankaratrite; 5, benmoreites; 6, Panzi pyroclastic series; 7, phonolites and trachytes). The dashed lines are normal faults.

been dated between 7 and 10 Ma (Pasteels *et al.* 1989). The middle and upper series are found at Bukavu. The middle series, of Mio-Pliocene age (Kampunzu *et al.* 1983; Pasteels *et al.* 1989), is intimately related to the rift faulting. The upper series dates from Pleistocene time and to the last century. The distribution of these deposits at Bukavu is shown in Figure 2.

Each lava series is believed to consist of numerous individual flows (UNESCO 2002). The complex geometry of the present lava layers is the result of successive rifting and eruptive episodes. Weathering and erosion, as well as normal faulting, occur between successive lava flow series, explaining the repeated occurrence of palaeorelief, contact metamorphism, smectite shrink–swell clay layers, probable shrink–swell vertisol palaeo-soils, and alluvial–colluvial clastic deposits.

In the absence of solid geological data, geomorphological evidence has often been used to locate active faults, believed to be the origin of the numerous escarpments crossing Bukavu (Fig. 3). However, there are doubts about geomorphological interpretations with regard to the exact locations of

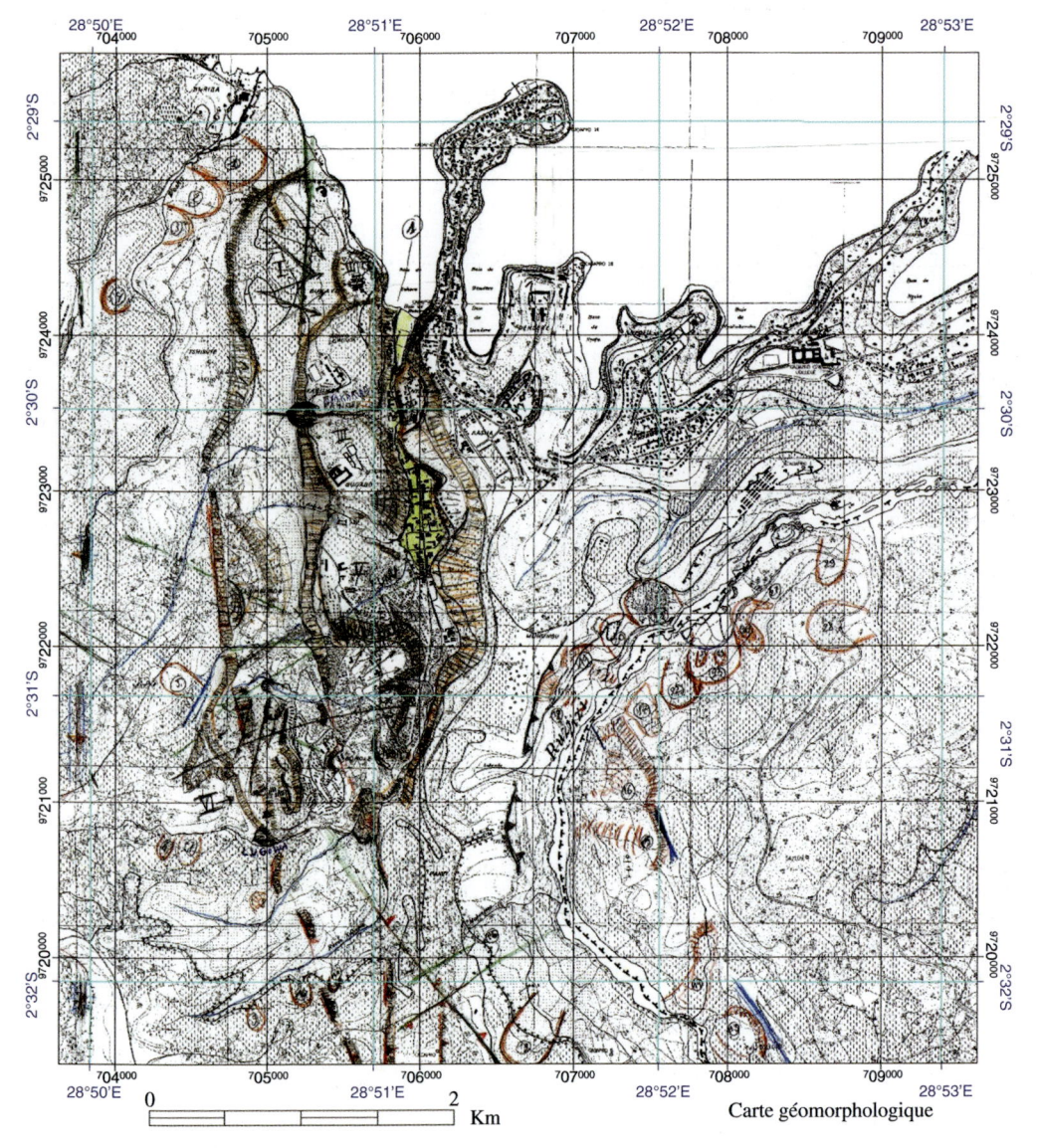

Fig. 3. Geomorphological map of Bukavu. BUK, Bukavu; BUG, Bugabo; ZA, Camp Zaïre; NY, Nyagongo; SA, Camp Saio; (1), (2), etc. are escarpments; 1, 2, etc. are landslides outside the micro-graben; I–VI are landslides inside the micro-graben.

the faults. Different geological maps indicate many different fault trajectories (Lambert 1981; Kampunzu *et al.* 1983; Mweze, cited by Munyololo *et al.* 1999).

Geomorphology

Bukavu lies on the west bank of the Ruzizi gorge, in a rolling landscape of convex and elongated hills, developed on the weathered lavas of Panzi–Muhungu–Dendere (1550–1650 m above sea level (a.s.l.)). This landscape is interrupted in the west by an asymmetrical north–south-trending corridor, the so-called 'Bukavu micro-graben' (UNESCO 2002). This starts at the Bay of Bukavu, gradually disappears south of Boholo, and is 1–1.5 km wide and about 4.5 km long. The Kawa River flows through the eastern side of the valley, bordered on the west by three successive north–south-trending escarpments that give rise to

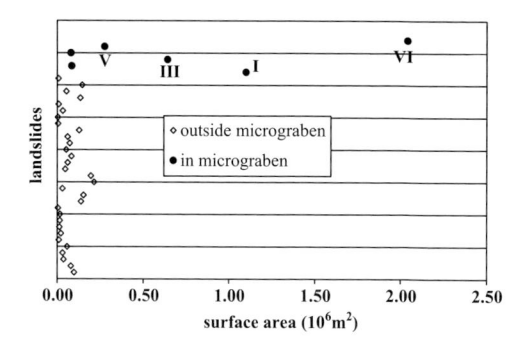

Fig. 4. Numbers of landslides by surface area in Bukavu district. Landslides (I), (III), (V) and (VI), located within the micro-graben or at its boundaries, are clearly larger than the others.

crest-lines at ±1550 m (Kabumba–Bugabo–Boholo), 1700–1750 m (Tshimbunda) and 1800–2000 m a.s.l (Karhale). For convenience, the term 'micro-graben' is used in this paper to indicate the asymmetric valley of the Kawa River (Fig. 3).

Recent investigations have aimed to determine the origin and activity of mass movements that have detrimentally affected, and are expected to continue to affect the rapidly urbanizing area in and around Bukavu. These investigations were accomplished by aerial photograph interpretations, ground reconnaissance, comparison with similar phenomena and investigations carried out in adjacent Rwanda, and the application of theoretical landslide models developed in the USA. This information may prove useful for urban planning and for the general understanding of mass movements in the Bukavu area, and perhaps elsewhere in tectonically active humid tropical countries.

The interpretation of stereoscopic aerial photographs led to the recognition of large landslides spatially linked to the Bakuvu micro-graben (II, III, IV and VI in Fig. 4). About 15% of the surface area of the Bakuvu district is covered by visible landslides, many of them clearly active, especially those in the micro-graben. First, there is an enormous scar, affecting the SE flank of Tshibuye Hill (I in Fig. 4), which is affected by a second Bukavu slide and partially affected by the Kabumba–Bugabu incision (II and III, respectively, in Fig. 4). Another landslide belt is in the valley of the lower Funu River (V in Fig. 4). This depression is bounded at its upper end by a huge crown-like fissure, a sharply expressed escarpment on the aerial photograph, which it is interpreted as an active landslide.

Stereoscopic interpretation of aerial photographs also shows the presence of numerous landslide scars outside the Bakuvu micro-graben. In many cases the lobe still partially fills the scar (Dikau et al. 1996). These have been numbered from 1 to 31. These stereoscopic study also reveals that none of these 31 landslides is located on or in the vicinity of visible active faults. Only landslides 13 and 15 have their headscarp in contact with a fault. Landslides 1–31 are clearly different from groups I–VI with respect to their width/length proportions and their size. The average surface of a landslide outside the micro-graben is ±6.5 ha, whereas the mean size of the landslides in the micro-graben is nearly 85 ha (Fig. 4). The landslides in the Bukavu micro-graben also appear to be relatively wide and short, suggesting some structural influence (Fig. 5).

In Figure 6, most Bukavu landslides fit within the topographic threshold envelope for North America (Chorley et al. 1984; Montgomery &

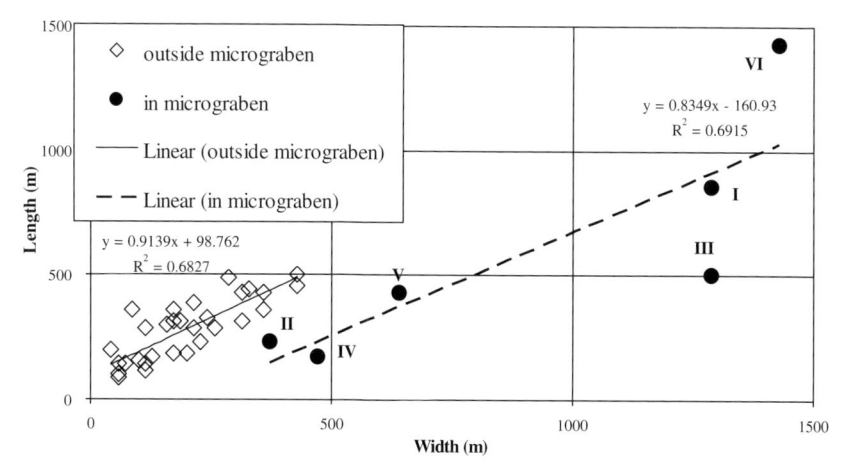

Fig. 5. Shape of the Bukavu landslides. The linear regression lines (length/width) show that the landslides in the micro-graben are significantly wider than the others.

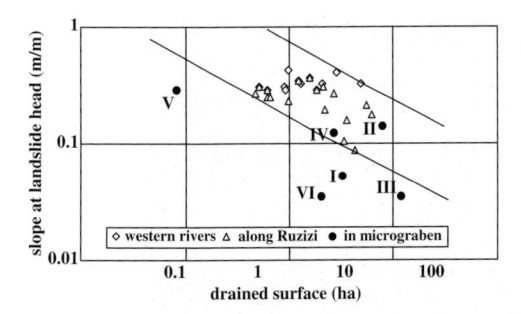

Fig. 6. Topographic thresholds of Montgomery & Dietrich (1994).

Dietrich 1994) and Rwanda (Moeyersons 2003), as delimited by the two lines. Landslides to the west of Bukavu are located in the valleys of the Wesha, Funu and Lugowa, as well as along Kahuma Gully. Landslides (I), (III), (V) and (VI) fall clearly below the Montgomery–Dietrich topographic threshold envelope. Their movement is probably not induced by hydrostatic pressures alone, but by such pressures in combination with tectonic and/or seismic activity. The validity of the topographic threshold criterion is supported by the fact that landslides (I), (III), (V) and (VI) are the only ones crossed or bounded by active faults.

Remote sensing

Data selection

This study used topographic maps from 1954, together with aerial photographs from 1959 and 1973. The main use of recent remote sensing data was for mapping the current city with regard to housing density, infrastructure, vegetation cover and slope instability features. This led to the following criteria for the selection of remote sensing data: (1) acquisition as recent as possible; (2) high spatial resolution (pixel size between 1 and 3 m); (3) multispectral data, allowing the rapid recognition of key features (soils, vegetation, buildings, roads, rivers) with simple image processing. In 2001, the

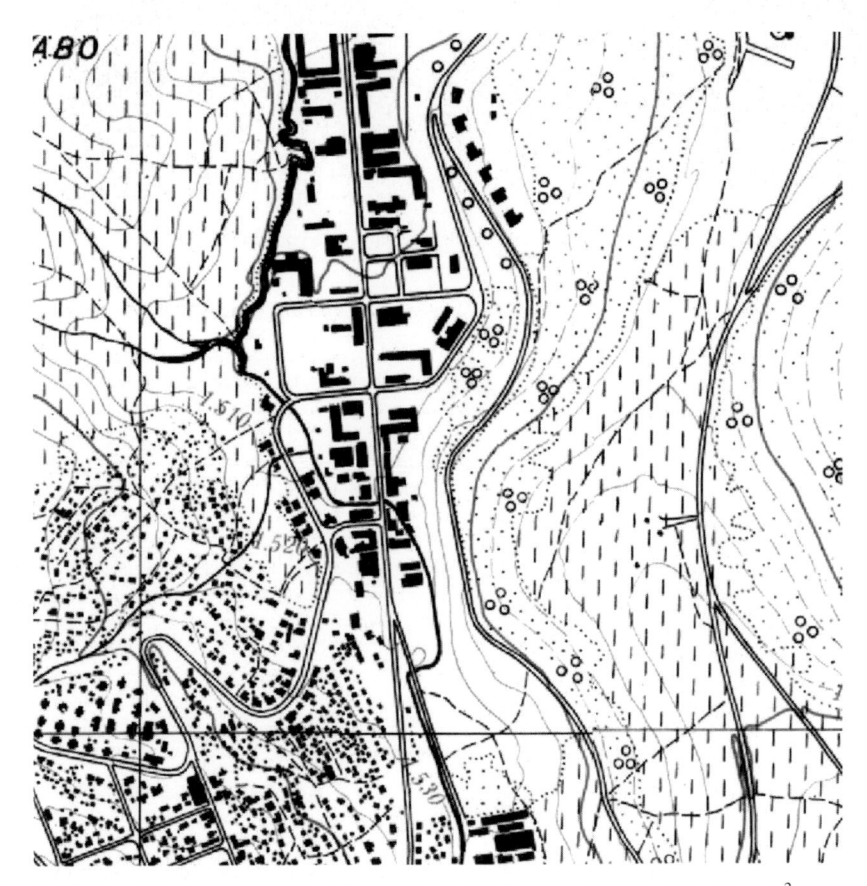

Fig. 7. The industrial area in 1954, surrounded by grass, bush and trees. The area shown is $c.$ 1 km^2 (IGCB 1957).

Fig. 8. The same area as in Figure 6, seen on the 2001 IKONOS image. The area shown is c. 1 km².

only data corresponding to these characteristics came from the IKONOS sensor. A scene of 11 km × 11 km was acquired, with a resolution of 4 m in the visible and near-infrared (VNIR) range and 1 m in the panchromatic channel.

Geometric corrections

The relief and the differences in altitude in the area are very important, with slopes dropping more than 500 m in a few kilometres, resulting in high values of parallax displacements in the image. Moreover, the angle of incidence of the IKONOS image acquisition was 28°18′, makes the geometric corrections more difficult. In these conditions, it becomes impossible to register the image to a cartographic projection system using simple warping with a polynomial transformation on x, y co-ordinates. The image was therefore orthorectified using geodetic control points from the Belgian Royal Africa Museum archive and a digital elevation model (DEM) derived from contour lines of the 1:10 000 scale 1954 topographic map.

Image processing and interpretation

The first step in the image processing was sharpening the 4 m pixel multispectral image by fusion with the 1 m pixel panchromatic channel, producing multispectral imagery with 1 m pixels. A true colour image at 1 m resolution was used as the 'base map' for the following GIS analyses and synthesis. Simple indices were computed to extract useful information for the project: the Normalized Difference Vegetation Index (NDVI) for vegetation cover and a brightness index for quantifying the growth of urban areas. A visual comparison between the 1954 map and the IKONOS image directly reveals the huge change in urbanization (Figs 7–10). Over much of Bukavu, the grass and bush areas of 1954 have been replaced by very high-density housing.

Field survey

Rapid geomorphological surveys were carried out to provide the GIS with updated information.

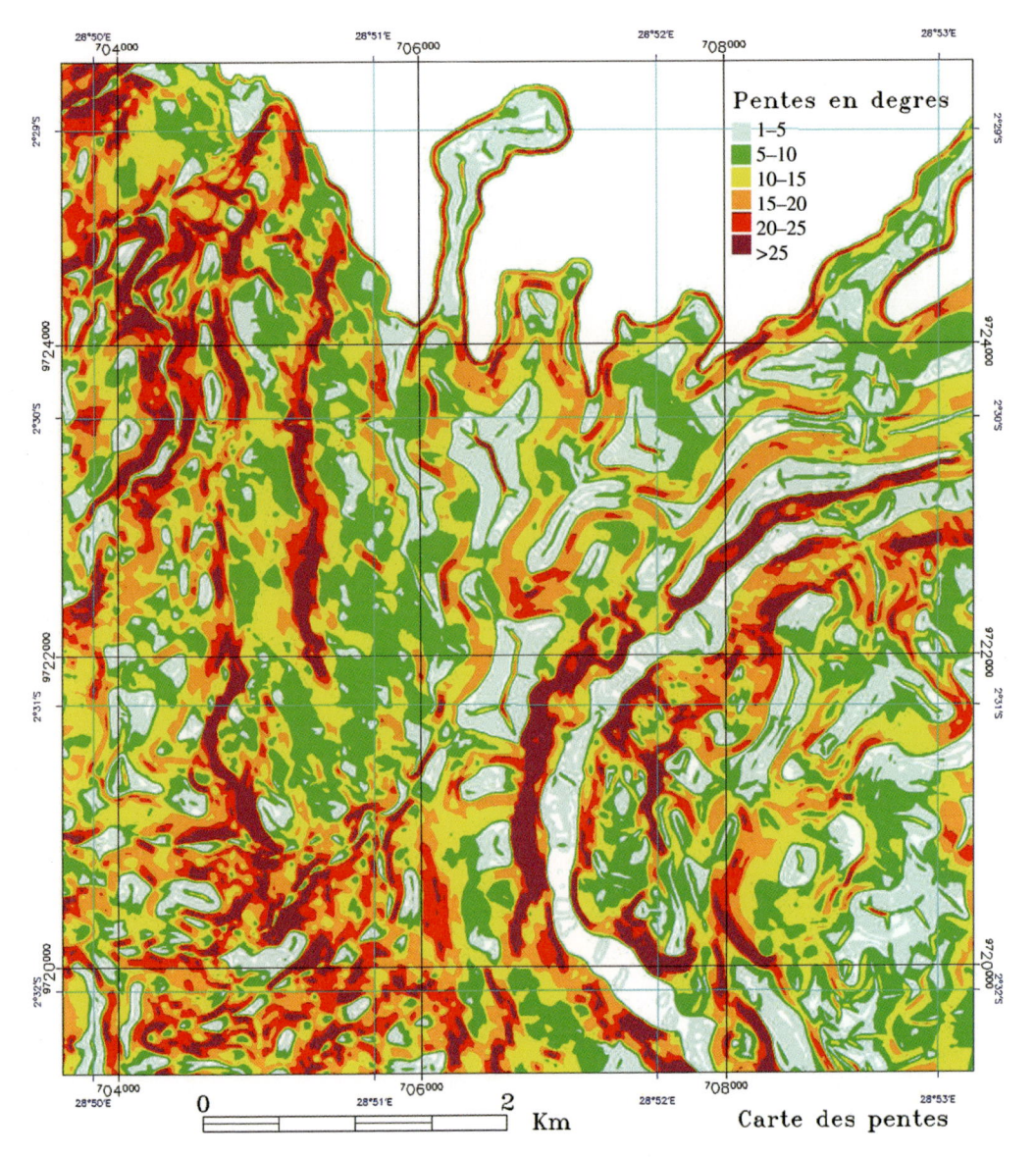

Fig. 9. Bukavu slope steepness map.

Archive information on Bukavu buildings and infrastructure was also digitized to complete the information set. The main topics covered were: outcrop locations and descriptions; locations of springs; the river network; biological pollution of drinkable water; street maps; the water distribution network and state of degradation; the sewerage network and state of degradation; slope instability (landslides, debris flows, mud flows); damage to buildings and infrastructure. Black and white image maps, based on the panchromatic IKONOS image, were edited at 1/5000 scale in a easy-to-use booklet format and used to locate the field observations. All of the field results were digitized and added to the Bakavu GIS database.

GIS synthesis

The GIS layers built during the Bukavu project allow the fast production of on-demand maps, at any required scale, according to user needs. The

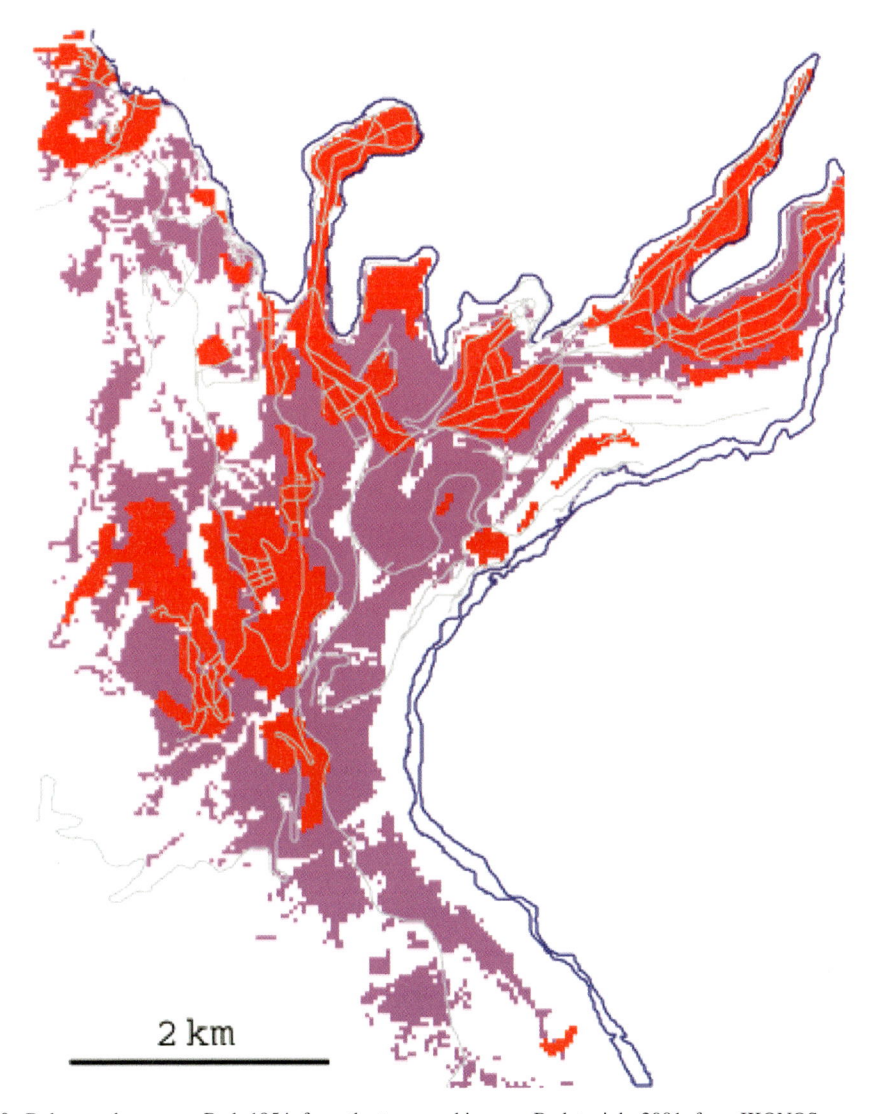

Fig. 10. Bukavu urban areas. Red: 1954, from the topographic map. Red + pink: 2001, from IKONOS.

GIS-generated maps are easily understood, even for non-specialists, because of the IKONOS image background. Furthermore, the GIS-based maps can be rapidly updated when new information is available. Figure 11 shows an example of the many possible data layer combinations.

Conclusion

A geomorphological approach to assessing geo-hazards in Bukavu, based on airphoto-based mapping of landslide morphologies and investigations into the geomorphological thresholds of the landslides, allowed a good understanding of local slope instability mechanisms and triggering factors, in the context of regional geological structures and seismic activity.

An orthorectified IKONOS satellite image served as a GIS base map for further spatial data. Combined with a DEM, the IKONOS image was very useful for highlighting the huge increase in urbanization, particularly on the steep to very steep slopes. Field observations and archive compilations formed the basis of GIS thematic layers, providing a flexible database for the production of on-demand maps by specific users.

This research was supported by the 'Géologie Urbaine de Bukavu. Interaction entre la stabilité du sol et la pression

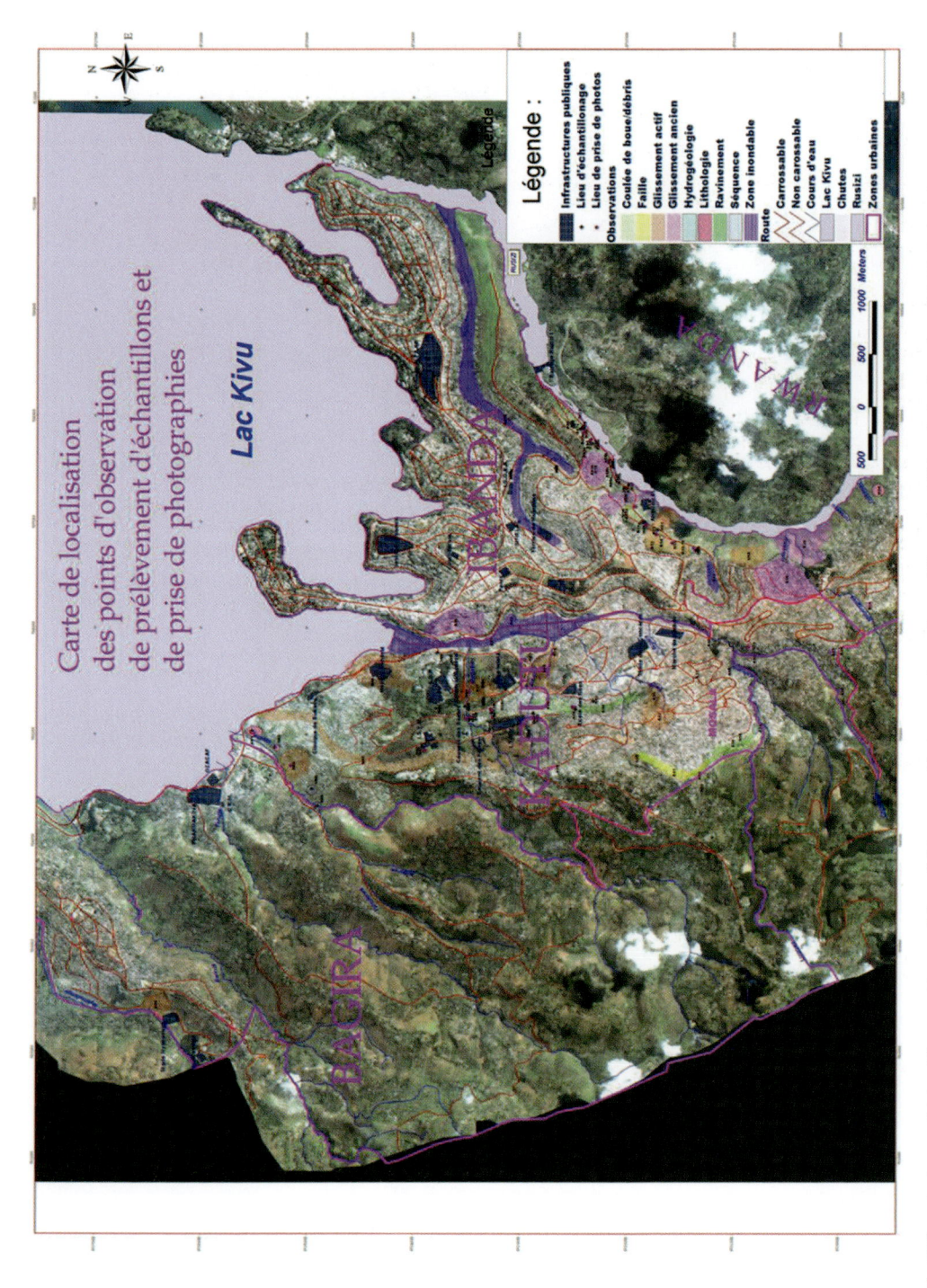

Fig. 11. GIS vector layers obtained from archives and from field surveys (river network, roads, mud flows, landslides, faults, location of field observations), with the 2001 IKONOS image as background.

démographique' project, contract SC/RP205.580.1, between the Royal Museum of Central Africa, Belgium, and UNESCO, as part of the Geological Networking (GEONET) Programme. The orthorectification of the IKONOS image and the map of settlements in 1954 and 2001 were computed at the 'SURFACES' laboratory of Liège University, Belgium.

References

CAHEN, L. 1954. *La Géologie du Congo Belge*, Vaillant–Carmanne, Liège.

CHORLEY, R. J., SCHUMM, S. A. & SUGDEN, D. E. 1984. *Geomorphology*. Methuen, London.

DIKAU, R., BRUNSDEN, D., SCHROTT, L. & IBSEN, M. L. 1996. *Landslide Recognition*. Wiley, Chichester.

IGCB 1957. *Carte topographique de Bukavu–Shangugu, 1:10 000*. Imprimerie de l'IGCB, Kinshasa.

ILUNGA, L. 1978. L'érosion dans la ville de Bukavu. *Antennes du Centre de Recherches Universitaires du Kivu*, **2**, 277–299.

ILUNGA, L. 1989. Problèmes géologiques d'aménagement dans la zone de Kadutu (Ville de Bukavu, Zaïre). *Cahiers du Centre de Recherches Universitaires du Kivu, Nouvelle Série*, **24**, 77–101.

ILUNGA, L. 1991. Morphologie, volcanisme et sédimentation dans le rift du Sud-Kivu. *Bulletin de la Société Géographique de Liège*, **27**, 209–228.

ISCHEBECK, O., KABAZIMYA, K. & VILIMUMBALO, S. 1984. Erosion à Bukavu. *Proceedings, Seminar, 23–28 January 1984, Bukavu*. ISP, Bukavu.

KAMPUNZU, A. B., VELLUTINI, P. J., CARON, J. P. H., LUBULA, R. T., KANINIKA, M. & RUMVEGERI, B. T. 1983. Le volcanisme et l'évolution structurale du Sud-Kivu, Zaïre; un modèle d'interprétation géodynamique du volcanisme distensif intracontinental. *Bulletin des Centres de Recherche Exploration–Production Elf-Aquitaine*, **7**, 257–271.

LAMBE, T. & WHITMAN, R. 1979. *Soil Mechanics*, SI version. New York.

LAMBERT, R. 1981. Carte géomorphologique. *In*: CHAMAA, M.-S., BIDOU, E. & BOUREAU, P.-Y. ET AL. (eds) *Atlas de la ville de Bukavu*. Centre de Recherches Universitaires du Kivu, Bukavu, sheet 10.

MOEYERSONS, J. 1989. *La nature de l'érosion des versants au Rwanda*. Annales du Musée, Royal de l'Afrique Central, Tervuren, Série Sciences Economiques, **19**.

MOEYERSONS, J. 2003. The topographic thresholds of hillslope incisions in South-Western Rwanda. *Catena*, **50**, 381–400.

MOEYERSONS, J., TREFOIS, PH., LAVREAU, J. ET AL. 2004. A geomorphological assessment of landslide origin at Bukavu, Democratic Republic of the Congo. *Engineering Geology*, **72**(1–2), 73–87.

MONTGOMERY, D. R. & DIETRICH, W. E. 1994. Landscape dissection and drainage area–slope thresholds. *In*: KIRKBY, M. J. (ed.) *Process Models and Theoretical Geomorphology*. Wiley, Chichester, 221–245.

MUNYOLOLO, Y., WAFULA, M., KASEREKA, M. ET AL. 1999. Récrudescence des glissements de terrain suite à la réactivation séismique du bassin du Lac Kivu. Région de Bukavu (Rép. Dém. Congo). Musée Royal de l'Afrique Central, Tervuren, Département de Géologie et Minerologie, Rapport Annuel, 1997–1998, 285–298.

PASTEELS, P., VILLENEUVE, P., DE PAEPE, P. & KLERCKX, J. 1989. Timing of the volcanism of the southern Kivu province: implications for the evolution of the western branch of the East African Rift system. *Earth and Planetary Science Letters*, **94**, 353–363.

UNESCO 2002. *Géologie Urbaine de Bukavu: interaction entre la stabilité du sol et la pression démographique*. Rapport Musée Royale de l'Afrique Central, Tervuren.

Landslide susceptibility assessment for St. Thomas, Jamaica, using geographical information system and remote sensing methods

S. MILLER[1,2], N. HARRIS[2], L. WILLIAMS[2] & S. BHALAI[2]

[1]*Department of Geography, University College Chester, Parkgate Road, Chester CH1 4BJ, UK (e-mail: s.miller@chester.ac.uk)*

[2]*Mines and Geology Division, Ministry of Lands and the Environment, Hope Gardens, Kingston 6, Jamaica*

Abstract: The St. Thomas district of Jamaica is prone to slope failure, which has resulted in extensive damage and in some cases loss of life. To reduce the effect from landslides, there was an urgent need to map and assess areas that may be prone to future failure. The interpretation of aerial photographs, together with geomorphological mapping and field surveys, was used to produce inventory maps of the landslides. The factors conditioning the slopes for failure were assessed and a weighting value was assigned to them. The weighting was achieved by using the principle of Bayesian conditional probability. The weighted factors were combined in a geographical information system (GIS) to produce a landslide susceptibility model for the study area. Comparison of the model with the existing landslides showed that 97% of the landslides fell within the high and very high susceptibility zones of the model. Comparison of the model with landslides that occurred during 2002, and that were not used in the construction of the model, shows that 83 of the 89 slides that occurred fell within the high and very high susceptibility zones. The landslide susceptibility model will be one of the first steps in assessing the risks that landslides pose to lives and new developments (housing, agriculture, physical infrastructure) in the region.

Landslides are a major natural hazard in Jamaica and have resulted in loss of life, major economic losses, social disruption and damage to public and private properties. The St. Thomas district has one of the highest densities of slope failures (Fig. 1) and is prone to large landslides, including the most catastrophic landslide recorded in Jamaica. Following heavy rainfall in June 1692 and the 6 June 1692 Port Royal earthquake, a landslide was triggered along the side of Mount Sinai overlooking the Yallahs valley. The landslide debris is estimated to be over 80×10^6 m^3: one village slipped down the mountain, another was buried, and a plantation was moved half a mile from its original location (Zans 1959). The site of the landslide is known as Judgement Cliff: it is one of the most prominent features in the local landscape (Fig. 2). The hurricanes and storms of 1951 (Charlie), 1963 (Flora), 1973 (Gilda), 1988 (Gilbert), 2001 (Iris), 2002 (Lily) and 2004 (Ivan) were instrumental in triggering widespread landslides throughout the district. During these events, lives were lost, houses destroyed, roads destroyed and blocked (Fig. 3), utility services interrupted and farms damaged or destroyed (ODPEM 2006). Although landslide damage in the district is a recurring problem, little has been done in terms of landslide hazard mapping to guide planning and rehabilitation. As is evident from the examination of aerial photographs and from speaking to local residents, roads and buildings have been blocked or destroyed at the same locations repeatedly.

The impetus for this research was the significant damage that resulted from slope instability in 2002. Between 22 and 30 May 2002, over 200 mm of rain fell as result of a tropical depression that passed over Jamaica (Metereological Office of Jamaica, unpubl. data). The rainfall triggered widespread landslide damage throughout St. Thomas. Significant damage was recorded at Trinity Ville, Friendship, Hillside, Windsor Forest, Mavis Bank, Bath, Friendship and Airy Castle (Fig. 4). This event highlighted the need to map these landslides and to create susceptibility–hazard maps to better guide planners. This paper covers the mapping and characterization of landslides within the area and the creation of a landslide susceptibility model using a geographical information system (GIS).

The study area

The St. Thomas district covers an area of *c.* 742 km^2 and is located on the eastern side of Jamaica (Fig. 4). There are *c.* 93 000 people living in the

From: TEEUW, R. M. (ed.) *Mapping Hazardous Terrain using Remote Sensing.* Geological Society, London, Special Publications, **283**, 77–91.
DOI: 10.1144/SP283.7   0305-8719/07/$15.00   © The Geological Society 2007.

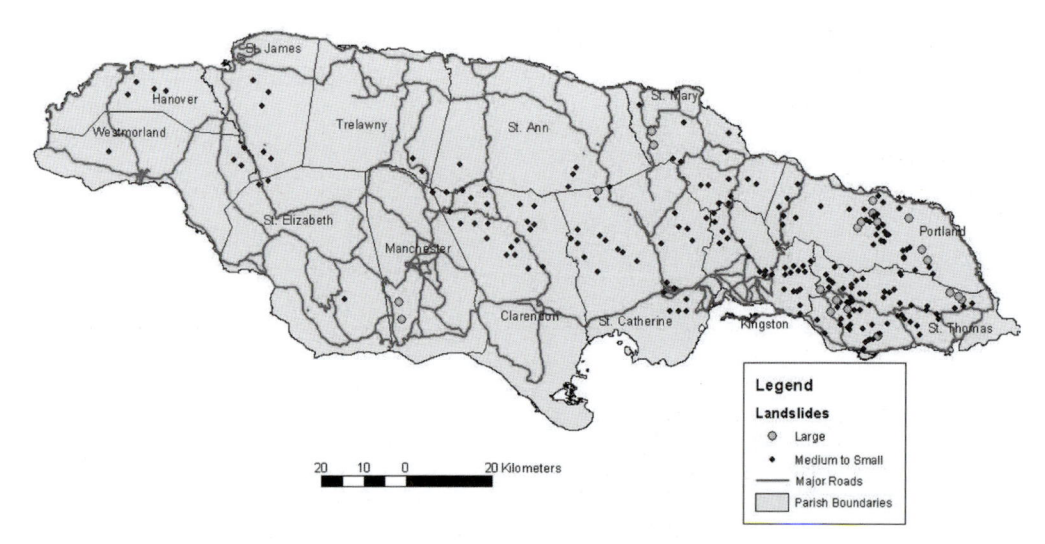

Fig. 1. Geographical distribution of major landslides in Jamaica (modified from Barker & McGregor 1995).

district and its major commercial activities are in manufacturing, mining and agriculture (Statistical Institute of Jamaica 2002). Local elevations range from sea level along the southern and eastern coasts, to 2256 m in the northern section, which is part of the Blue Mountains. The area is generally rugged and mountainous, with deeply incised and steep-sided valleys. There are three major rivers in the region, the Plantain Garden, Morant and Yallahs, all of which have tributaries in the Blue Mountains and drain into the Caribbean Sea. Unpublished data from the Meteorological Office of Jamaica indicate that the 30 year average rainfall for the area is 2288 mm a^{-1}, with October having the highest amount of rainfall.

The bedrock geology is composed mainly of Cretaceous schists, volcaniclastic rocks, marbles and volcanic rocks; Eocene–Miocene limestone, shale–sandstone sequences and conglomerates; and Pleistocene limestone and sandstone (Mines and Geology Division, Ministry of Mining and Energy, Jamaica (MGD) unpubl. data). Quaternary alluvium is mostly found in low-lying areas in the southern and eastern sections of the district. Rocks in the area are extensively faulted, with the major faults being the Plantain Garden fault, a section of the Blue Mountains fault and the Wagwater Fault system (MGD, unpubl. data).

As a result of the combination of the geology, steep slopes, deep chemical weathering and high volume of rainfall in the St. Thomas district, the land surface is highly susceptible to slope instability. Changing land-use, particularly for housing, quarrying and agricultural activities, further compounds the problem of slope instability. The

Fig. 2. Photograph of the Judgement Cliff landslides, looking NE.

Fig. 3. Landslide damage to road in St. Thomas district.

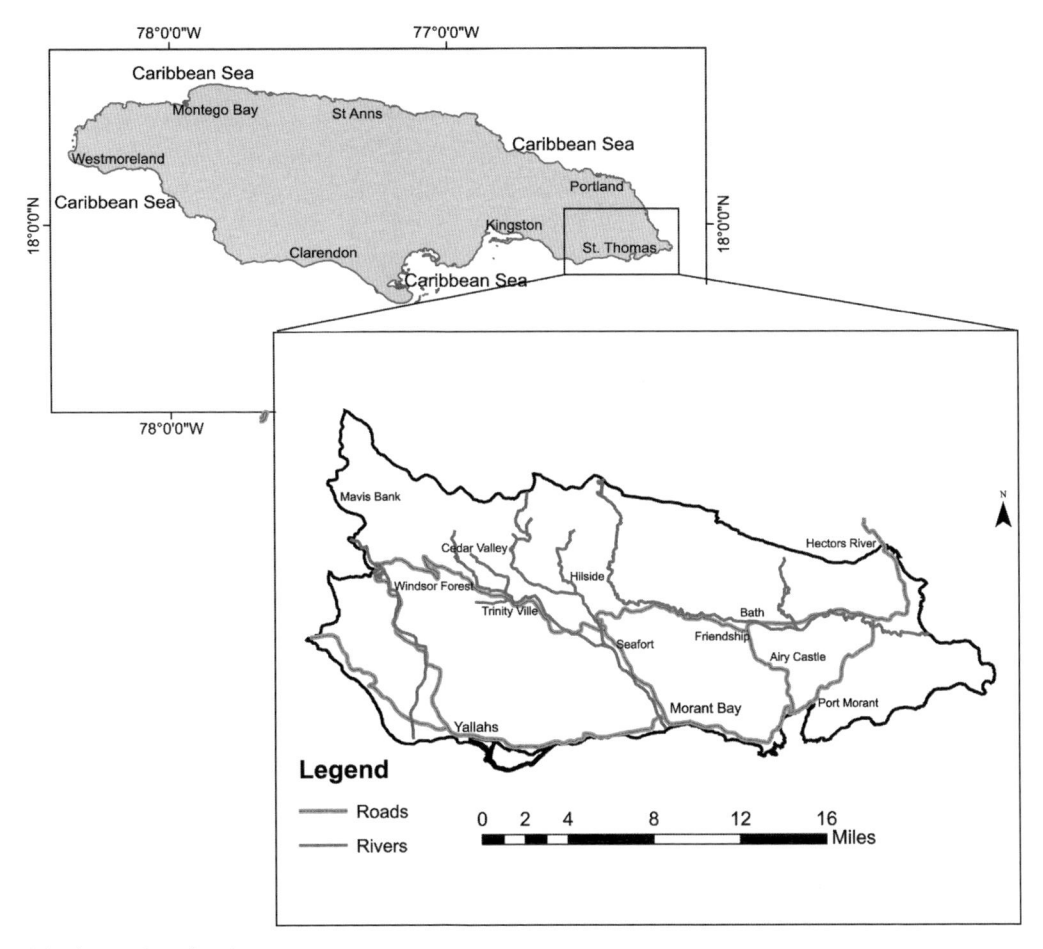

Fig. 4. Location of study area.

damages caused by landslides are of economic, societal and environmental concern. Although it is difficult to prevent most of these landslides from occurring, we can certainly mitigate their impact. To do this, we first need to know where they are or where they are likely to occur. A landslide susceptibility map will satisfy these criteria, and may be one of the first steps in looking at the risks landslides pose to lives and new developments.

Methods

To carry out landslide hazard assessment for St. Thomas, a combination of direct, indirect and non-deterministic methods were applied. Landslide occurrence data collected using direct methods (aerial photograph or satellite imagery interpretation and field mapping), were compared with a number of factors (geological, geomorphological and hydrological) that are believed to be contributing to slope instability, to determine statistically their relationship. The strength of the relations between the presence or absence of landslides and the factors chosen was used to determine the weighting given to each factor. The weighted factors were then combined to create the final susceptibility–hazard model. The statistical analysis was carried out using the Statistical Program for the Social Sciences (SPSS) and the results were exported to the GIS. To create the landslide susceptibility map in a GIS there are five general stages.

(1) Creation of a landslide digital inventory, using existing sources or remote sensing and fieldwork.
(2) Identification of the controlling parameters (factors) contributing to slope instability.
(3) Acquisition of base-maps (factor maps) in a compatible digital format or converting hard copy information into digital information. This stage

also involves the classification of variable maps into appropriate classes.

(4) Assigning weights to factor maps and the classes within them. In this study the Bayesian probability weight of evidence method was used. The various factor maps are reclassified to reflect these assigned weights.

(5) Combination of these weighted maps to produce a final susceptibility map.

Landslide data acquisition

Because of the size of the study area and the rugged terrain, aerial photographs at scales of between 1:12 000 and 1:40 000 were used to map the landslides, using airphoto interpretation and geomorphological mapping. 'Ground truth' was checked by walk-over surveys and traverses, with the results recorded on 1:12 500 topographical maps. Additional landslides, which were not identifiable on the aerial photographs, were mapped during fieldwork. Information recorded for these slides included; type of landslides, size, degree of activity, relative age and certainty in data mapping.

Landslide inventory map

The landslide inventory map was digitized using Autocad and imported into ArcGIS. The information collected (e.g. typology, size, activity, impact, mechanism of failure and impact of the slide) was incorporated into a database and then linked to the digitized landslides. The resulting inventory map shows the spatial location and distribution of landslides within the study area (Fig. 5).

Most of the landslides were translational slides (85%), followed by rotational slides (13%); the other 2% were flows and compound slides. The translational slides tended to be smaller than the rotational slides and predominantly occurred in weathered volcanic and volcaniclastic rocks. Translational slides also occur in shale–sandstone sequences, where the permeable and more competent sandstone slides over the less competent and impervious shale. There are two modes of occurrence of the rotational slides; those in weathered material (weathered volcanic and volcaniclastic rocks and conglomerates) and those in limestone. The landslides in the limestone, such as at Judgement Cliff, tended to be larger than those in weathered material. The slides in the study area ranged in size from 553 m^2 to 912 605 m^2 in area. In total, landslides debris covered c. 3.49% of the study area. Half of the landslides mapped were converted into a binary map (presence or absence of landslide), which was later used in the analysis to create the susceptibility map. The other half of the landslides were used to validate the model.

Selection of appropriate variable map

Preparatory factors may best be described as those factors that make the slope susceptible to slope movement, but do not initiate any form of movement. Eight preparatory factors were selected and

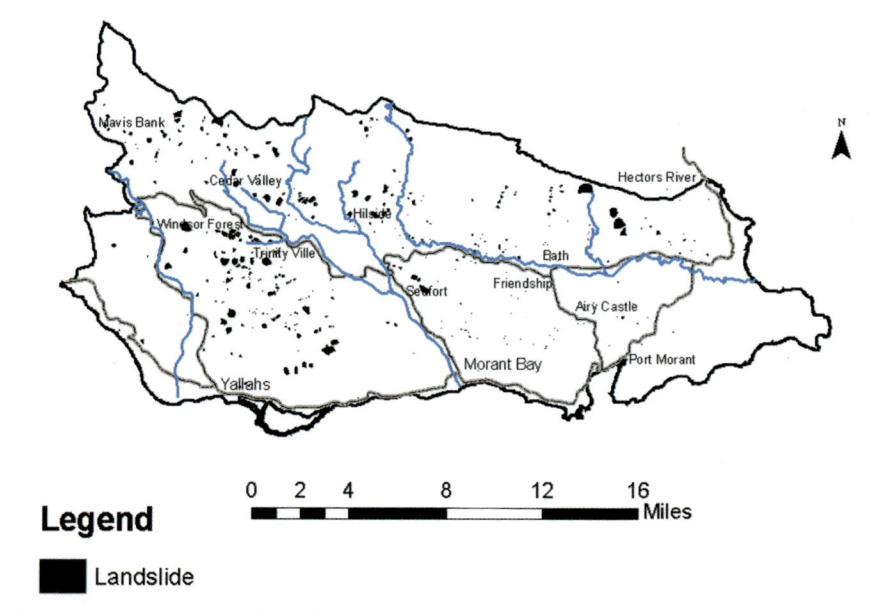

Fig. 5. Landslide inventory map of the study area.

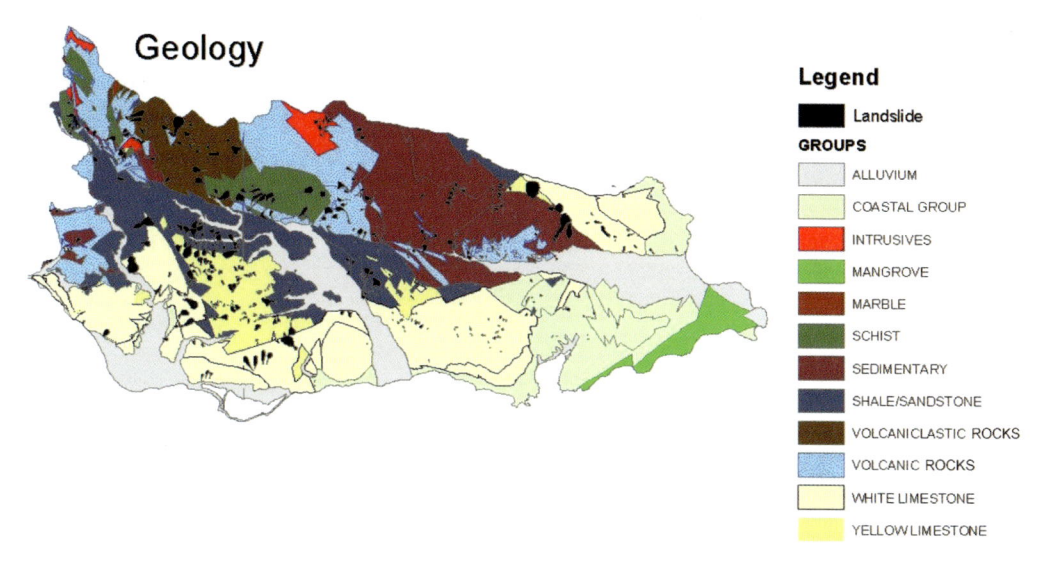

Fig. 6. Lithological map with landslides overlaid, study area (source: MGD unpubl. data).

used in the initial landslide susceptibility analysis: lithology; distance from fault lines; slope gradient; slope aspect; distance from fluvial channels; land use and rainfall.

Lithology. A lithological map (Fig. 6) was derived from the Mines and Geology Department (MGD) digital geological sheets (unpubl. data). This information was subsequently modified and the attributes table updated, based on additional information gleaned in the field. The landslide distribution map was compared with the lithological map and the density of landslides within each lithological grouping was calculated (see Fig. 7). The lithologies that are most susceptible to slope instability tend to be the yellow limestones, schists and the volcaniclastic rocks (Fig. 7). The volaniclastic rocks tend to be deeply weathered and are found in mountainous regions, with steep slopes. This combination of these factors may be the reason why this particular lithology is susceptible to slope failure. The schists, as they are foliated rocks, already have planes of weaknesses. Weathering of these rocks further makes them

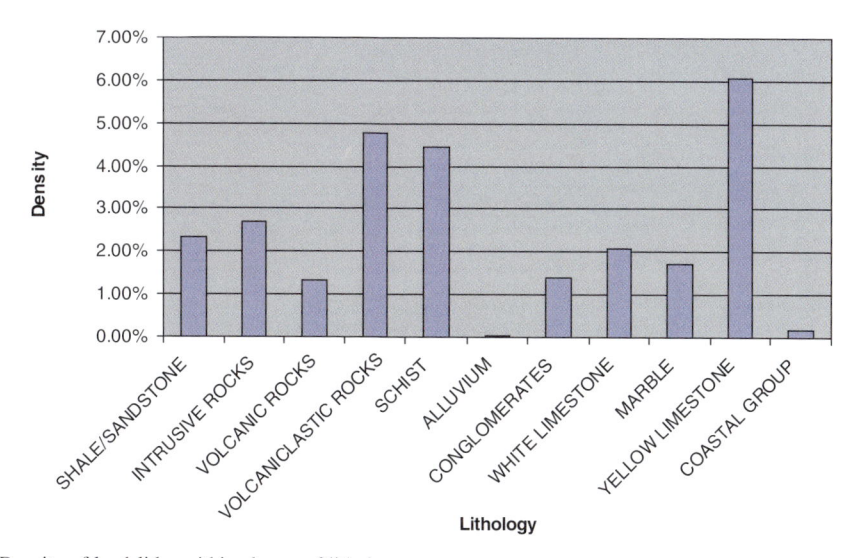

Fig. 7. Density of landslides within classes of lithology.

susceptible to slope failure, particularly where they form steep slopes. The most susceptible of the lithologies, based on the density of landslide within it, is the Yellow limestone, on which the landslide density is over 6%, which far exceeds the average of the study area (3.49%). The Yellow limestone is an impure limestone with bands of clay that act as an impervious layers, preventing the flow of water and increasing shear stress in the rock, which may contribute to slope failure. The clay bands also act as planes of weakness along which movement may occur.

Distance from fault lines. Fault lines were extracted from the MGD digital geological sheets (MGD, unpubl. data). A buffer was created around the faults using the 'Distance' command in ArcGIS. Distances created were in 100 m increments up to 700 m. When the landslide distribution map was compared with the distance from fault line, there is a clear indication that fault lines are affecting the susceptibility of rocks. The density of 3.7% within 100 m of fault lines exceeded the average landslide density of the study area (Fig. 8). There was a gradual decrease in the density of landslides as the distance from fault lines increased. However, beyond 100 m there was still a high number of landslides, which indicates that other factors may be playing a more prominent role.

Slope gradient. A digital elevation model (DEM) was created from contours provided by the Survey Department of Jamaica (SDoJ 2002). The contours were interpolated in Arc Info, using the TOPOGRID function. A 30 m grid DEM was created from this interpolation process. The slope gradient variable map was created from this DEM using the SLOPE surface function in ArcGIS. The slopes were then divided into 10° classes (Fig. 9).

The landslide distribution was compared with slope angle and density within each class of slope gradient calculated. Figure 9 shows an initial increase in the density of landslides with increased slope gradient up to 40°. After 40° there is a sudden decrease in density followed by a sudden increase in density for slopes over 60°. Further analysis indicated that landslides in weathered superficial material failed at a lower angle, compared with solid rocks. This is to be expected, as solid rocks generally have higher shear or cohesive strength than weathered or poorly lithified rocks and will resist slope movement at higher slope angles, and explains bimodal distribution of density (see Fig. 9): landslides in soils are failing at angles less than 40° and those in solid geology mostly at angles above 60°.

Slope aspect. The slope aspect map was created from the 30 m DEM using the ASPECT surface function in ArcGIS, with aspect divided into 45° classes. When the landslide distribution map was compared with the slope aspect map, higher densities were observed on slopes facing the east and west (see Fig. 10). There are two possible reasons for this: the Blue Mountains in the northern and eastern section strongly influence the pattern of rainfall in the area. Slopes facing west and NW receive more rainfall than other slopes, which may trigger landslides. On the other hand, western slopes receive less rainfall and are left bare for

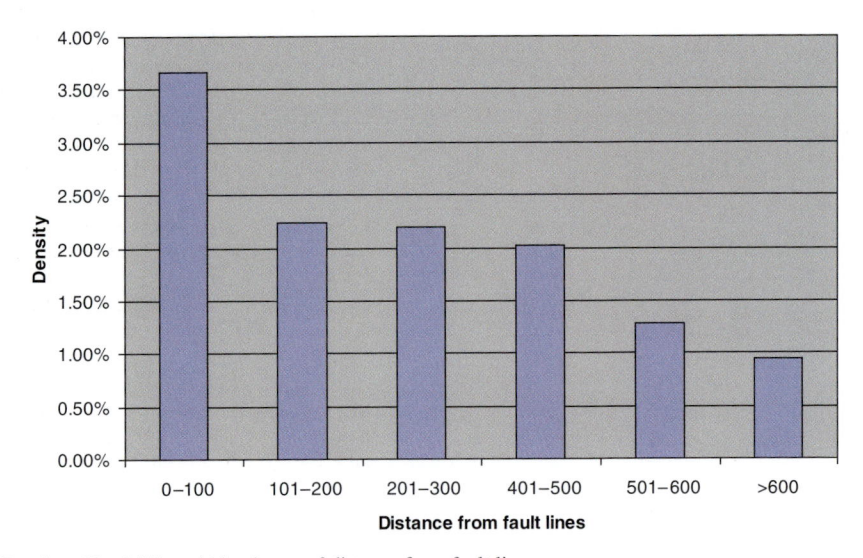

Fig. 8. Density of landslides within classes of distance from fault line.

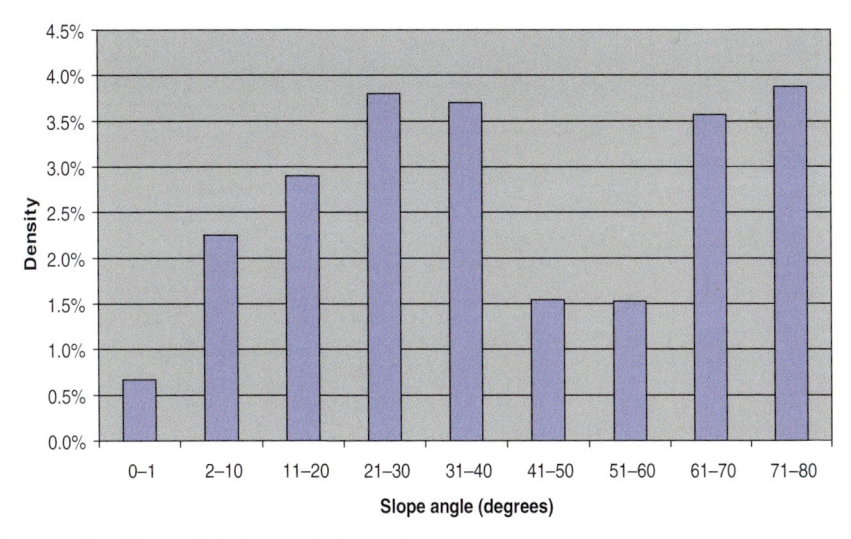

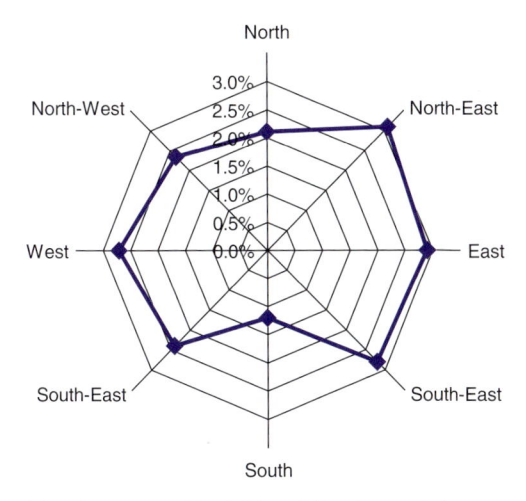

Fig. 9. Density of landslides within classes of slope angle.

longer periods of time. Sudden and intensive rainfall tends to trigger shallow landslides. There is little vegetation to help to hold the soil together or reduce the rate of infiltration during periods of intense rainfall. The soils rapidly become oversaturated, resulting in mudflows and/or translational slides. Examination of Figure 10 shows that none of the classes has a landslide density that exceeds the average for the study area; it appears that aspect does not play a significant role in making the slopes unstable.

Distance from fluvial channels. The river network was digitized from the Survey Department of Jamaica topographical maps (SDoJ 1984). Distance calculations were calculated using the Distance function in ArcGIS and the resulting map was divided into 100 m increments. If the rivers were undercutting the slopes and thus making them prone to failure, more landslides would be expected in close proximity to river channels than away from them. Figure 11 shows that this is generally true, although there is also a higher density of landslides in the zone 600 m or more from river channels. The presence of a high density of landslides within 100 m of a river indicates that river undercutting may be playing a role in some areas.

Land-use. The land-use pattern strongly influences the susceptibility of the land to slippage. Areas with thick vegetation are generally less prone to slope failure than areas that are devoid of vegetation. The vegetation decreases the rate at which meteoric water enters the soil. This allows for a gradual infiltration rate through the soil down to the bedrock. This prevents the build-up of neural pore water pressure in the soil, thereby reducing the risk of slope failure. The roots of the plants increases the shear strength of the soil as far down as the roots goes (Donatti & Turrini 2002). Based on the density of landslides, disturbed forest and fields are the most susceptible areas (see Fig. 12). These areas include areas cleared for agriculture (small fields and planted forests), quarrying, roads and housing. Arable land and urban areas have the lowest density of landslides: both are usually in low-lying areas on alluvium plains.

Rainfall. Most of the landslides were recorded after heavy rainfall, indicating that the slides are

Fig. 10. Density of landslides within classes of slope aspect.

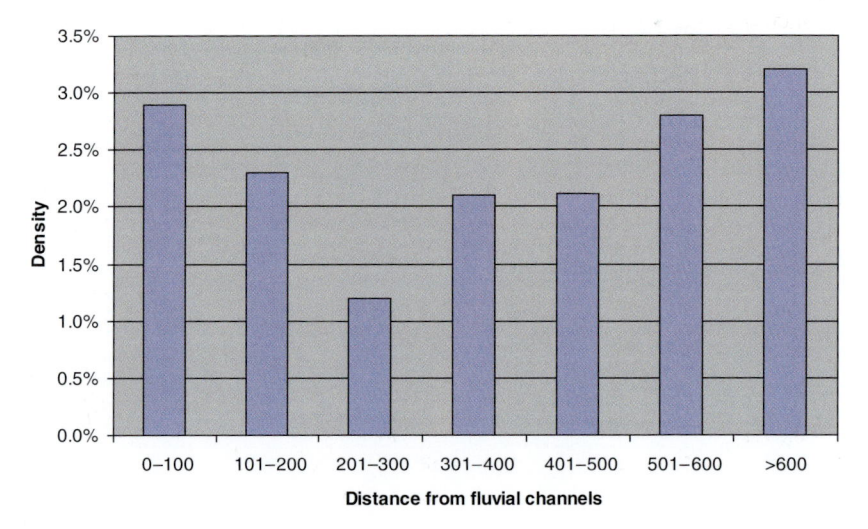

Fig. 11. Density of landslides within classes of distance from fluvial channels.

rain-induced, with rainfall acting as one of the triggering mechanisms. The rainfall data were obtained as rainfall measurements at selected measuring stations within the study area from the Meteorological Office of Jamaica (MOJ, unpubl. data). This data were first converted to point data with associated attributes and then interpolated in ArcGIS using the CONTOUR surface function. This resulted in a continuous surface for rainfall covering the entire study area.

The highest density of landslides was within areas of relatively low rainfall, with a gradual increase in landslide density where rainfall exceeds 2500 mm a^{-1} (Fig. 13). This result is opposite to what was expected. A possible explanation for this is that areas of low rainfall intensity have less vegetation cover and hence are left bare for longer periods of time. As these slopes are bare, during rainfall they are more prone to failure than those with a higher vegetation cover.

There was a need to identify systematically the relationship between variables and the landslides before the analysis could proceed and to decide which variables were most suitable to be used for the susceptible model creation. A backward stepwise discriminant analysis was used to determine

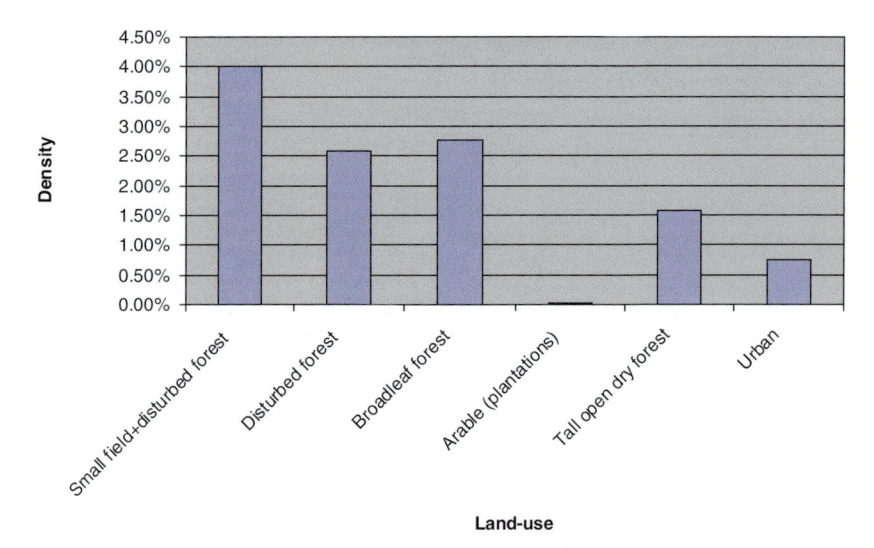

Fig. 12. Density of landslides within classes of land-use.

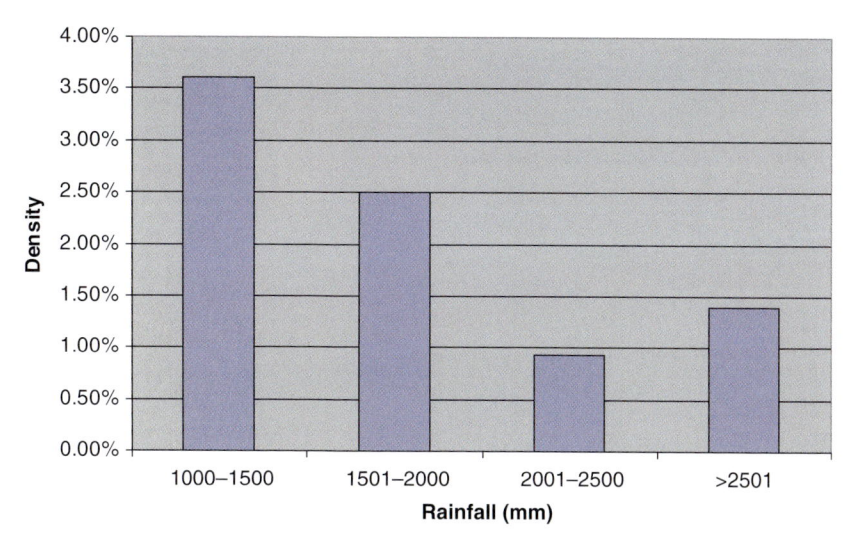

Fig. 13. Density of landslides within classes of rainfall.

the significant variables and those that would contribute least to the model. F-values were used to eliminate those variables that were contributing the least to the model. The F-value for a variable indicates its statistical significance in discrimination between groups, which is a measure of the extent to which a variable makes a unique contribution to the prediction of group membership. The smaller the F-value of the model, the better it is. Based on this discriminant analysis, aspect was removed first and then distance from stream-head. The variables left in the model were; geology, slope angle, distance from fault line, land-use and rainfall. A correlation between the variables and landslides also shows that aspect and distance from stream-head do not correlate with landslides (see Table 1). Removing these variables from the creation of the susceptibility map was therefore justified.

Slope instability modelling using Bayesian probability modelling

The fourth step in the creation of a susceptibility map is to assign a weighting to factors and classes within them. The weighting of the controlling variables can be done in a subjective way, or it may be done in a more objective manner using statistical or deterministic methods. As this model will later be used in hazard and risk analysis, probability maps will have to be created, so a subjective method is not recommended to assign the weighting. Deterministic and statistical methods are preferred methods to create probability maps. Deterministic methods are generally expensive and time consuming to implement, and are only practical if used on a local scale. On the other hand, statistical methods are suitable for use at a regional scale and are less time consuming than deterministic methods. The advantages of statistical methods are that the model from one study area can be extrapolated to other areas with similar characteristics. The Bayesian conditional probability statistical method is used to facilitate comparison between models. It has been used successfully by a number of researchers carrying out landslide susceptibility mapping (e.g. Chung & Fabbri 1999; Clerici *et al.* 2002; Lee *et al.* 2002; Pistocchi *et al.* 2002). Bayesian probability can measure the likelihood that an event will occur given the past events. This approach is based on the idea of prior and posterior probability. Bayes's Theorem begins with a statement of knowledge prior to performing the experiment, the prior probability. In the context of landslide assessment, the prior probability would be the likelihood of a landslide occurring anywhere in the study area. The average landslide density would be the prior probability in this study. Where there is evidence that a particular area is less or more prone to slope instability, this prior probability can be updated, giving the posterior probability.

The formulation of the Bayesian probability model outlined below was modified from the approaches of Bonham-Carter (1994) and Lee *et al.* (2002). In general terms, the probability of a landslide occurring for a particular binary pattern, for example a geological unit, is equal to the prior probability multiplied by the posterior probability.

Table 1. *Correlation between landslides and factors contributing to slope instability*

Kendall's tau_b

Correlations

		GEO	FAULT	RAINFALL	SLOPE	LANDUSE	stream_dist	ASPECT	landslide
GEO	Correlation coefficient	1.000	−0.114*	−0.205*	−0.039*	0.039*	0.112*	−0.019	0.338*
	Sig. (2-tailed)		0.000	0.000	0.006	0.010	0.000	0.169	0.000
	N	3016	3016	3016	3016	3016	3016	3016	3016
FAULT	Correlation coefficient	−0.114*	1.000	−0.022	−0.057*	0.028	0.061*	−0.017	−0.114*
	Sig. (2-tailed)	0.000	0	0.148	0.000	0.069	0.000	0.226	0.000
	N	3016	3016	3016	3016	3016	3016	3016	3016
RAINFALL	Correlation coefficient	−0.205*	−0.022	1.000	−0.149*	−0.070*	0.120*	0.024	−0.242*
	Sig. (2-tailed)	0.000	0.148		0.000	0.000	0.000	0.102	0.000
	N	3016	3016	3016	3016	3016	3016	3016	3016
SLOPE	Correlation coefficient	−0.039*	−0.057*	−0.149*	1.000	−0.123*	−0.160*	0.039*	0.271*
	Sig. (2-tailed)	0.006	0.000	0.000		0.000	0.000	0.005	0.000
	N	3016	3016	3016	3016	3016	3016	3016	3016
LANDUSE	Correlation coefficient	0.039*	0.028	−0.070*	−0.123*	1.000	−0.120*	−0.005	−0.109*
	Sig. (2-tailed)	0.010	0.069	0.000	0.000		0.000	0.760	0.000
	N	3016	3016	3016	3016	3016	3016	3016	3016
stream_dist	Correlation coefficient	0.112*	0.061*	0.120*	−0.160*	−0.120*	1.000	−0.003	0.030
	Sig. (2-tailed)	0.000	0.000	0.000	0.000	0.000		0.808	0.065
	N	3016	3016	3016	3016	3016	3016	3016	3016
ASPECT	Correlation coefficient	−0.019	−0.017	0.024	0.039*	−0.005	−0.003	1.000	−0.027
	Sig. (2-tailed)	0.169	0.226	0.102	0.005	0.760	0.808		0.085
	N	3016	3016	3016	3016	3016	3016	3016	3016
landslide	Correlation coefficient	0.338*	−0.114*	−0.242*	0.271*	−0.109*	0.030	−0.27	1.000
	Sig. (2-tailed)	0.000	0.000	0.000	0.000	0.000	0.065	0.085	
	N	3016	3016	3016	3016	3016	3016	3016	3016

*Correlation is significant at the 201 level (two-tailed).
Sig., significance

In this case the prior probability may be expressed as

$$P\{D\} = \frac{N\{D\}}{N\{T\}} \quad (1)$$

where $N\{D\}$ is the number of cells containing landslides and $N\{T\}$ is the total number of cells in the study area.

Posterior probability was calculated for each variable. The calculation is similar for all the variables (patterns), therefore the geology map will be used as an example from here on. If the binary predictor pattern, B(geology), occupying $N\{B\}$ unit cells occurs in the region, and a number of known landslides occurs preferentially within the pattern, i.e. $N\{D \cap B\}$, then the favourability of locating an occurrence, given the presence of a predictor and the absence of a pattern, can be expressed by the conditional probabilities

$$P\{D|B\} = \frac{P\{D \cap B\}}{P\{B\}} = P\{D\}\frac{P\{B|D\}}{P\{B\}} \quad (2)$$

and

$$P\{D|\overline{B}\} = \frac{P\{D \cap \overline{B}\}}{P\{\overline{B}\}} = P\{D\}\frac{P\{\overline{B}|D\}}{P\{\overline{B}\}}. \quad (3)$$

The posterior probabilities of an occurrence, given the presence and absence of the predictor pattern, are denoted by $P\{D|B\}$ and $P\{D \setminus \overline{B}\}$, respectively; $P\{B|D\}$ and $P\{\overline{B}|D\}$ are the posterior probabilities of being inside and outside the predictor pattern B, respectively, given the presence of an occurrence D. Also, $P\{B\}$ and $P\{\overline{B}\}$ are the prior probabilities of being inside and outside the predictor pattern, respectively. The same model can be expressed in an odds-type formulation where the odds, O, are defined as $O = P/(1 - P)$. Expressed as odds, equations (2) and (3) become

$$O\{D|B\} = O\{D\}\frac{P\{B|D\}}{P\{B|\overline{D}\}} \quad (4)$$

and

$$O\{D|\overline{B}\} = O\{D\}\frac{P\{\overline{B}|D\}}{P\{\overline{B}|\overline{D}\}} \quad (5)$$

where $O(D|B)$ and $O\{D\setminus\overline{B}\}$ are the posterior odds of an occurrence given the presence and the absence of a binary predictor pattern, respectively, and $O(D)$ is the prior odds of an occurrence. The weights for the binary predictor pattern are defined as

$$W^+ = \log_e \frac{P\{B|D\}}{P\{B|\overline{D}\}} \quad (6)$$

and

$$W^- = \log_e \frac{P\{\overline{B}|D\}}{P\{\overline{B}|\overline{D}\}} \quad (7)$$

where W^+ and W^- are the weights of evidence when a binary predictor pattern is present and it is these values that are used to assign the weighting to each class within a factor. W^+ and W^- were calculated for each class within each of the variable maps (geology, slope gradient, land use, rainfall and distance from fault line; see Table 2). The use of Bayesian theorem requires that binary maps must be created. Each variable map was converted to a binary map based on the contrast values. In most cases the threshold or 'cut-off' value (see Table 3) was obtained where the contrast was at a maximum and/or positive and changed to negative for the continuous variables and for the categorical data where the W^+ values were positive. The variable map was reclassified into a binary map using the W^+ and W^- values at the maximum and/or positive contrast or the highest positive W^+ value. The steps to combine these maps are as follows:

(1) Calculate the prior logit, which = log(prior probability/1 − prior probability).
(2) Calculate the posterior logit, which = the prior logit + the sum of all the weighted binary maps.
(3) Convert to posterior odds, which = exp(posterior logit).
(4) Convert to posterior probability, which = posterior odds/(1 + posterior odds).

The combination of the maps was done in ArcGIS using the Spatial analysis, Raster calculator. Figure 14 represents the final susceptibility map with associated hazard classes for the study area.

Model result

Based on visual inspection, the very high susceptibility zone is in general agreement with the distribution of known landslides within the study area. High and very high susceptibility zones are generally associated with mountainous regions, with moderate to steep slopes and areas with deeply weathered rock and/or Yellow limestone and interbedded shale–sandstone sequences. Areas calculated as low susceptibility are mostly low-lying, where alluvium and coastal limestone are present. The success of the landslide susceptibility models is measured by their ability to

Table 2. *Result from Bayesian analysis for individual factors, showing weighted values*

Variable	W⁺	W⁻	Contrast	Variable	W⁺	W⁻	Contrast
Slope (degrees)				*Distance from fault lines (m)*			
0–10	−0.5210	0.3251	−0.8461	>100	0.0748	−0.0326	0.1074
11–20	0.0789	−0.0189	0.0977	101–200	−0.2011	0.0760	−0.2771
21–30	0.5431	−0.1412	0.6843	200–300	−0.2210	0.0653	−0.2863
31–40	0.3732	−0.0539	0.4272	301–400	−0.1475	0.0198	−0.1673
41–50	0.3077	−0.0160	0.3236	401–500	0.2373	−0.0136	0.2509
51–60	−0.4086	0.0049	−0.4135	>500		−0.1357	
61–70	0.7163	−0.0089	0.7252	*Geology (lithology)*			
71–80	0.6747	−0.0047	0.6795	Shale–sandstone	0.1788	−0.0266	0.2054
Rainfall (mm)				Intrusive rocks	0.4242	−0.0067	0.4309
1000–1500	1.0658	−0.1198	1.1856	Volcanic rocks	−0.0285	0.0035	−0.0320
1501–2000	−0.0370	0.0229	−0.0599	Volcaniclastic rocks	0.5978	−0.0376	0.6353
2001–2500	−0.2850	0.1219	−0.4069	Schist	0.7477	−0.0404	0.7881
>2500	−0.0350	0.0092	−0.0442	Alluvium	−2.9678	0.1534	−3.1211
Land-use				Conglomerates	0.1024	−0.0174	0.1199
Urban	−1.2430	0.0241	−1.2671	White limestone	−0.1227	0.0310	−0.1537
Disturb–forest	0.3992	−0.1003	0.4995	Marble	0.5860	−0.0008	0.5869
Field	0.0469	−0.0375	0.0844	Yellow limestone	1.3199	−0.1489	1.4688
Forest	−0.1416	0.0650	−0.2065	Coastal group	−0.8176	0.0656	−0.8832
Arable							

accurately define existing areas of landsliding, which can be validated in two ways (Chung-Jo & Fabbri 2003): (1) comparison between predicted results and the occurrence of past landslides (using the other half of mapped landslides not used to create model); (2) comparison between predicted results and the occurrence of 'future landslides' (using landslides that occurred after the date of the initial data used to construct the model); in this case, the 'future landslides' is a set of landslides mapped along roads after the heavy rainfall of June 2002.

The Bayesian model was first compared with the remaining half of the landslides (those landslides not used in the model creation). The comparison shows that 97% of the landslides fell within the high and very high zones of the Bayesian model, approximately 2% in the moderate susceptibility zone and the remaining 1% in the low and very low susceptibility zones (see Table 4). When the model was compared with the 'future landslides', 83 of the 88 landslides fell within the high and very high susceptibility zone, with five in the moderate zone and none in the low or very low

Table 3. *The cut-off value used for constructing binary maps from multiclass maps*

Variable	Present			Absent		
	Class	W⁺		Class	W⁻	
Distance from fault lines	≤100 m	0.0748		>101 m	−0.0326	
Rainfall	≤1500 mm	1.0658		>1500 mm	−0.0029	
Slope	11–50° and 61–80°	0.3077		≥11° and 51–60°	−0.160	
Land-use	Disturbed broadleaf, fields; fields and disturbed broadleaf; broadleaf forest	0.3992		Built-up areas; dry tall forest; plantation; wetland	−0.1003	
Geology	Schist; Yellow limestone; white limestone; volcaniclastic rocks; shale–sandstone	0.7477		Coastal limestone; marble, intrusive rocks; volcanic rocks; alluvium	−0.0404	

Prior probability = 0.186.

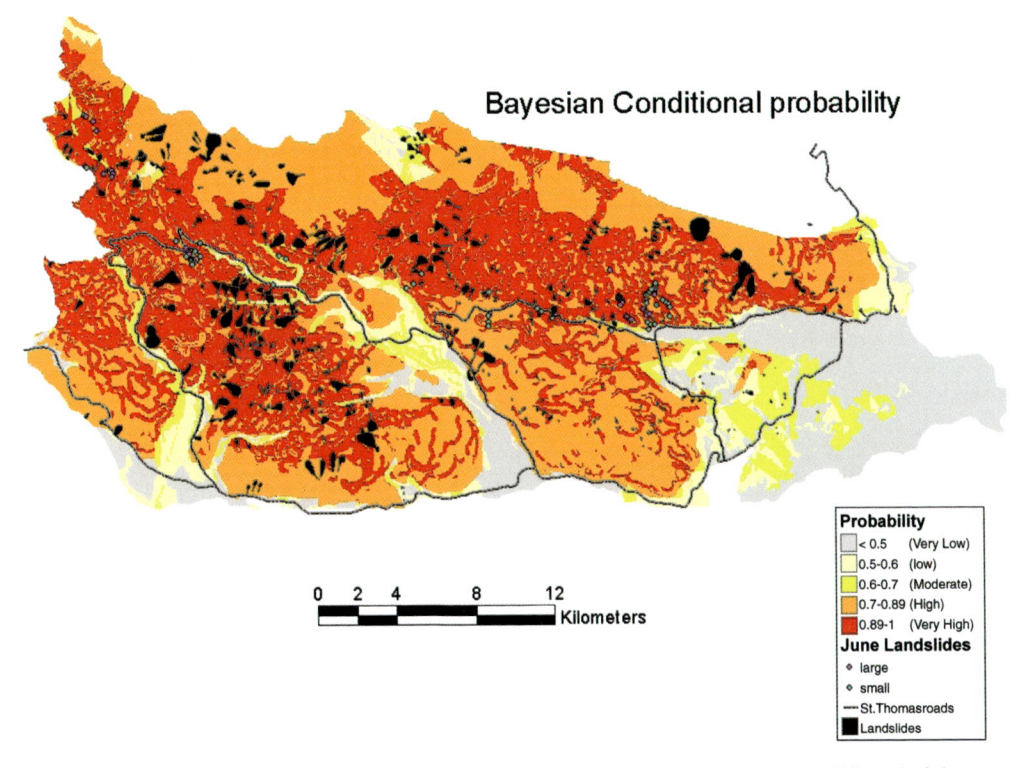

Fig. 14. Landslide susceptibility map for St. Thomas, created using Bayesian conditional probability principles.

susceptibility zone. Based on this, it can be concluded that the model has performed well and may be used as a good predictive tool for slope instability in the area.

A χ^2 test was carried out to ascertain that the high rate of success was not a result of the high percentage of area classified within the high and very high susceptibility zones. An increase in area may increase the chance of a landslide falling within one these zones. The χ^2 test serves as a test of randomness and it was used to determine which of the following hypotheses stands.

Null hypothesis (H_0). The landslides are randomly distributed and the high number of landslides occurring within the high and very high susceptibility zones is a result of chance.

The alternative hypothesis (H_1). The landslides are not randomly distributed and are not occurring

Table 4. *Number of landslides within the respective susceptibility zones*

Susceptibility zone	Area(km^2)	Landslides ('future slides': June 2002) Small	Landslides ('future slides': June 2002) Large	Landslides (%) – remaining half of slides
Very low	117.8	0	0	0.2
Low	56.9	0	0	0.9
Moderate	51.3	1	4	1.9
High	307.7	19	15	39.7
Very high	283.7	29	20	57.3

A total of 49 small landslides (width <10 m) and 39 large landslides (width >10 m) were mapped.

Table 5. *Results from χ^2 test for landslides occurring by chance within susceptibility zones*

Susceptibility category	Percentage of area	Observed	Expected	d	d^2	d^2/e
Very low	14.4	2	7.4371	5.4371	29.5623	3.9750
Low	7.0	1	3.7882	2.7882	7.77388	2.0521
Moderate	6.3	4	3.4140	−0.5860	0.3434	0.1006
High	37.6	12	20.4790	8.4790	71.8941	3.5106
Very high	34.7	35	18.8817	−16.118	259.800	13.7594
		Total = 54				$\chi^2 = 23.3977$

by chance in the high and very high susceptibility zone.

Eighteen 2 km² quadrats, representing 5% of the study area, were placed randomly over the susceptibility map created. This was achieved by using a moving window in the GIS. The slides observed within each susceptible zone were counted and totalled (see Table 5). If the landslides were occurring by chance within the susceptibility zones, then the number of landslides would be expected to correspond to the proportion of area covered by each zone (see Table 5). The χ^2 test is based on the equation

$$\chi^2 = \sum \frac{d^2}{e} \qquad (8)$$

where, d^2 is the difference between the observed and expected frequency for each category and e is the expected frequency for each category. The critical value of χ^2 at the selected 0.01 significance level, with four degrees of freedom ($n - 1$ categories), would be 13.28. As the χ^2 test had a value of 23.39, the null hypothesis is rejected and the alternative hypothesis accepted. It can be concluded that the susceptibility map correctly delineates areas of slope instability and potential instability.

Conclusions

From the mapping and susceptibility mapping it is evident that the St. Thomas area is highly susceptible to landslides. The use of airphoto interpretation and field mapping was successful in locating the existing landslides. By identifying possible factors that are conditioning the slope for failure and applying Bayesian conditional probability theorem, it was possible to create a landslide susceptibility model for the study area. This model was able to successfully delineate areas of known slope instability and to predict areas that may be prone to future slope failure. A significant proportion of the high and very high susceptibility zones indicated by all the models are outside the urbanized areas, which is reassuring. The susceptibility maps created will serve as a guide to planners, environmental managers and policy makers, allowing them to better prepare and implement plans for development in areas prone to slope failure.

I would like to acknowledge the contribution of the Jamaica Mines and Geology Division and members of staff who provided the resources that made this project possible.

References

BARKER, D. & MCGREGOR, F. M. (eds) 1995. *Environment and Development in the Caribbean: Geographical Perspectives*. The Press, University of the West Indies, Barbados, Jamaica, Trinidad & Tobago, 146–169.
BONHAM-CARTER, G. F. 1994. *Geographic Information Systems for Geoscientists: Modelling with GIS*. Pergamon, Oxford.
CHUNG-JO, F. & FABBRI, A. 2003. Validation of spatial prediction models for landslides hazard mapping. *Natural Hazards*, **30**, 451–472.
CLERICI, A., PEREGO, S., TELLINI, C. & VESCOVI, P. 2002. A procedure for landslide susceptibility zonation by the conditional analysis method. *Geomorphology*, **48**, 349–364.
DONATTI, L. & TURRINI, M. 2002. An objective method to rank the importance of factors predisposing to landslides with GIS methodology: application to an area of the Apennines (Valnerina; Perugia, Italy). *Engineering Geology*, **23**, 277–289.
LEE, S., CHOI, J. & MIN, K. 2002. Landslide susceptibility analysis and verification using the Bayesian probability model. *Environmental Geology*, **43**, 120–131.
ODPEM 2006. *ODPEM National Disaster Catalogue*. Office of Disaster Preparedness and Emergency Planning, Kingston, Jamaica (Online database, January 2006) http://www.odpem.org.jm/.
PISTOCCHI, A., LUZI, L. & NAPOLITANO, P. 2002. The use of predictive modeling techniques for optimal exploitation of spatial databases: a case

study in landslide hazard mapping with expert system-like methods. *Environmental Geology*, **41**, 765–775.

SDoJ 1984. *1:50 000 Topographical metric sheets 18 and 19*. Survey Department of Jamaica, Government of Jamaica, Kingston.

Statistical Institute of Jamaica (STATIN) 2002. *Population Census, Jamaica Country Report, Volume 1*, STATIN, Kingston.

ZANS, V. A. 1959. Judgment Cliff Landslide in the Yallahs valley. *Geonotes (Journal of the Geological Society of Jamaica)*, **II**(2(7)), 43–48.

Applications of remote sensing for geohazard mapping in coastal and riverine environments

R. M. TEEUW

Geohazards Research Centre, School of Earth and Environmental Sciences, University of Portsmouth, Portsmouth, UK (e-mail: richard.teeuw@port.ac.uk)

Abstract: This paper reviews various types of remote sensing that can be applied to mapping and monitoring riverine and coastal geohazards. It focuses on flooding, ground instability, erosion and sedimentation, mainly drawing on examples from the UK. Both airborne and space-borne systems are examined, with assessments of the merits, limitations and relative costs of each system. Remote sensing offers a wide range of useful techniques for mapping coastal and riverine features, particularly where there are access problems associated with sites prone to flooding or slope instability.

Some 20% of the world's population live within 30 km of the sea and these populations are growing at about twice the overall global rate (Nicholls *et al.* 1997; Haslett 2000). Cliffs account for about 75% of the world's coastline (Bird 2000): cliff retreat poses significant threats to many communities, especially in the face of global warming, sea-level rise and increased storminess (Houghton *et al.* 2001; Lee 2001). Floods are the most costly type of natural hazards, accounting for 31% of the economic losses (Munich Re 1997). In the UK, about 5 million people and 2 million properties are in areas with flood risk (Webster 2006). The region most at risk is the Thames floodplain, with *c*. 3 million people, 1.4 million residential properties and 100 000 commercial properties in the immediate vicinity of the river (Galy & Sanders 2002). In 2001, the UK government allocated £564 million (*c*. US$ 1billion) for flood defence and coastal protection schemes (DEFRA 2001). The effective implementation of such schemes requires a good understanding of the processes and landforms associated with coastal and riverine settings; remote sensing provides many ways of mapping and monitoring those environments.

Coastal and riverine environments contain many features that are difficult to map, sample or monitor using conventional fieldwork surveys, even when access is assisted by the use of a boat. Access is particularly difficult, sometimes dangerous, in marshland and mudflats, especially in tidal areas. Cliffs are inherently dangerous, with both steep slopes and slope instability making fieldwork difficult and hazardous. Furthermore, many of the landforms in coastal and riverine settings are frequently changing, requiring frequent repeat surveys to understand their dynamics.

Remotely sensed data can be rapidly collected at a range of spatial scales, from the airborne detection of detailed features, such as the dimensions of trees on a floodplain, to satellite observation of regional features, such as the morphology of a major drainage basin. A wide range of temporal scales are also possible, from observation balloons providing continuous 24 h coverage, to daily coverage from aeroplanes or weather satellites, and fortnightly to monthly coverage from Earth observation satellites, such as Landsat. This paper reviews coastal and riverine applications of remote sensing, primarily drawing on examples from the UK. Three main sets of geohazard are examined: (1) flooding; (2) ground instability; (3) erosion and sedimentation.

Remote sensing applications

In remote sensing, observed features interact in characteristic ways with electromagnetic radiation. Table 1 illustrates how some of those interactions can be used to map coastal and riverine features. Summaries of remote sensing applied to riverine and/or coastal zones have been given by Van Zuidam (1993), Ritchie & Rango (1996), Cracknell (1999), Schultz & Engman (2000), Leuven *et al.* (2002), Mertes (2002) and Schmugge *et al.* (2002). Phinn *et al.* (2000) provided case studies examining the selection of optimum remote sensing systems for monitoring and managing coastal environments.

Airborne remote sensing

For detailed mapping of riverine features, aerial photography remains the most widely used and well-tested method for mapping land cover and geomorphology, as well as the generation of detailed contour maps via photogrammetry. Aerial photography has been a mainstay of land cover mapping since the 1950s: the characteristic textures, tones and patterns displayed by different

Table 1. *The electromagnetic spectrum: coastal and riverine applications*

Wavelength	Mapping uses
Ultraviolet (UV)	Sensitive to areas of snow cover
Blue (B)	Distinguishing soil from vegetation and deciduous from coniferous trees; bathymetric mapping, to *c.* 15 m depth
Green (G)	Plant vigour assessment; bathymetric mapping, to *c.* 10 m depth
Red (R)	Vegetation discrimination; bathymetric mapping, to *c.* 5 m depth
Blue, Green, Red laser altimetry (LiDAR)	Detects vegetation canopy height and ground elevation with sub-metre precision; bathymetric mapping to *c.* 30 m depth
Near Infrared (NIR)	Land–water boundaries; penetrates thin cloud; discriminates soil and vegetation moisture contents
Middle Infrared (MIR)	Discriminates soils and sediments via mineralogy variations.
Thermal Infrared (TIR)	Evapotranspiration and soil moisture variations; pollution mapping; penetrates thin cloud
Microwave radar	Sensitive to surface roughness and soil moisture; penetrates cloud; interferometry detects millimetre-scale elevation changes, can produce 3D elevation models

vegetation types are used to map geo-ecological zones, each with a distinctive underlying geology, soil type and hydrology. Better discrimination of vegetation types and land–water boundaries is possible using infrared film. Digital aerial photography has recently begun to rival the detail provided by conventional (analogue) aerial photography. Using digital photography eliminates the loss of detail and time associated with scanning aerial photographs. Furthermore, the images can be readily imported into a geographical information system (GIS), facilitating spatial analysis and map production. Digital-format aerial photography also allows automated photogrammetry, which facilitates the generation of high-density digital elevation models (DEMs) (Chandler 2001).

Aerial photography is limited to visible and near infrared (NIR) wavelengths: however, multispectral digital sensors have been developed that allow the detection of middle infrared (MIR) and far/thermal infrared (TIR) wavelengths. The 11-band Airborne Thematic Mapper (ATM) can detect in the visible, NIR, MIR and TIR: it has been used extensively in the UK to map fluvial and coastal features (Fuller *et al.* 1995; Milton *et al.* 1995; Bryant & Gilvear 1999; Rainey *et al.* 2000; Winterbottom 2000). The Compact Airborne Spectrographic Imager (CASI) is a hyperspectral sensor that records reflectance in the visible to NIR wavelengths (400–900 nm) in up to 288 bands, at 1.8 nm intervals (Babey & Anger 1989). There are trade-offs with CASI hyperspectral data: the greater the number of spectral bands, the fewer the number of pixels that can be scanned and vice versa. For instance, programming CASI to detect in 200 spectral bands will result in a pixel size of *c.* 10 m, whereas adjusting the scanner to use *c.* 1 m pixels limits the number of spectral bands to no more than 18. CASI has been used extensively by UK researchers to map vegetation and landform assemblages along river valleys, estuaries and coastal zones (e.g. Fuller *et al.* 1995, 1996; Thomson *et al.* 1998a,b; Eastwood *et al.* 1999; Hunter *et al.* 2000; Hunter & Power 2002; Lucas *et al.* 2002). Another hyperspectral sensor, HyMap, which detects in 144 bands from visible to MIR wavelengths, has been used simultaneously with airborne radar to map UK wetlands (Denniss 1999).

Until recently, airborne radar was not widely used for the mapping of terrain in riverine and coastal settings, because of complex, time-consuming image processing issues and the relatively distorted nature of the imagery. However, improvements in microwave radar technology, involving multiple-frequency and variable polarity systems, have greatly improved the range of features that can be detected (Kasischke *et al.* 1997). By far the biggest improvement has been with airborne radar, with Intermap Technologies Inc. surveying the whole of the UK using X-band interferometric synthetic aperture radar (InSAR), producing DEMs with 5 m spacing and an accuracy of ± 1 m (Dowman *et al.* 2003). The DEM, marketed as NextMap UK, was commissioned by Norwich Union Insurance, allowing them to obtain elevation maps of floodplains and coastal lowlands that are more detailed than those of the national mapping agency. The NextMap DEM data have been used extensively by the British Geological Survey to map geomorphological and geological features across the UK: part of this dataset is shown in Figure 1. The NextMap DEM has been shaded and a colour-ramp applied. Abandoned meanders and river terraces are visible, along with subtle topographic variations within the floodplain which may influence the pattern of over-bank flows during floods.

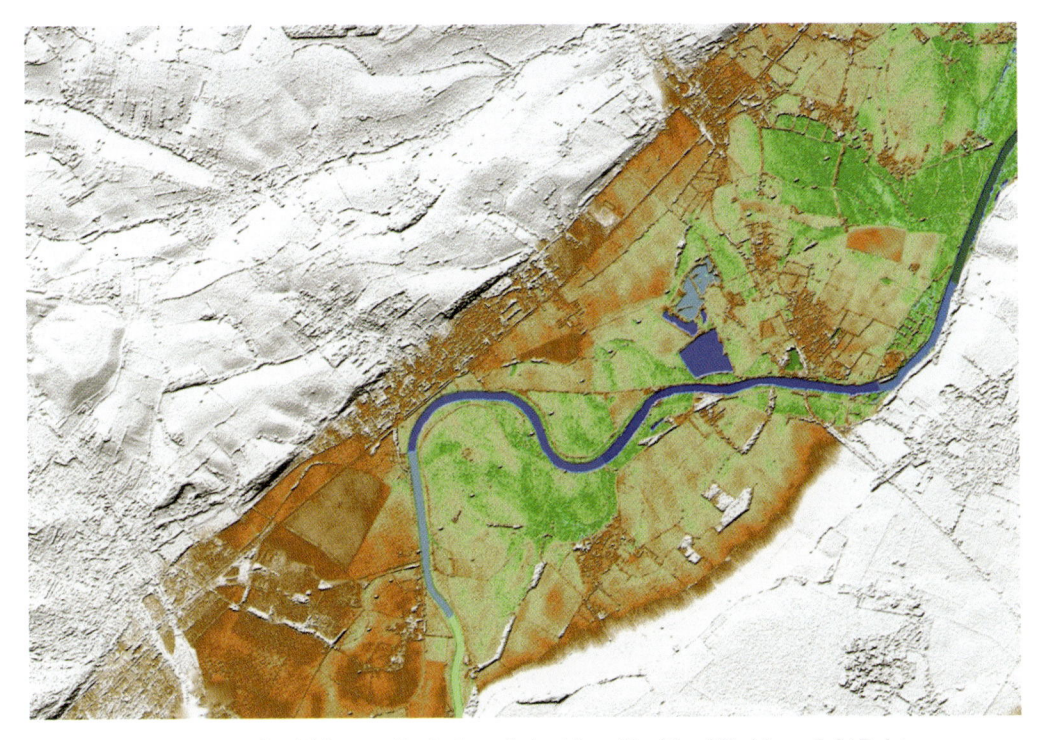

Fig. 1. DEM of the Trent floodplain, near Nottingham, derived from NextMap UK airborne InSAR data. Image courtesy of the British Geological Survey, data from Intermap Technologies.

Laser altimetry, or light direction and ranging (LiDAR), involves the scanning of the land surface using a laser: the longer it takes for the laser beam to be reflected back from a target to the sensor, the further away that target is (Wehr & Lohr 1999). Improvements in accuracy, mainly involving the use of higher-frequency lasers (producing a denser set of 'hits' on the ground, from which average elevations are calculated), have led to the latest LiDAR systems having 'footprints' of 0.3–1.0 m and elevation intervals of 0.1–0.5 m. A particularly useful feature of LiDAR is its ability to detect signal reflectances from the top surface of vegetation cover, as well as from the underlying ground. This allows the direct production of two types of DEM: a digital surface model (DSM), showing the elevation of vegetation canopies, and a digital terrain model (DTM), which shows only the ground surface, including that under vegetation. Bathymetric mapping to depths of *c.* 20 m is also possible using LiDAR; however, clear water is needed for the best results, and Charlton *et al.* (2003) found errors associated with aquatic vegetation and deep water.

Space-borne remote sensing

Earth observation satellites, such as SPOT, IRS, ASTER and Landsat ETM, have panchromatic pixels in the 5–15 m range, as well as multispectral (visible and IR) pixels in the 10–90 m range. This level of detail is sufficient for the detection of landform assemblages with characteristic sets of geo-ecological features, such as a saltmarsh, fluvial floodplain or coastal dune fields (Donoghue & Shennan 1987; Muller *et al.* 1993; Seaman & Fletcher 2000; Bourcier *et al.* 2002). SPOT and ASTER can provide stereopair images that allow 3D viewing and the production of DEMs with 10–15 m contour intervals, useful for both geomorphological and hydrological analyses (Hirano *et al.* 2003; Kormus *et al.* 2004). The detection of relatively small landforms, such as tidal creeks or dune blow-outs, requires very high-resolution satellites, such as IKONOS and Quickbird, with their 0.5–1.0 m panchromatic pixels and 2–4 m multispectral pixels.

A limitation of satellite remote sensing is that most of the sensors do not have the operational flexibility to obtain images at the ideal times for mapping and monitoring, as a result of two factors beyond the control of the operators: (1) cloud cover obscuring target areas; (2) tidal conditions. Radar remote sensing is useful for overviews of catchments during flood events and has been used for monitoring flooding on the large river systems, notably the Rhine, Rhône and Mississippi, when cloud cover precluded the use of other satellite remote sensing systems (e.g. Blyth 1994). Imaging radar systems are sensitive to variations in soil moisture content, which is useful when assessing runoff potential and susceptibility to flooding, and to variations in surface roughness, which can be used to map land cover types.

Radar interferometry (InSAR) uses data from two antenna, viewing a given part of the Earth's surface from different look-angles, to produce a DEM. InSAR has been used by the Space Shuttle Radar Topographic Mission (SRTM) to produce DEM coverage of the Earth's land surface between 80°N and 70°S. SRTM DEMs have 30 m pixels over US territory and 90 m pixels elsewhere, with vertical accuracies of ± 16 m (SRTM mission, JPL website 2005).

Differential synthetic aperture radar (DifSAR) can detect centimetre-scale changes in surface elevation. DifSAR mostly utilizes data from the European Space Agency's ERS series of satellites, as their orbits are more precisely determined than those of other radar satellites; furthermore, there is an extensive archive of ERS coverage, dating back to the early 1990s. A recent variant, persistent scatterer interferometry (PSI), uses at least 30 satellite radar images of a given area to reduce errors associated with the technique, resulting in millimetre-scale quantification of ground movement (for more details, see Riedmann & Haynes 2007).

Coastal and riverine geohazards

Flooding

The spatial and temporal resolutions of key remote sensing systems, in relation to flood events, are illustrated in Figure 2. Remote sensing applications with regard to flooding fall into two categories: (1) early warning and monitoring; (2) mapping and monitoring of flood defence structures and land cover.

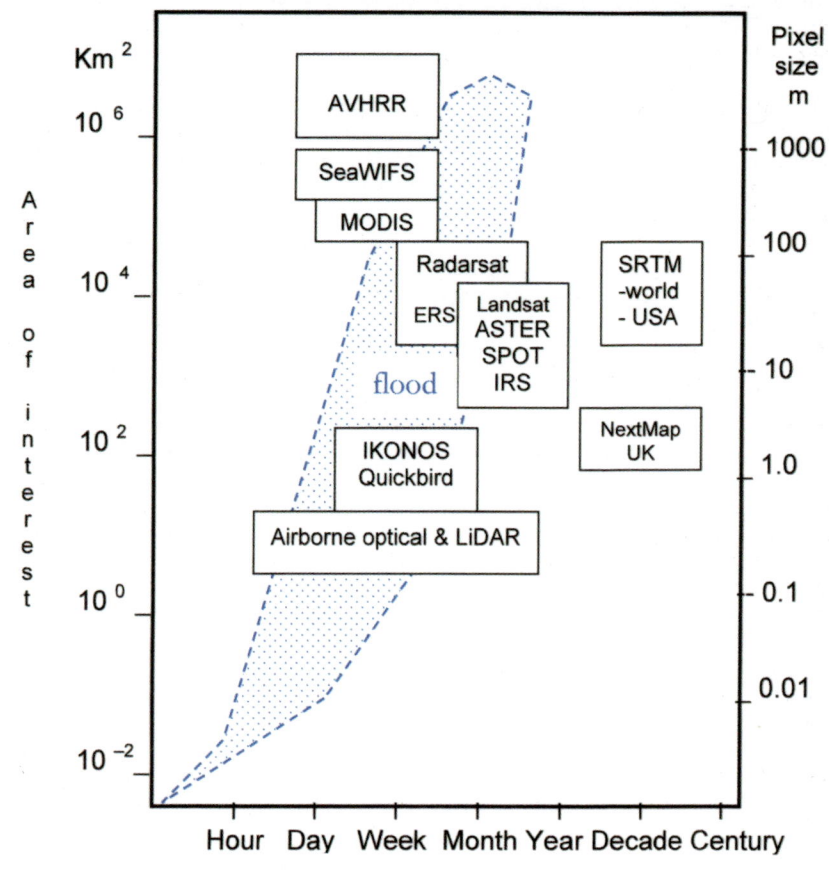

Fig. 2. Spatial and temporal resolutions of key remote sensing systems, in relation to flood events of varying magnitudes. The right ordinate scale represents the finest pixel size of the various sensors. The horizontal axis shows the frequency of data acquisition and also represents flood duration (based on Hirschboeck 1995) for drainage basins at scales shown on the left ordinate. Modified after Mertes (2002).

Early warning and monitoring. Remote sensing plays key roles in the early prediction, detection, delineation, monitoring and modelling of floods. On a continental scale, this involves the tracking of regional weather systems and storm clouds using weather satellites. On a regional scale, the freely available SRTM data have been used for the mapping and modelling of inundation zones; a useful innovation in countries with limited map coverage, particularly those susceptible to tsunami impacts (e.g. Theilen-Willige 2006).

Optical satellite sensors, such as Landsat and SPOT, cannot penetrate through cloud. This limits their usage for flood mapping, although a rare exception was the monitoring of the November 2000 Gloucester flooding using SPOT imagery (Sarti *et al.* 2001). Radar's ability to operate at night and to 'see' through cloud makes it particularly useful for mapping and monitoring major floods. Imaging radar also allows assessments of catchment soil moisture conditions, river levels, flooded areas and floodwater volumes, as well as the probable rate at which floodwater will move down-river, allowing early warnings to be given to vulnerable sites (e.g. Koblinsky *et al.* 1993; Goodrich *et al.* 1994; Nico *et al.* 2000; Horritt & Bates 2002).

The detailed mapping of floodplains involves the quantification of features such as flood compartments (areas where flood water would be contained by raised land features or structures such as dykes) and the rate at which floodwaters travel over a flood plain, involving evaluations of surface roughness related to vegetation types and built structures. In the light of these requirements for hydraulic modelling, detailed mapping and monitoring are needed of features with hydro-geomorphological significance on floodplains. Similar detailed mapping and monitoring is needed in coastal zones that are susceptible to flooding, with the added imperative of modelling the impacts of storm surges in the light of sea-level rise and global warming. The UK Environment Agency (EA) use photogrammetry on 1:3000 scale colour aerial photographs, to produce floodplain maps with 15–25 cm contour intervals. Improvements in LiDAR, which allow levels of mapping that rival photogrammetry, as well as LiDAR's ability to quantify the height and volume of objects on a floodplain, such as buildings and trees, will dramatically improve floodplain hydrodynamic modelling (e.g. Cobby *et al.* 2001; Hollaus *et al.* 2005; Mason *et al.* 2003; Lane *et al.* 2003; Straatsma & Middelkoop 2006).

The NextMap UK airborne InSAR DEM dataset has been used by insurers to map floodplain boundaries, allowing them to check the flood risk of a given building, based on its postcode location (Sanders *et al.* 2005). Of particular interest has been the Thames floodplain, which has the highest economic risk in the UK (DEFRA 2001). The EA has shown the benefits of multi-sensor aerial surveys, typically collecting LiDAR data simultaneously with CASI data. The LiDAR–CASI datasets can readily be processed within a GIS, yielding DEM data for flood modelling and land cover data that can be converted into a map showing the potential cost of flooding, allowing an evaluation of flood risk (Fig. 3 and Table 2).

Mapping and modelling of flood defence structures. The main application of remote sensing in the flood defence sector is in the production of GIS inventories of flood defence structures (e.g. flood walls, embankments, sluices) for facilities management and hydrological modelling. Mapping is carried out using 1:3000 scale airphotos and photogrammetry, recently augmented by detailed LiDAR surveys, producing maps with 15–25 cm contours. Future UK flood defences will involve new defence structures, with the removal of some flood walls and the 'managed retreat' of flood-defended areas, along with the creation of flood detention areas elsewhere. Maps showing land cover types and detailed (centimetre-scale) elevation will be needed, along with evaluations of potential flood damage costs, and for this purpose simultaneous LiDAR–CASI surveys appear be the most effective solution, as illustrated by the Environment Agency in Figure 3 and Table 2.

Floodplain encroachment is a severe problem: a walk-over survey along the River Lee, north of London, found that 90% of the floodplain structures (fences, walls, sheds, footbridges, etc.) had been constructed without official permission (Day & Fenner 1996). Features such as walls and fences are difficult to detect, requiring detailed aerial photography, expensive photogrammetric processing and time-consuming airphoto interpretation mapping. Recent developments have involved the use of (1) CASI hyperspectral and multispectral imagery to map areas of aquatic weed infestation and locate previously unknown discharge pipes (Thomson *et al.* 1998a,b; Hunter *et al.* 2000); (2) LiDAR, to rapidly identify features constructed on floodplains or affecting river channels (Webster 2006).

Ground instability

The mechanisms controlling coastal slope instability are driven by numerous environmental factors, notably the types of cliff material, groundwater conditions, wave impacts and human activity, all of which vary over space and time, and all of which can be detected, with varying degrees of success, by remote sensing. A detailed review of the types of remote sensing systems that can be used to detect ground instability features is

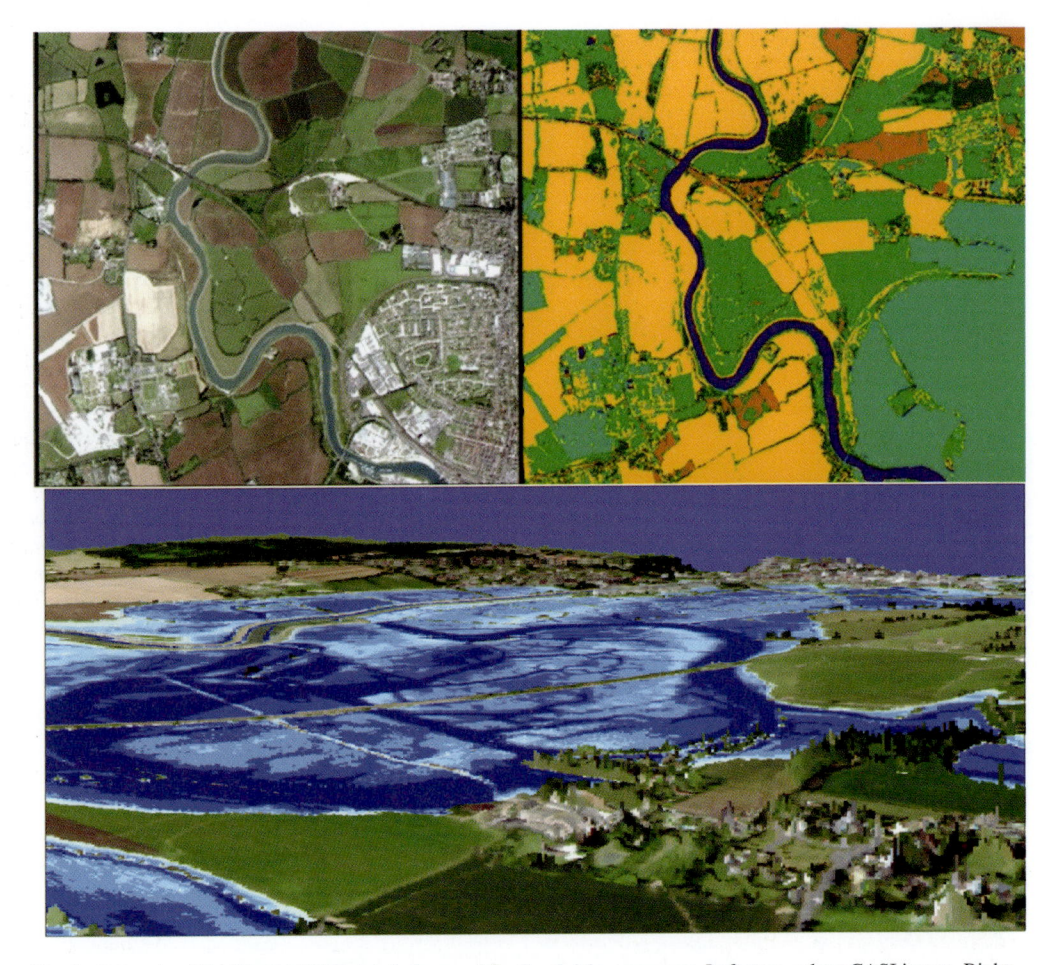

Fig. 3. Example of CASI and LiDAR data being used for flood risk assessment. Left, true-colour CASI image. Right, classified CASI image (yellow, arable; brown, rough grassland; turquoise, urban; light green, pasture; dark green, woodland). Bottom, 3D perspective view: true-colour CASI image draped over LiDAR DEM; the blue tones represent a simulated flood event. *Source*: Environment Agency.

beyond the scope of this review; instead the reader is referred to the studies by Mantovani *et al.* (1996) and Metternicht *et al.* (2005). Soeters & van Westen (1994) summarized various ways in which remote sensing can be used in conjunction with GIS to recognize and analyse zones of slope instability.

Geomorphological mapping, based on airphoto interpretation and photogrammetric analysis, is a well-established technique for mapping cliffs,

Table 2. *The cost of inundation for various land cover types derived from the CASI image displayed in Fig. 3*

Class	Area (ha)	Housing equivalent ha^{-1}	Total housing equivalents
Arable	280.11	0.5320	149.0
Pasture	229.17	0.0260	6.0
Rough grass	53.18	0.0106	0.6
Forest	40.49	0.0002	0.1
Urban	10.19	15.0000	152.8
Total	613.14	15.5698	308.5

Source: Environment Agency.

particularly where steep and unstable slopes make sites too dangerous for walk-over surveys. Cliffs still pose major problems for photogrammetric surveys, as a result of perspective variations and near-vertical slopes that create shadows and strong contrasts: these factors make it difficult to collect ground control points, which limits accuracies and the production of automated stereo-matches (Chandler & Moore 1989).

LiDAR, both airborne and terrestrial, has recently been applied to mapping and monitoring cliffs, with encouraging results. Figure 4 is an example of airborne LiDAR applied to quantifying retreat at Black Ven, a soft-rock cliff on the Dorset coast of England, the differencing between data collected a year apart highlighting zones of sediment transfer from the cliff to the beach. Adams & Chandler (2002) carried out a comparative study of Black Ven DEMs produced by LiDAR and photogrammetry: the former gave root mean square (RMS) errors of 0.26 m, the latter of 0.43 m. Both systems gave centimetre-scale underestimates of the elevation; the greater error from the photogrammetry was due to the very variable slope angles of the cliff. Recently, terrestrial LiDAR has been used to quantify the morphological features of cliffs: millimetre-accuracy DEMs, readily usable by most geotechnical slope stability software, can now be produced from ranges of up to 2 km (Gibson *et al.* 2003; Nagahara *et al.* 2004).

Variations in mineral spectral responses provide a number of ways to detect areas susceptible to slope instability, using visible and IR sensors (e.g. Ben-Dor *et al.* 2002; Kariuki *et al.* 2003). At Black Ven, shrink–swell clays were detected by spectrometry studies of landslide debris; furthermore, waterlogged areas prone to failure could be highlighted using a soil moisture index (Gibson 2005). Scaling-up from laboratory spectrometry to satellite imagery, Bourguingnon *et al.* (2007) determined the spectra of shrink–swell clays and went on to use ASTER multispectral imagery to map high concentrations of those clays and target areas of potential ground instability. Pre-dawn TIR imagery has been used to detect buried structures

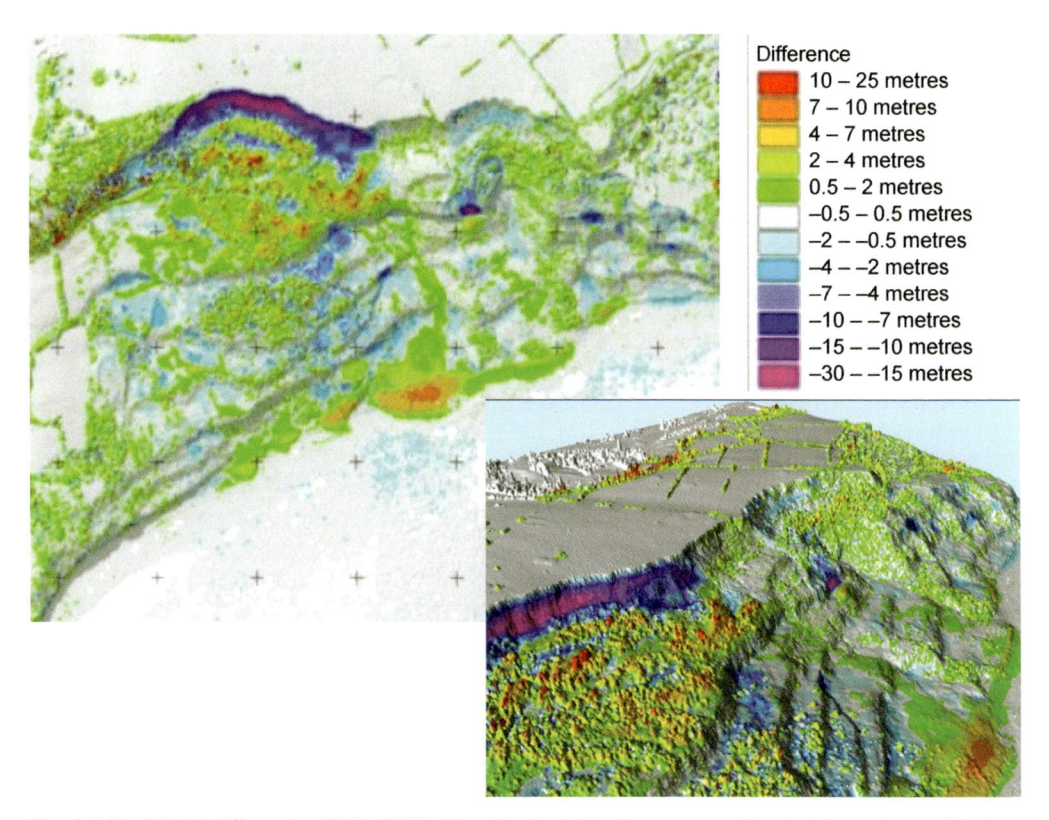

Fig. 4. LiDAR DEM differencing (2002–2001): Black Ven landslide, Dorset coast of England. Inset shows a 3D view of the landslide. *Source*: Environment Agency.

or leaking pipelines, but it is potentially useful for detecting cliff fracture zones and waterlogged soils (e.g. Loughlin 1990; Davidson & Watson 1995; Pickerill & Malthus 1998).

Annual clay shrinking and swelling in the London basin can cause up to 50 mm of vertical movement, resulting in widespread structural damage to buildings and infrastructure, with water mains being particularly affected. Satellite InSAR provides a regional overview of ground movements resulting from subsidence and/or the action of shrink–swell clays (Boyle *et al.* 2000). Persistent scatterer interferometry (PS-InSAR), allows the detection of millimetre-scale ground movements. PS-InSAR has highlighted localized subsidence, of up to 50 mm per year, under sections of the Thames flood embankments (Fig. 5).

Erosion and sedimentation

Despite the high degree of floodplain encroachment and engineering along much of the UK's river system and coast, river channels and the tidal foreshore are still very active terrains, both in terms of geomorphology (landforms: their component materials and formative processes) and sedimentology (deposits: their formative processes and resulting forms).

Archives of aerial photography suitable for stereoscopic viewing and photogrammetric analysis extend back to the late 1940s in the UK and cover the past 50 years in many other parts of the world. Although largely limited to analogue panchromatic airphotos, archive airphotos are a valuable data source for mapping and monitoring land cover and geomorphological features. The time range of the

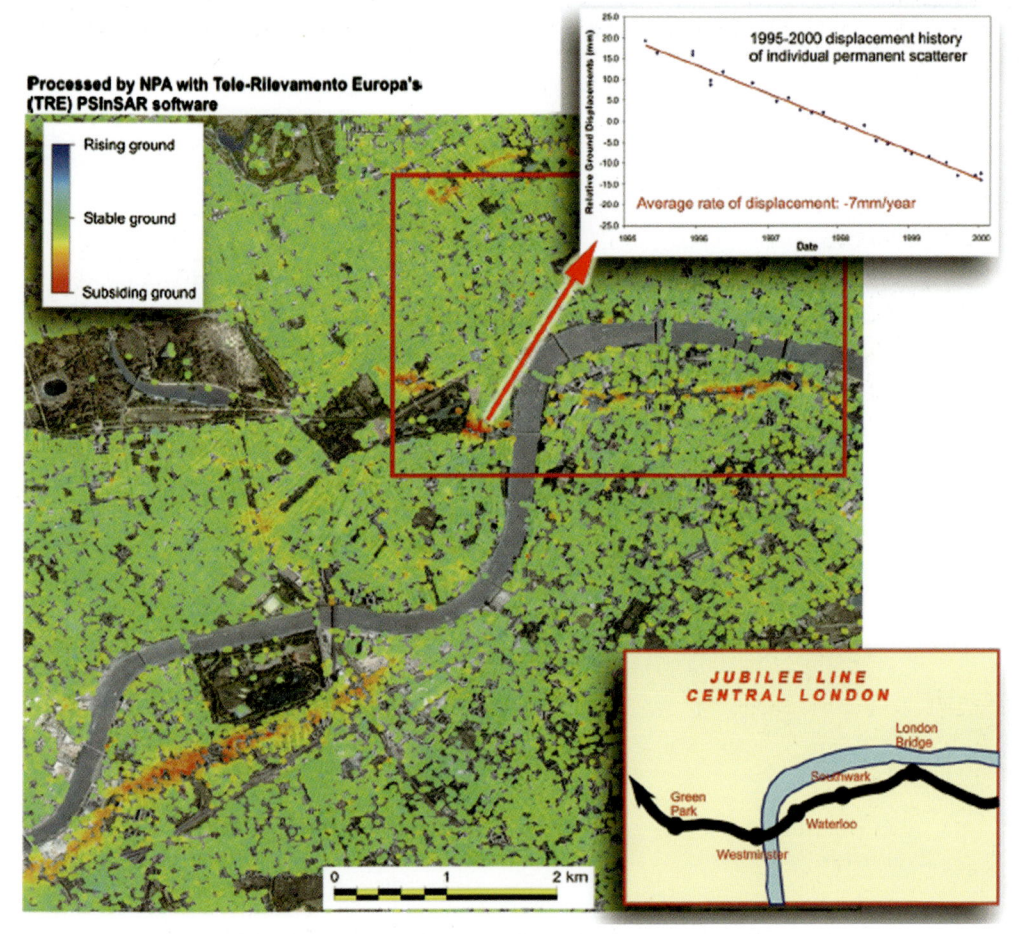

Fig. 5. PS-InSAR image of London, showing areas of pronounced subsidence, as a result of a new underground railway (see inset), under sections of the Thames flood embankments. *Source*: NPA Group.

UK airphoto archives is particularly useful in the light of the large changes in population density and land cover types that have occurred across many drainage basins and floodplains over the past 60 years. Archive aerial photography has been used extensively in the UK to map and quantify changes in riverine and coastal landforms (e.g. Winterbottom & Gilvear 1997; Lee & Brunsden 2001; Brasington et al. 2003). Repeated close-range terrestrial photogrammetric surveys have been used to map spatio-temporal changes along gravel-bed river banks and channels, allowing the identification of locations where individual clasts had been removed (Pyle et al. 1997; Butler et al. 1998). Figure 6 shows meander migration along a reach of the Afon Trannon, a tributary of the River Severn, derived from a set of archive aerial photography dating back to 1948. At a number of locations along the Trannon, digital photogrammetric analysis of bank erosion allowed estimates to be made of the volumes of sediment transported over time, highlighting the accelerated erosion and ensuing river channel instability that resulted from upper catchment afforestation in the late 1940s (Mount et al. 2003).

Airborne digital videography is used by the EA for qualitative assessments of flooded areas. Livingstone et al. (1999) have applied this approach to visualizing landforms along the UK's Norfolk coast. In the USA videography has been used from fixed locations to observe wave activity and changes in beach morphology, as part of a coastal zone management system (Davidson et al. 2004). The Airborne Thematic Mapper has been used to determine the bathymetry of channels along the River Tay in Scotland. Surveys before and after a major flood allowed the mapping and quantification of areas of erosion and deposition, along with sediment volume changes (Winterbottom & Gilvear 1997; Bryant & Gilvear 1999). Airborne LiDAR has proved to be a very effective technique for accurately mapping, monitoring and quantifying the morphology and dynamics of coastal landforms, thus helping to determine zones of erosion and deposition along beaches, sand dunes and cliffs, as illustrated in Figure 4.

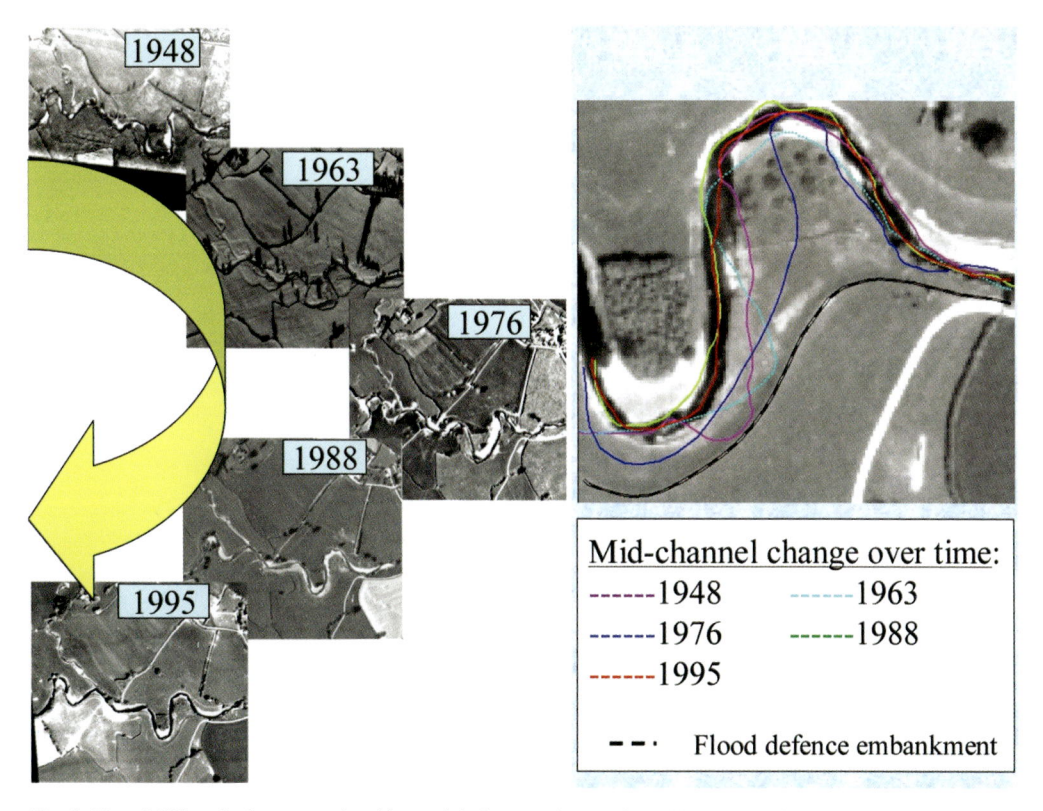

Fig. 6. Use of GIS and othocorrected archive aerial photography (1948–2000) to quantify erosion along the Afon Trannon, Wales.

The alternately wet and dry sediments of intertidal zones are difficult to map using remote sensing (Bryant et al. 1996). Imaging radar is useful for differentiating between coarse-grained and fine-grained sediments, as a result of differences in roughness, but radar is of limited use once the sediments are wet. Rainey et al. (2000) used an airborne multispectral scanner to examine sandy and clayey sediments on mudflats: the different spectral responses caused by wet and dry sediments were greater than the differences in spectral response caused by variations in sediment particle size. Recent research by Deronde et al. (2004) has shown how the spectra of a given sandy sediment when dry can be matched with the spectra of the same sediment when wet, allowing the spectral mapping of intertidal sediment types.

Pilling (1989) used low-tide summer Landsat TM imagery to map changes in the morphology of tidal sandbanks in the Mersey estuary with estimated accuracies of over 90%. Landsat TM's IR bands were also used to illustrate tidal sediment fluxes during environmental impact studies for proposed barrages on the Mersey and Severn estuaries. Middelkoop (2002) sampled suspended sediment loads in the Rhine during major floods and has been able to calibrate Landsat TM images of those flood events, allowing better estimations of both suspended sediment loads and zones of deposition.

Cost considerations

The various costs associated with remotely sensed data are summarized in Table 3. For regional mapping, spaceborne sensors are clearly the best value, particularly where archive Landsat and SRTM data can be utilized. Monitoring features at regional scales using satellite sensors is less straightforward and generally more expensive. SPOT, IKONOS and Quickbird can be programmed to capture daily repeat coverage, but data from these high resolution satellites are expensive. Prior to the systematic airborne IfSAR survey of Britain by Intermap Technologies, the collection of such data was limited to small one-off surveys, resulting in relatively expensive data. At £7 km^{-2} (Dowman 2004) the NextMap UK DEM is excellent value, ideal for rapid assessments of floodplain topography and flood hazard. Where detailed (10–20 cm contour) mapping of floodplains is required, high-cost photogrammetry and LiDAR are needed. LiDAR is proving to be better value, as it can rapidly produce DSMs, showing dimensions of buildings and vegetation, and DTMs, penetrating vegetation cover to record ground elevations, something that is not possible with photogrammetry.

The main cost incurred during airborne remote sensing is the flying time, so running different remote sensing systems simultaneously during a single flight (e.g. CASI + LiDAR + thermal scanning) is clearly a more cost-effective strategy than separate flights for each system. Running CASI simultaneously with LiDAR on survey flights appears to be a particularly effective strategy. The CASI data allow the spectral mapping of land cover types, particularly vegetation. The LiDAR data allow the mapping of micro-relief and surface roughness, as well as providing data on vegetation height. Brown (2004), examining salt marshes and sand dunes with statistical and neural network classifiers, has shown how vegetation classifications are improved by inputting CASI and LiDAR data layers.

Table 3. *Comparative costs of various types of remote sensing for riverine and coastal applications*

Type of remote sensing	£ km^{-2}	If more detail required	£ km^{-2}
Airborne data			
Stereoscopic airphotos	100–400	Photogrammetry	400–2000
ATM	140–250	Major processing	250–350
CASI	140–280	Major processing	280–400
LiDAR (>1 m posting)	150–300	Sub-metre posting	300–620
IfSAR DEM	3–300	Sub-metre posting	300–600
Thermal	100–150	Major processing	150–200
Spaceborne data			
Landsat	Free–0.02	Advanced processing	0.02–8
ASTER	0.01	With processing + DEM	0.01–12
SPOT	0.3	With processing + DEM	0.6–16
ERS imaging radar	0.3	Advanced processing	0.3–8
ERS InSAR	40	PS-InSAR	40–220
SRTM IfSAR DEM	Free		

Fuller et al. (1996, 1997), Environment Agency (1998), Gebhardt (1998), NRT Associates (2000), and Airborne costs compiled from Dowman (2004).

Conclusion

Aerial photography remains the most widely used system of remote sensing, allowing geomorphological interpretive mapping, as well as analytical photogrammetry and the production of detailed contour maps. However, the time and levels of expertise required for effective airphoto-based mapping are relatively high and expensive. Airborne multispectral and hyperspectral remote sensing cannot yet match aerial photography for detail, but offers possible 'short-cuts' for geohazard detection, such as the automated detection of shrink–swell clays, waterlogged soils and vegetation communities associated with frequent flooding. The poor spatial resolution of multispectral satellite remote sensing systems, relative to airborne systems, is becoming less of a problem, with some new satellite sensors having 0.6 m pixels. The trade-off is that satellite systems with 5–15 m pixels provide a useful low-cost multispectral synoptic overview of large regions.

LiDAR requires complex computer processing, involving vast amounts of data, resulting in relatively expensive datasets. However, the detailed elevation models provided by LiDAR are proving to be invaluable for the mapping and modelling of fluvial and coastal terrains. A major cost involved with airborne remote sensing is for the survey flight: multi-sensor systems, particularly LiDAR–CASI, have been relatively accurate and cost-effective for mapping areas at risk of flooding.

Three types of microwave radar remote sensing have been successfully applied to mapping geohazards in riverine and coastal settings. Imaging radar's ability to detect areas of water and variations in soil moisture has been particularly useful in flood management: new developments with imaging radar involve multifrequency and multi-polarizing sensors. Radar interferometry (InSAR) has made rapid improvements, allowing the production of low-cost DEMs, from both aircraft (e.g. NextMap) and the Space Shuttle (SRTM). Persistent scatterer InSAR, using repeat images from the ERS satellite, has allowed millimetre-scale monitoring of ground deformation, highlighting previously unknown subsidence hazards.

Archives of remotely sensed imagery, some spanning the past half-century, are proving to be valuable sources of information about the extent of land cover changes, particularly urbanization, and associated problems with flooding, erosion, slope instability and sedimentation.

Many thanks go to the British Geological Survey, the Environment Agency and NPA Group for providing images and giving permission to reproduce them in this paper.

References

ADAMS, J. C. & CHANDLER, J. H. 2002. Evaluation of LiDAR and medium-scale photogrammetry for detecting soft-cliff coastal change. *Photogrammetric Record*, **17**, 405–418.

BABEY, S. K. & ANGER, C. D. 1989. A Compact Airborne Spectrographic Imager (CASI). *Proceedings of the IEEE International Geoscience and Remote Sensing Symposium (IGARSS'89), Vancouver, Canada, 10–14 July 1989*. IEEE Press, New York, 1028–1031.

BEN-DOR, E., PATKIN, K., BANIN, A. & KARNIELI, A. 2002. Mapping several soil properties using DAIS-7915 hyperspectral scanner data—a case study over clayey soils in Israel. *International Journal of Remote Sensing*, **23**, 1043–1062.

BIRD, E. C. F. 2000. *Coastal Geomorphology: an Introduction*. Wiley, Chichester.

BLYTH, K. 1994. The use of satellite radar for monitoring fluvial and coastal flooding. *In*: WADGE, G. (ed.) *Natural Hazards and Remote Sensing. Proceedings of UK IDNDR Conference*. Royal Society, London, 59–63.

BOURCIER, A., POUDEVIGNE, I. & TEEUW, R. M. 2002. The use of SPOT imagery as a tool for ecological analysis of river floodplains: a case study in the Seine valley. *In*: LEUVEN, R. S. E. W., POUDEVIGNE, I. & TEEUW, R. M. (eds) *Application of Geographic Information Systems and Remote Sensing in River Studies*. Bakhuys, Leiden, 63–74.

BOURGUIGNON, A., DELPONT, G., CHEVREL, S. & CHABRILLAT, S. (2007) Detection and mapping of shrink–swell clays in SW France, using ASTER imagery. *In*: TEEUW, R. M. (ed.) *Remote Sensing of Hazardous Terrain*. Geological Society, London, Special Publications, **283**, 117–124.

BOYLE, J., STOW, R. & WRIGHT, P. 2000. *InSAR imaging of London surface movement for structural damage management and water resource conservation*. Report for BNSC Link programme, project R4/019.

BRASINGTON, J., LANGHAM, J. & RUMSBY, B. 2003. Methodological sensitivity of morphometric estimates of coarse fluvial sediment transport. *Geomorphology*, **53**, 299–316.

BROWN, K. 2004. Increasing classification accuracy of coastal habitats using integrated airborne remote sensing. *EARSeL eProceedings*, **3**, 34–42.

BRYANT, R. G. & GILVEAR, D. J. 1999. Quantifying geomorphic and riparian land cover changes either side of a large flood event, using airborne remote sensing: River Tat, Scotland. *Geomorphology*, **29**, 307–321.

BRYANT, R, TYLER, A., GILVEAR, D., MCDONALD, P., TEASDALE, I., BROWN, J. & FERRIER, G. 1996. A preliminary investigation into the spectral characteristics of inter-tidal estuarine sediments. *International Journal of Remote Sensing*, **17**, 405–412.

BUTLER, J., LANE, S. N. & CHANDLER, J. H. 1998. DEM quality assessment for surface roughness characterisation using close-range photogrammetry. *Photogrammetric Record*, **16**, 271–291.

CHANDLER, J. H. 2001. Terrain measurements using automated digital photogrammetry. *In*: GRIFFITHS, J. S.

(ed.) *Land Surface Evaluation for Engineering Practice*. Geological Society, London, Engineering Geology Special Publications, **18**, 13–18.

CHANDLER, J. H. & MOORE, R. 1989. Analytical photogrammetry: a method for monitoring slope instability. *Quarterly Journal of Engineering Geology*, **22**, 97–110.

CHARLTON, M. E., LARGE, A. R. G. & FULLER, I. C. 2003. Application of airborne LiDAR in river environments: the River Coquet, Northumberland, UK. *Earth Surface Processes and Landforms*, **28**, 299–306.

COBBY, D. M., MASON, D. C. & DAVENPORT, I. J. 2001. Image processing of airborne scanning laser altimetry data for improved river flood modelling. *ISPRS Journal of Photogrammetry and Remote Sensing*, **56**, 121–138.

CRACKNELL, A. P. 1999. Remote sensing in estuaries and coastal zones—an update. *International Journal of Remote Sensing*, **20**, 485–496.

DAVIDSON, D. A. & WATSON, A. I. 1995. Spatial variability in soil moisture as predicted from airborne thematic mapper (ATM) data. *Earth Surface Processes and Landforms*, **20**, 219–230.

DAVIDSON, M. A., AARNINKHOF, S. G. J., VAN KONINGSVELD, M. & HOLMAN, R. A. 2004. Developing coastal video monitoring systems in support of coastal zone management. *Journal of Coastal Research*, **SI 39**, 1–14.

DAY, R. A. & FENNER, R. A. 1996. The effectiveness of the Land-Drainage Consent system. *CIWEM Journal*, **10**, 111–117.

DEFRA 2001. *National appraisal of assets at risk from flooding and coastal erosion, including the potential impact of climate change. Final Report, July 2001*. Flood Management Division, DEFRA, London.

DENNISS, A. 1999. Moving into the hyperspectral age. *Mapping Awareness*, **13**, 47–49.

DERONDE, B, HOUTHUYS, R., STERCKX, S., DEBRUYN, W. & FRANSAER, D. 2004. Sand dynamics along the Belgian coast, based on airborne hyperspectral data and LiDAR data. *EARSeL eProceedings*, **3**, 26–33.

DONOGHUE, D. N. M. & SHENNAN, I. 1987. A preliminary assessment of Landsat TM imagery for mapping vegetation and sediment distribution in the Wash estuary. *International Journal of Remote Sensing*, **8**, 1101–1108.

DOWMAN, I. 2004. *Integration of LiDAR and IFSAR for mapping*. Invited Paper Commission II, WG2, ISPRS, Istanbul.

DOWMAN, I., BALAN, P., RENNER, K. & FISCHER, P. 2003. *An evaluation of NextMap terrain data in the context of UK national datasets*. Report to Getmapping, Dept. of Geomatic Engineering, UCL.

EASTWOOD, J. A., YATES, M. G., THOMSON, A. G. & FULLER, R. M. 1999. The reliability of vegetation indices for monitoring salt marsh vegetation cover. *International Journal of Remote Sensing*, **18**, 391–3907.

ENVIRONMENT AGENCY 1998. *Airborne Light Detection and Range feasibility study*. Environment Agency, R&D Technical Report, **E43**.

FULLER, R. M., THOMSON, A. G., EASTWOOD, J. A., YATES, M., SPARKS, T. & WARMAN, E. 1995. *Further development of airborne remote sensing techniques: cover classification in intertidal zones and river corridors. Final Report: Part 1*. National Rivers Authority, R&D Note **472**.

FULLER, R. M., THOMSON, A. G. & EASTWOOD, J. A. 1996. *Strategic remote sensing of coastal and estuarine environment types for flood defence and conservation: feasibility study. Part 1*. Environment Agency, R&D Technical Report, W47.

GALY, H. M. & SANDERS, R. 2002. Using Synthetic Aperture Radar imagery for flood modelling. *Transactions in GIS*, **6**, 31–42.

GEBHARDT, M. 1998. *Airborne Synthetic Aperture Radar (SAR) Feasibility Study*. Environment Agency, R&D Technical Report, **E37**.

GIBSON, A. D. 2005. *The spectral characterization of landslide debris*. PhD thesis, University of Portsmouth.

GIBSON, A. D., FORSTER, A. F., POULTON, C., ROWLANDS, K., JONES, L. J., HOBBS, P. R. N. & WHITWORTH, M. C. Z. 2003. An integrated method for terrestrial laser-scanning subtle landslide features and their geomorphological setting. *In*: APLIN, P. & MATHER, P. M. (eds) *Proceedings RSPSoc 2003*. Remote Sensing & Photogrammetry Society, Nottingham.

GOODRICH, D. C., SCHMUGGE, T. J., JACKSON, C. L., UNKRICH, T. O., PARRY, R., BACH, L. B. & AMER, S. A. 1994. Runoff simulation sensitivity to remotely sensed initial soil water content. *Water Resources Research*, **30**, 1393–1405.

HASLETT, S. K. 2000. *Coastal Systems*. Routledge, London.

HIRANO, A., WELCH, R. & LANG, H. 2003. Mapping from ASTER stereo image data: DEM validation and accuracy assessment. *ISPRS Journal of Photogrammetry and Remote Sensing*, **57**, 356–370.

HIRSCHENBROEK, K. K. 1995. Flood hydroclimatology. *In*: BAKER, V. R., KOCHEL, R. C. & PATTON, P. C. (eds) *Flood Geomorphology*. Wiley, New York, 27–49.

HOLLAUS, M., WAGNER, W. & KRAUS, K. 2005. Airborne laser scanning and usefulness for hydrological models. *Advances in Geosciences*, **5**, 57–63.

HORRITT, M. S. & BATES, P. D. 2002. Evaluation of 1D and 2D numerical models for predicting river flood inundation. *Journal of Hydrology*, **268**, 87–99.

HOUGHTON, J. T., DING, Y., GRIGGS, D. J., NOUGER, M., VAN DER LINDEN, P. J. & XIAOSU, D. (eds) 2001. *Climate Change: The Scientific Basis: Contribution of Working Group 1 to the Third Assessment Report of the International Panel on Climate Change*. Cambridge University Press, Cambridge.

HUNTER, E., DALTON, F. & POWER, C. 2000. Thames Estuary marshland habitat discrimination using multispatial resolution remotely sensed data. *In: Proceedings RSS 2000 Conference*. Remote Sensing Society, Nottingham, 655–663.

HUNTER, E. L. & POWER, C. H. 2002. An assessment of two classification methods for mapping Thames Estuary intertidal habitats using CASI data. *International Journal of Remote Sensing*, **23**, 2989–3008.

KARIUKI, P. C., VAN DER MEER, F. D. & SIDERIUS, W. 2003. Classification of soils based on engineering indices and spectral data. *International Journal of Remote Sensing*, **12**, 2567–2574.

KASISCHKE, E. S., MELACK, J. M. & DOBSON, M. C. 1997. The use of imaging radars for ecological applications – a review. *Remote Sensing of Environment*, **59**, 141–156.

KOBLINSKY, C. J., CLARKE, R. T., BRENNER, A. C. & FREY, H. 1993. Measurement of river level variations with satellite altimetry. *Water Resources Research*, **29**, 1839–1848.

KORMUS, W., ALAMUS, R., RUIZ, A. & TALAYA, J. 2004. Assessment of DEM accuracy derived from SPOT 5 High Resolution Stereoscopic imagery. *In*: *Proceedings ISPRS Conference, Istanbul.*

LANE, S. N., JAMES, T. D., PRITCHARD, H. & SAUNDERS, M. 2003. Photographic and laser altimetric reconstruction of water levels for extreme flood event analysis. *Photogrammetric Record*, **16**, 793–821.

LEE, E. M. 2001. Living with natural hazards: the costs and management framework. *In*: HIGGETT, D. L. & LEE, E. M. (eds) *Geomorphological Processes and Landscape Change.* Blackwell, Oxford, 237–268.

LEE, E. M. & BRUNSDEN, D. 2001. Sediment budget analysis for coastal management, west Dorset. *In*: GRIFFITHS, J. S. (ed.) *Land Surface Evaluation for Engineering Practice.* Geological Society, London, Engineering Geology Special Publications, **18**, 13–18.

LEUVEN, R. S. E. W., POUDEVIGNE, I. & TEEUW, R. M. (eds) 2002. *Application of Geographic Information Systems and Remote Sensing in River Studies.* Bakhuys, Leiden.

LIVINGSTONE, D., RAPER, J. & MCCARTHY, T. 1999. Integrating aerial videography with terrain modelling: an application for coastal geomorphology. *Geomorphology*, **29**, 77–92.

LOUGHLIN, W. P. 1990. Geological exploration in the western United States by use of airborne scanner imagery. *In*: *Remote Sensing an Operational Technology for the Mining and Petroleum Industries.* IMM (Institute of Mining & Metallurgy), London, 223–241.

LUCAS, N. S., SHANMUGAM, S. & BARNSLEY, M. 2002. Sub-pixel habitat mapping of a coastal dune ecosystem. *Applied Geography*, **22**, 253–270.

MANTOVANI, F., SOETERS, R. & VAN WESTERN, C. 1996. Remote sensing techniques for landslide studies and hazard zonation in Europe. *Geomorphology*, **15**, 213–225.

MASON, D. C., COBBY, D. M., HORRIT, M. S. & BATES, P. 2003. Floodplain friction parameterization in two-dimensional river flood models using vegetation heights derived from airborne scanning laser altimetry. *Hydrological Processes*, **17**, 1711–1732.

MERTES, L. A. K. 2002. Remote sensing of riverine landscapes. *Freshwater Ecology*, **47**, 799–816.

METTERNICHT, G., HURNI, L. & GOGU, R. 2005. Remote sensing of landslides: an analysis of potential contribution to geo-spatial systems for hazard assessment in mountainous environments. *Remote Sensing of Environment*, **98**, 284–303.

MIDDELKOOP, H. 2002. Application of remote sensing and GIS-based modelling in the analysis of floodplain sedimentation. *In*: LEUVEN, R. S. E. W., POUDEVIGNE, I. & TEEUW, R. M. (eds) *Application of Geographic Information Systems and Remote Sensing in River Studies.* Bakhuys, Leiden, 95–117.

MILTON, E. J., GILVEAR, D. J. & HOPPER, I. D. 1995. Investigating change in fluvial systems using remotely sensed data. *In*: GURNELL, A. M. & PETTS, G. E. (eds) *Changing River Channels.* Wiley, Chichester, 277–298.

MOUNT, N. J., LOUIS, J., TEEUW, R. M., ZUKOWSKYJ, P. M. & STOTT, T. 2003. Estimation of error in bank-full width comparisons from temporally sequenced raw and corrected aerial photographs. *Geomorphology*, **56**, 65–77.

MULLER, E., DECAMPS, H. & DOBSON, M. K. 1993. Contribution to space remote sensing to river studies. *Freshwater Biology*, **29**, 301–312.

MUNICH, Re 1997. *Flooding and Insurance.* Munich Re, Munich.

NAGAHARA, S., MULLIGAN, K. R. & XIONG, W. 2004. Use of a three-dimensional laser scanner to digitally capture the topography of sand dunes in high spatial resolution. *Earth Surface Processes and Landforms*, **29**, 391–398.

NICHOLLS, R. J., HOOZEMANS, F. & MARCHAND, M. 1997. Impacts of sea level rise on coastal areas. *In*: *Climate Change and Its Impacts – A Global Perspective.* Meteorological Office, Bracknell, 14–15.

NICO, G., PAPPALEPORE, M., PASQUARIELLO, G., RECIFE, A. & SAMARELLI, S. 2000. Comparison of SAR amplitude vs. coherence flood detection methods—a GIS application. *International Journal of Remote Sensing*, **21**, 1619–1631.

NRT Associates 2000. *Potential use of CASI for monitoring.* Phase II and III reports to the Environment Agency (Thames Region).

PHINN, S. R., MENGES, C., HILL, G. J. E. & STANFORD, M. 2000. Optimising remote sensing for monitoring, modeling and managing coastal environments. *Remote Sensing of Environment*, **73**, 117–132.

PICKERILL, J. M. & MALTHUS, T. J. 1998. Leak detection from rural aqueducts using airborne remote sensing techniques. *International Journal of Remote Sensing*, **19**, 2427–2433.

PILLING, I. 1989. *Sandbank Mapping in the Mersey Estuary.* SPI Division Working Paper, SP(89) WP35, National Remote Sensing Centre, Farnborough, UK.

PYLE, C. J., RICHARDS, K. S. & CHANDLER, J. H. 1997. Digital photogrammetric modelling of river bank erosion. *Photogrammetric Record*, **15**, 753–763.

RAINEY, M. P., TYLER, A. N., BRYANT, R. G., GILVEAR, D. J. & MCDONALD, P. 2000. The influence of surface and interstitial moisture on the spectral characteristics of intertidal sediments: implications for airborne image acquisition and processing. *International Journal of Remote Sensing*, **21**, 3025–3038.

RIEDMANN, M. & HAYNES, M. 2007. Developments in synthetic aperture radar interferometry for monitoring geohazards. *In*: TEEUW, R. M. (ed.) *Remote Sensing of Hazardous Terrain.* Geological Society, London, Special Publications, **283**, 126–133.

RITCHIE, J. C. & RANGO, A. 1996. Remote sensing applications to hydrology—introduction. *Hydrological Sciences*, **41**, 429–431.

SANDERS, R., SHAW, F., MACKAY, H., GALY, H. & FOOTE, M. 2005. National flood modelling for insurance purposes: using IFSAR for flood risk estimation in Europe. *Hydrology and Earth System Sciences*, **9**, 449–456.

SARTI, F., INGLADA, J., LANDRY, R. & PULZ, T. 2001. Risk management using remote sensing data: moving from scientific to operational applications. *In*: Proceedings 10th BSR Symposium, Brazil, 23–27 April, http://www.treemail.nl/download/sartiol.pdf.

SCHMUGGE, T. J., KUSTAS, W. P., RITCHIE, J. C., JACKSON, T. J. & RANGO, A. 2002. Remote sensing in hydrology. *Advances in Water Resources*, **25**, 1367–1385.

SCHULTZ, G. A. & ENGMAN, E. T. 2000. *Remote Sensing in Hydrology and Water Management*. Springer, Berlin.

SEAMAN, E. J. & FLETCHER, P. A. 2000. *An evaluation of satellite sensing to monitor changes in coastal landscape*. Environment Agency, R&D Technical Report, E81.

SOETERS, R & VAN WESTEN, C. J. 1994. Slope instability: the role of remote sensing and GIS in recognition, analysis and zonation. *In*: WADGE, G. (ed.) *Natural Hazards and Remote Sensing. Proceedings of UK IDNDR Conference*. Royal Society, London, 44–50.

SRTM MISSION, JPL WEBSITE 2005. World Wide Web Address: http://www2.jpl.nasa.gov/srtm/.

STRAATSMA, M. W. & MIDDELKOOP, H. 2006. Airborne laser scanning as a tool for lowland floodplain vegetation monitoring. *Hydrobiologia*, **565**, 87–103.

THEILEN-WILLIGE, B. 2006. Tsunami risk site selection in Greece, based on remote sensing and GIS methods. *Science of Tsunami Hazards*, **24**, 35–48.

THOMSON, A. G., FULLER, T. H., SPARKS, T. H., YATES, M. G. & EASTWOOD, J. A. 1998a. Ground and airborne radiometry over intertidal surfaces: waveband selection for cover classification. *International Journal of Remote Sensing*, **19**, 1189–1205.

THOMSON, A. G., FULLER, R. M. & EASTWOOD, J. A. 1998b. Supervised versus unsupervised methods for classification of coasts and river corridors from airborne remote sensing. *International Journal of Remote Sensing*, **19**, 3423–3431.

VAN ZUIDAM, R. A. 1993. Review of remote sensing applications in coastal zone studies. *In*: VAN GELDEREN, J. L., VAN ZUIDAM & POHL, C. (eds) *Proceedings International Symposium: 'Operationalisation of Remote Sensing'*, ITC, Enschede, **7**, 1–15.

WEBSTER, N. 2006. The use of elevation models for flood modelling by the Environment Agency. *In: Elevation Models for Geoscience. GRSG-GIG conference*. Geological Society, London, 2 (abstract).

WEHR, A. & LOHR, U. 1999. Airborne laser scanning—an introduction and overview. *ISPRS Journal of Photogrammetry and Remote Sensing*, **54**, 68–82.

WINTERBOTTOM, S. J. 2000. Medium and short-term channel planform changes on the Rivers Tay and Tummel, Scotland. *Geomorphology*, **34**, 195–208.

WINTERBOTTOM, S. J. & GILVEAR, D. J. 1997. Quantification of channel bed morphology in gravel-bed rivers, using airborne multi-spectral imagery and aerial photography. *Regulated Rivers: Research & Management*, **13**, 489–499.

Application of hyperspectral remote sensing data in the monitoring of the environmental impact of hazardous waste derived from abandoned mine sites

G. FERRIER[1], B. RUMSBY[1] & R. POPE[2]

[1]*Department of Geography, University of Hull, Hull HU6 7RX, UK*
(e-mail: g.ferrier@hull.ac.uk)

[2]*Division of Geographical Sciences, University of Derby, Kedleston Road, Derby DE22 1GB, UK*

Abstract: The release of hazardous mine waste from abandoned gold mining areas is a major global environmental problem. The complexity of the processes, the scale of the mine sites, problems with accessibility and the lack of site information mean that field-based survey methods are often too costly and provide incomplete coverage. The results of a study of an abandoned gold mine at Rodalquilar in southern Spain have demonstrated the potential of airborne-mounted hyperspectral remote sensing instruments to resolve the distributions of mine waste and secondary iron species on the mine site and in adjacent rivers. The advantages of using higher spatial resolution hyperspectral data in identifying secondary iron species and resolving geomorphological settings of mine waste accumulations were also demonstrated. Integration of laboratory-derived correlations of secondary iron species, cyanide and heavy metals concentrations with the results of the remote sensing study allowed the identification of locations of hazardous materials over the study area and a more accurate understanding of the environmental status of the Rodalquilar area.

Acid mine drainage emanating from abandoned mines can adversely affect the quality of drinking water and the health of riparian ecosystems (Toy & Hadley 1987). The large number and scale of abandoned mine sites, the rapid changes in tailings and river channel form, and the difficulty of gaining access to relevant areas because of the steep topography and instability of waste piles mean that field-based geomorphological and geochemical mapping are often prohibitively time-consuming, costly and incomplete. A technique to carry out a rapid and accurate mapping of the mineralogy of mine waste affected areas is therefore required to identify the locations of the most significant sources of acid mine drainage discharge (AMD) and mine waste accumulation. The objectives of this study were to (1) investigate the utility of spectral remote sensing data to map Fe-bearing mineral indicators of acidic mine waste, (2) investigate the connection between spectrally detectable minerals, and heavy metal and cyanide concentrations, and (3) resolve the geomorphological controls on the accumulation of the hazardous mine waste.

Geology and mining history of the Rodalquilar mining district

The Rodalquilar mining district is located in an area of late Tertiary calc-alkaline volcanic and low-grade metamorphic rocks in southeastern Spain (Fig. 1). Extensive hydrothermal alteration of the volcanic host rocks has resulted in formation of hydrothermal alteration zones from high to low intensity in the sequence: silica–alunite–kaolinite–illite–smectite–chlorite. Associated with this mineral alteration are high sulphidation gold deposits and low sulphidation base metal deposits, hosted by rhyolitic ignimbrite deposits and domes (Rytuba *et al.* 1990). In the final stage of mining, between 1943 and 1966 (Rytuba *et al.* 1990), the gold was extracted from finely ground ore (<63 μm) using cyanide leachate and the waste material was deposited onto a tailings dump (Fig. 1). The tailings dump contains 1.440×10^6 m^3 of mine waste extending across two shallow valleys covering an area of 122 365 m^2. The tailings dump material has a distinctive light red colour, is weakly cemented, displays sub-centimetre layering and comprises predominantly quartz with subordinate kaolinite and alunite, with traces of jarosite (Rytuba *et al.* 1990).

Rodalquilar has a typical Mediterranean semi-arid climate. Thirty year data (1960–1990) indicate that average annual rainfall for the area is 202 ($\pm$100) mm with minimum and maximum annual rates of 102 and 549 mm, respectively. During the year, the majority of rainfall is concentrated within the winter months (November–January) as moderate to intense events lasting 1 or 2 days. The absence of significant vegetation cover on the

From: TEEUW, R. M. (ed.) *Mapping Hazardous Terrain using Remote Sensing.* Geological Society, London, Special Publications, **283**, 107–116.
DOI: 10.1144/SP283.9   0305-8719/07/$15.00   © The Geological Society 2007.

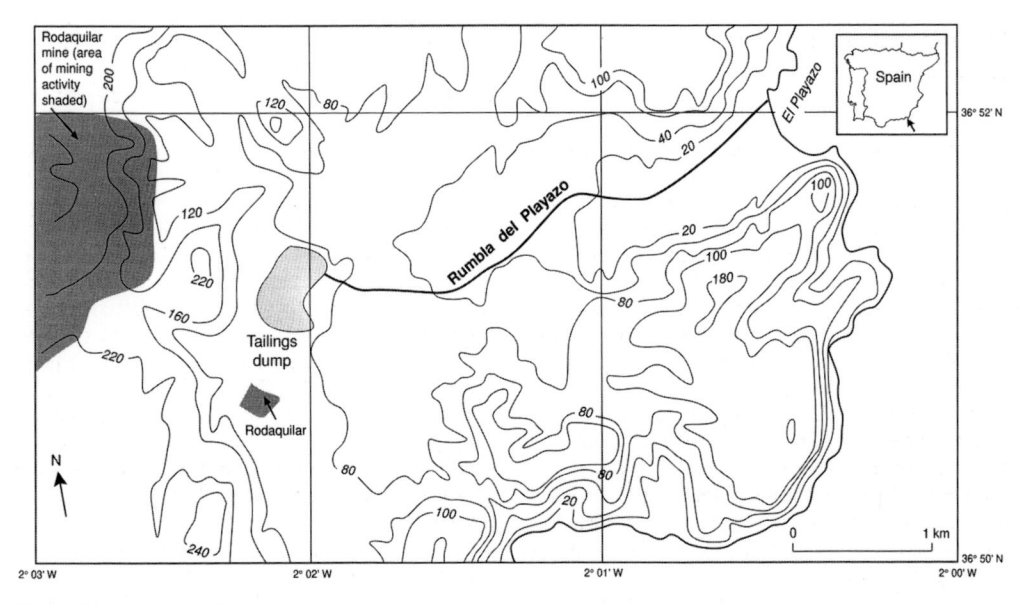

Fig. 1. Location map of the study area.

tailings dump means that the very fine-grained, non-cohesive mine waste will be readily loosened by rainsplash, making it highly vulnerable to localized transfer by overland flow, and rilling and gullying processes (Evans *et al.* 2000). Once the mine waste has been delivered to ephemeral drainage systems, it will be dispersed under the flashy discharge generated by these brief rainfall events, and transferred through, or deposited within, channels and floodplains alongside uncontaminated river sediment (Hudson-Edwards *et al.* 2003). This combination of episodic flow events and periodic failures in containment structures has produced a complex sequence of erosional and depositional features within the Rodalquilar riverbed.

Environmental impacts of mine waste

As meteoric water passes through mine waste it is acidified by sulphide oxidation and then partly neutralized by hydrolysis reactions with aluminosilicates and other minerals present in the waste piles. This leads to the accumulation of Fe sulphates, oxyhydroxides and oxides in a spatial and temporal sequence that represents the buffering of the acidic solution as it moves away from its source (Swayze *et al.* 2000).

The distribution of these secondary minerals about a source of active pyrite oxidation forms a spatial pattern in which copiapite and jarosite are relatively abundant near or at the source, and are surrounded by goethite and hematite (Montero *et al.* 2005). Hematite accumulates away from the sources of acidity after forming in a pH-dependent process that may involve the dehydration and transformation of earlier precipitates, such as those of goethite and ferrihydrite (Alpers *et al.* 1994). Trace elements can become aggregated in these iron oxide and oxyhydroxide minerals and/or mineraloids either through adsorption on the mineral surface or incorporation into the mineral structure (Schwertmann & Taylor 1977; Ferris *et al.* 1989).

The cyanide heap–leach technique is a commonly used method for the extraction of gold from low-grade deposits and remnant tailings dumps. The cyanide leachate generally forms iron-cyanide complexes in Fe-rich environments (Meeussen *et al.* 1995). These complexes still pose an environmental risk, as they can undergo rapid photolysis in the presence of sunlight, or react to changing pH conditions and/or redox potential to yield the free cyanide form. Cyanide readily combines with most major and trace metals producing a wide variety of toxic, cyanide-related compounds and reacts readily with other carbon-based matter, including living organisms. While leachate operations are active, solutions are kept at alkaline pH levels because metal extraction is more efficient at lower pH levels. Once an operation is abandoned acid mine drainage develops and the cyanide complexes decompose more readily. The presence of acid mine drainage, and high concentrations of toxic heavy metals and a range of cyanide complexes make mine waste from abandoned gold mines extremely hazardous to both humans and many other organisms, particularly aquatic ones.

Mapping mine waste with hyperspectral remote sensing data

Direct mapping of pyrite using hyperspectral remote sensing data would be the ideal method of detecting sources of acidic mine waste. The low reflectance of pyrite, saturated Fe absorptions and coating by secondary Fe minerals hamper its direct spectral detection (Swayze et al. 2000). The secondary iron minerals deposited from acid mine drainage runoff have unique reflectance characteristics in the visible and near-IR portion of the spectrum, making them highly amenable to detection and mapping using hyperspectral remote sensing data (Swayze et al. 2000).

Two airborne remote sensing datasets of the study area have been acquired. Airborne Visible Infrared Imaging Spectroradiometer (AVIRIS) data were acquired in July 1991, and Compact Airborne Spectrometer (CASI) and Airborne Thematic Mapper (ATM) data were acquired in May 2001 (Table 1). Field spectral measurements were acquired coincident with both airborne data acquisitions using Geophysical and Environmental Research Corporation (GER) 3700 and 1500 spectroradiometers (Table 1). Prior to each field measurement, reference spectra from a calibrated Spectralon tablet were collected to convert final measurements to absolute percent reflectance. For each spectral sample location, three replicate spectra were recorded under clear skies around local noon. The field of view of the field spectroradiometers was set at $8°$ and the sensor head located 1 m above the target.

The airborne hyperspectral remote sensing data were calibrated to apparent surface reflectance using the empirical line technique utilizing the field spectra acquired coincident with the airborne data acquisitions (Ferrier 1997). The Minimum Noise Fraction (MNF) algorithm (Green et al. 1988) was applied to the remote sensing dataset to identify the number of spectrally distinct surface materials (endmembers) present. The purest pixel(s) of each endmember was then identified using principal component analysis and the pixel purity index algorithm (Boardmann et al. 1995). The proportions of each spectral endmember for each image pixel were derived using the Matched Filtering and Linear Spectral Unmixing techniques. Matched Filtering (Harsanyi & Chang 1994) performs a partial unmixing to find the abundances of user-defined endmembers. It provides a rapid means of detecting specific minerals based on matches to specific library or image endmember spectrum and does not require knowledge of all the endmembers within an image scene. Linear Spectral Unmixing is a means of determining the relative abundances of materials depicted in multispectral imagery based on the materials' spectral characteristics. The reflectance at each pixel of the image is assumed to be a linear combination of the reflectance of each material present within the pixel (Settle & Drake 1993). Two spectral analysis techniques were employed to identify the endmembers. The spectral angle mapper (SAM) is a pixel-based supervised classification technique that measures the similarity of an image pixel reflectance spectrum to a reference spectrum from either a spectral library or a field spectrum (Kruse et al. 1993). The lower the spectral angle between two spectra, the more similar they are. The second technique applied was cross-correlogram spectral matching (Van der Meer & Bakker 1997). A cross-correlogram is constructed by calculating the cross-correlation coefficient between a test spectrum (a pixel spectrum) and a reference spectrum (a laboratory or endmember pixel) at different match positions.

Field mapping and laboratory analysis

A mapping and sampling survey was undertaken in April 2001 to acquire a representative set of sediment samples from the tailings dump and river. X-ray fluorescence (XRF) and X-ray diffraction (XRD) analyses were carried out on all the samples. The high concentration of heavy minerals in the samples meant that the results of XRD analyses were unusable. A sequence of selective dissolution techniques were used instead. The chromous chloride reduction method was applied to identify the pyrite concentration. The dithionite extraction method (van Oorschots et al. 2001) was applied, followed by the ammonium oxalate extraction method (van Oorschots et al. 2001). The analysis was carried out using a Varianspectr AA-10

Table 1. *Spectral and spatial specifications of available remote sensing data*

Sensor	Spectral resolution (nm)	Spectral range (nm)	Spatial resolution	Acquisition date
AVIRIS	10	440–2500	20 m^2	1991
CASI	10	400–1100	3 m^2	2001
GER1500	1.5	300–1100	Point	2001
GER3700	1.5	300–2500	Point	1991
ATM	100	400–14000	3 m^2	2001

atomic absorption X-ray spectrometry. The samples were sent to a commercial laboratory for analyses of total and free cyanide concentrations.

A digital elevation model (DEM) of the study area was created from 1:9000 scale topographic using a geographical information system (GIS) (ArcGIS, version 9.1). High-resolution colour aerial photography and the results of the analysis of the airborne remote sensing data were orthorectified to the DEM. The integration of the 3D dataset within ArcGIS allowed quantitative spatial analysis of the multi-spatial resolution dataset and provided a means of assessing the influence of topography and distance from source locations on geomorphological and geochemical processes. Analysis using ArcGIS also allowed volumes of mine waste accumulations at specific locations to be calculated by subtracting the palaeo-topography of the valleys and river channels from current topography.

Spectral mineral map

The spectral profile of the tailings dump material (Fig. 2) has an absorption edge at 0.54 μm, a reflectance shoulder at 0.63 μm, a local reflectivity maximum at 0.74 μm, a band minimum at 0.85 μm, and another reflectance shoulder at 1.04 μm. These features correspond very strongly to the hematite (OH-1A) reference spectrum (Grove et al. 1992). The mean spectrum from a large homogeneous area on the tailings dump was calculated from the AVIRIS data and used as the endmember in the matched filtering analysis. The match filter results are presented as a grey-scale image with values ranging from zero (white) to one (black) giving an estimate of the degree of match to the reference spectrum (where 1.0 is a perfect match). The contour lines (see Fig. 1) are also in black and underlie the match filter results.

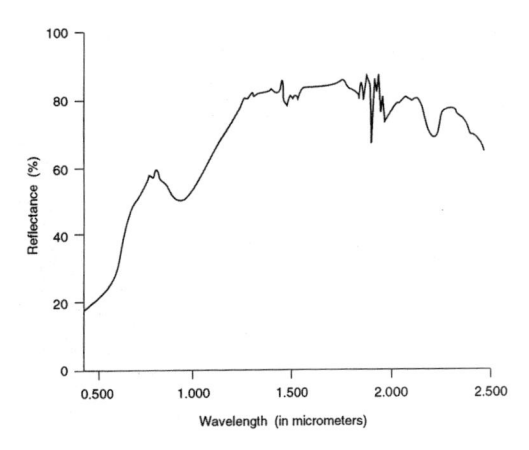

Fig. 2. Field spectrum of tailings dump material.

The match scores range from 0.2 to 0.9, vary rapidly on a pixel-to-pixel basis, and clearly show the tailings dump, the mine workings and a concentrated trail of mine waste extending down the river to 500 m from the beach, where the mine waste becomes much more dispersed (Fig. 3). Linear unmixing analysis of the AVIRIS data identified the distribution of three spectral endmembers having absorption features diagnostic of ferruginous material. One endmember is concentrated in the tailings dump (Fig. 4), a second is found within the mine, and the third is found dispersed along the river. SAM analyses of the AVIRIS-derived spectra found that the tailings dump endmember was very similar to hematite (0.039). The best match for the mine endmember was with jarosite (0.052), whereas the stream-bed endmember has a best match with goethite (0.042). The results of the cross-correlation spectral matching analyses were similar to those of the SAM analyses except for the mine endmember (Table 2).

Linear unmixing and matched filtering analysis of the CASI data showed a very similar distribution to the AVIRIS-derived results, with hematite-rich material concentrated on and around the tailings dump and along the upper reaches of the river. The results of the CASI match filter results for the eastern section of the study area are displayed overlying an aerial photograph draped over the DEM with contours (Fig. 5). The results clearly show a marked change in the stream sediments composition around 3000 m from the tailings dump, with the hematite concentration reducing and the ferrihydrite concentration increasing significantly (Fig. 5). The analysis of the CASI data identified concentrations of mine waste and particular secondary iron species at the geomorphological settings identified in field mapping.

Dispersion of the mine waste

The locations where mine waste enters the main drainage channel display a pronounced reddening of sediments and a marked reduction in sediment grain size (from dominantly gravel to fine sands and finer) as a result of the introduction of fine-grained mine waste and the deposition of secondary iron species on the surface of the stream-bed sediment. Within the main river channel, a 1 m deep flood plain deposit of very finely stratified tailings material has preferentially formed in three main zones: first, at the junction where tributaries join the main channel; second, immediately adjacent to the inner bank of gently meandering channel sections; third, on mid-channel bars that are restricted to wider channel sections. The presence of multiple very finely stratified layers indicates that the tailings banks formed as a

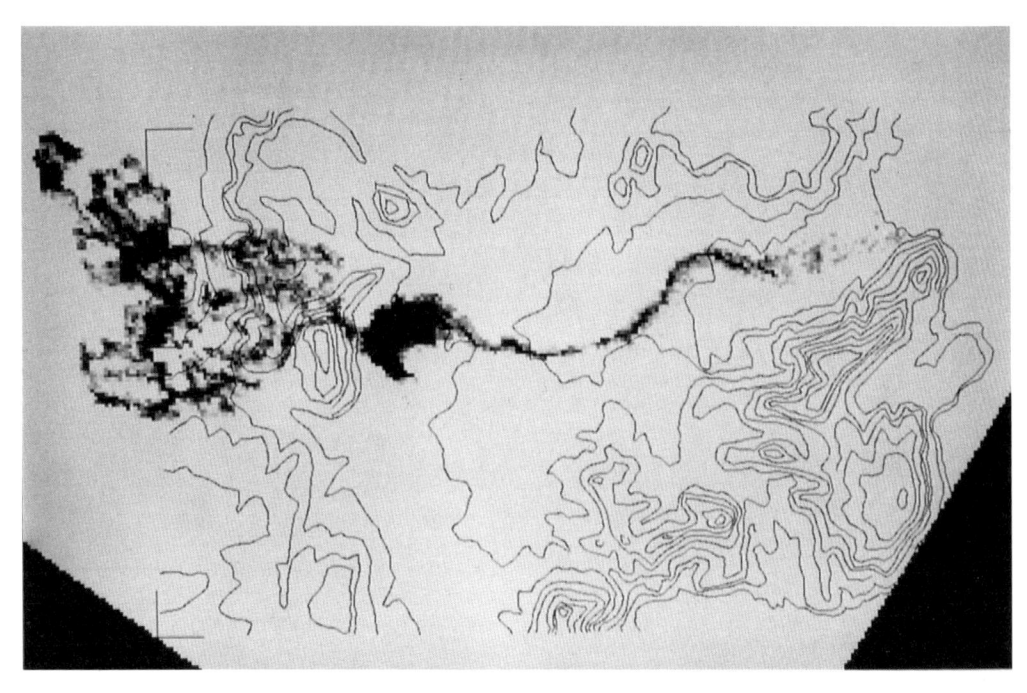

Fig. 3. Results of Matched Filter Analysis of AVIRIS data, showing distribution of mine waste material.

Fig. 4. Results of Linear Unmixing Analysis, showing distribution of hematite and jarosite endmembers.

Table 2. *Results of cross-correlation spectral match analyses*

Endmember	Best match	Hematite (skewness)	Goethite (skewness)	Jarosite (skewness)
Tailings endmember	0.986 (hematite)	0.995	0.987	0.970
Streambed endmember	0.96 (goethite)	0.968	0.992	0.985
Mine endmember	0.779 (jarosite)	0.983	0.989	0.955

result of several depositional events. Deposition within each of the three specific locations clearly reflects the dynamics of ephemeral flows (stream velocity and competence) coupled with variations in the channel geometry. With the remediation efforts effectively reducing the input of tailings material to the main channel, ephemeral flows have progressively eroded the tailings, with scour culminating in bank collapse. The tailings released during phases of bank collapse have undergone variable amounts of transport downstream. Where the channel widens tailings are redeposited as interstitial fills between gravels, whereas in narrower sections tailings deposits form drapes above coarse sands. Tailings that have been transported to the coast zone are currently accumulating within a broad 5 m by 20 m lagoonal-type depression adjacent to the beach.

The introduction of mine waste has also dramatically increased concentrations of heavy metals within the Rodalquilar river. Concentrations of Sb and As in the stream sediments above the location where the tailings material enters the stream are relatively low, whereas immediately downstream of the tailings dump there is a dramatic increase in both elements. Sb and As show a gradual reduction in concentration downstream of the tailings dump to *c.* 3000 m downstream, where the concentration has reduced to background values found above the tailings dump (Figs 6 and 7). At *c.* 4000 m downstream, however, there is a significant increase in Sb and As. The highly elevated concentrations of these insoluble heavy metals along the river indicate that a significant amount of the mine waste material has been transported along the length of the river in particulate form.

Fig. 5. Results of Matched Filter results of CASI data for the eastern section of the study area displayed overlying an aerial photograph draped over the DEM with contours, showing distribution and relative concentration of ferrihydrite.

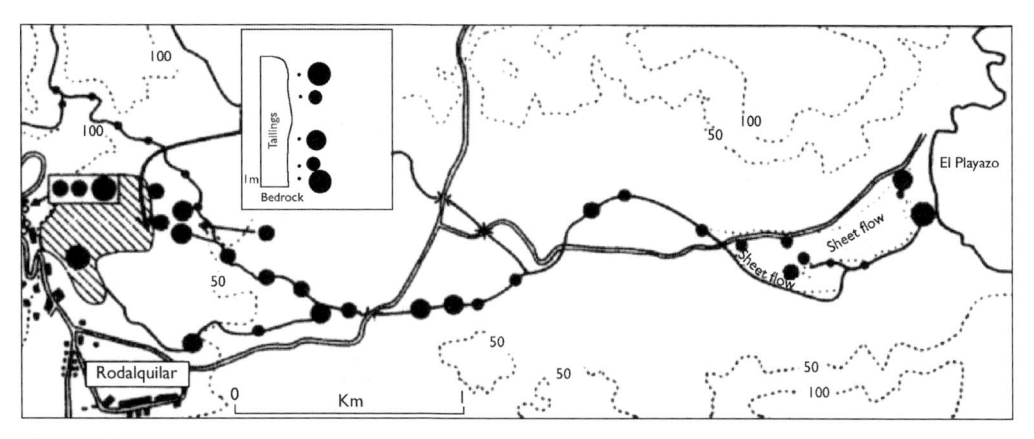

Fig. 6. Distribution of Sb in stream sediments and tailings (after Wray 1998).

- ● ≥ 0.0 ppm
- ● ≥ 14.9 ppm (25%)
- ● ≥ 29.7 ppm (50%)
- ● ≥ 39.2 ppm (75%)
- ● ≥ 50.1 ppm (90%)

Laboratory analyses of tailings material and stream sediments showed good correlations between the concentrations of specific secondary iron species, total cyanide and particular heavy metals. Although the number of samples is limited, the pyrite concentration can be seen to vary markedly downstream from the tailings dump, with a significant increase at 1350 m and a less marked increase at 3800 m downstream (Fig. 8).

The distributions of pyrite and ferrihydrite are highly correlated ($R^2 = 0.806$), suggesting that the ferrihydrite is most probably a product of the *in situ* weathering of pyrite. The ferrihydrite concentration is relatively constant from the tailings dump downstream to within 500 m of the beach, where there is a significant increase. Dithionite extracted iron (Fe$_d$) has a poor general correlation with both pyrite and ferrihydrite (0.158) and shows a pronounced gradient, with high concentrations near the tailings dump and along the upper reaches of the river but very low concentrations in the lower reaches down to the beach. Good correlations were found between the total cyanide concentration and both the ferrihydrite ($R^2 = 0.823$) and the Fe$_d$ ($R^2 = 0.904$) concentrations, indicating that these secondary Fe minerals are associated with mine waste. Poor correlations were found between total cyanide and both total iron ($R^2 = 0.006$) and pyrite ($R^2 = 0.102$). The concentration of total cyanide is extremely high

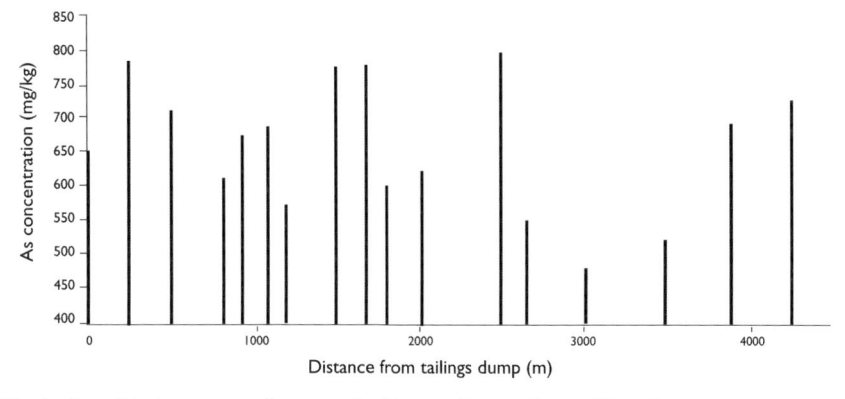

Fig. 7. Distribution of As in stream sediments and tailings v. distance from tailings dump (m).

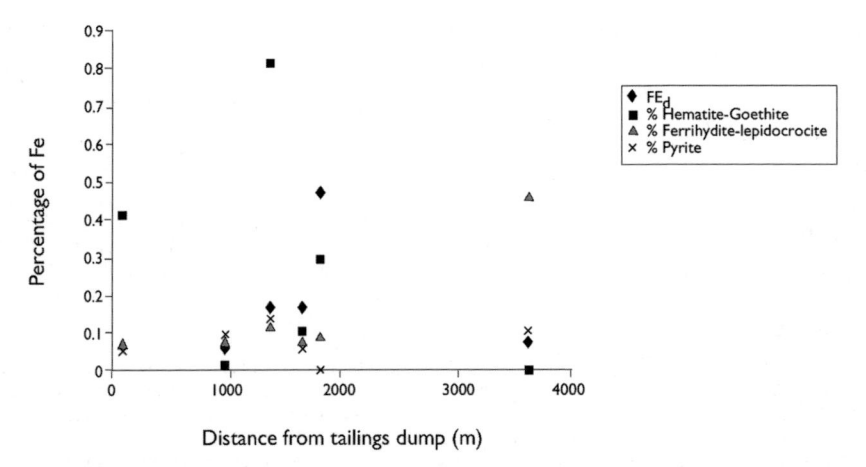

Fig. 8. Distribution of dithionite, chromium chloride and ammonium oxalate extracted iron in stream sediments v. distance from tailings dump (m).

immediately adjacent to the tailings dump, then is much lower along most of the river, finally rising significantly in the depression adjacent to the beach (Fig. 9).

Ferrihydrite is a poorly ordered, unstable, hydrous form of iron oxide formed in environments where high levels of Fe(III) are made available by the rapid oxidation of Fe(II). Hematite is a secondary iron mineral that is representative of formation over a longer time period within more stable water chemistry and physical settings (Schwertmann & Taylor 1977). This suggests that the mine waste in the stream characterized by high hematite concentrations underwent initial weathering on or adjacent to the tailings dump prior to being transferred into the stream. Subsequent erosion has resulted in this weathered, hematitic-rich mine waste being transported down the stream to the edge of the coastal zone (3500 m downstream). A limited development of ferrihydrite has occurred at one locality immediately adjacent to the tailings dump and in the area immediately adjacent to the coastal zone. This distribution of secondary iron minerals corresponds well to models of the accumulation of secondary Fe minerals in Fe sulphide-rich mine environments (Ferris et al. 1989; Montero et al. 2005). In all these models a clear paragenetic relationship between pyrite, ferrihydrite and hematite is demonstrated, with pH and biotic oxidation being critical factors and ferrihydrite being an intermediate, unstable phase.

At Rodalquilar, pyrite occurs naturally only in the mineralized zone of the gold deposit. The distribution of unweathered, pyrite-rich mine waste along the length of the stream to the coast zone, particularly concentrated at depositional zones such as

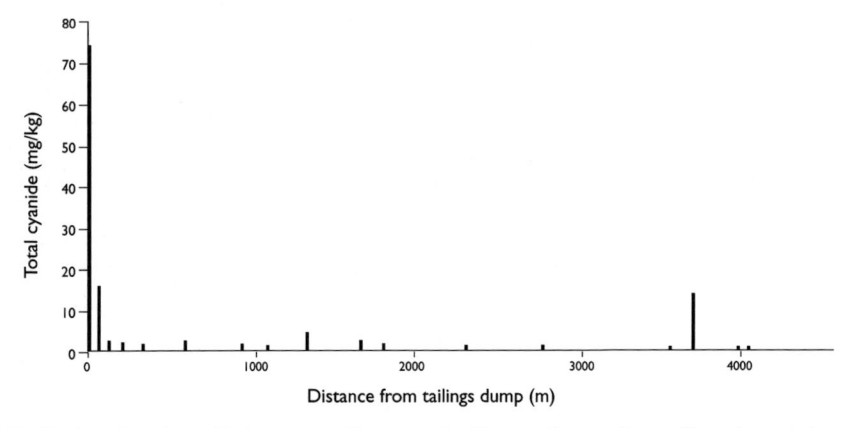

Fig. 9. Distribution of total cyanide in stream sediments and tailings v. distance from tailings dump (m).

point bars, strongly suggests that shortly after the abandonment of the mine, a major flood event transported a significant volume of unweathered, pyrite-rich mine waste into and along the length of the stream, depositing a significant volume in the lagoon area.

Conclusions

The results of the remote sensing study supported by field mapping and laboratory analyses have shown that the Valle de Rodalquilar has been severely affected by the escape of mine waste from the tailings dump. Both airborne hyperspectral remote sensing datasets clearly identified the distribution of significant amounts of mine waste material along the length of the river and that the mine waste is very hematitic-rich apart from at two localities, one immediately adjacent to the tailings dump and the other adjacent to the beach. The results from the analysis of the CASI data demonstrated the additional capability provided by high spatial resolution hyperspectral instruments to resolve the geomorphological setting of accumulations of mine waste. The ability of spectral remote sensing data to identify secondary iron species supported by laboratory-derived iron species–cyanide–heavy metal correlations offers the capability of predicting the distributions of hazardous mine waste over large areas.

Although this study has demonstrated the utility of spectral remote sensing in mapping the environmental impact of hazardous mine waste at the landscape scale, the limitations and sources of error in the analysis of airborne remote sensing data must be fully appreciated to avoid misinterpretation of the results. The spatial and spectral limitations of the airborne instruments, coupled with the errors introduced in geometric and atmospheric correction processing, can modify the target spectra at the airborne sensor. The effects of intimate mixing of particulates, presence of vegetation and many other environmental factors can also dramatically affect the spectral response.

The authors would like to acknowledge the assistance provided by NERC ARSF in providing the CASI and ATM datasets, and NERC FSF for providing loans of the field spectroradiometers.

References

ALPERS, C. N., BLOWES, D. W., NORDSTROM, D. K. & JAMOBOR, J. L. 1994. Secondary minerals and acid mine-water chemistry. *In*: JAMBOR, J. L. & BLOWES, D. W. (eds) *Environmental Geochemistry of Sulfide Mine-Wastes*. Mineral Association of Canada, Short Courses, **22**, 247–270.

BOARDMANN, J. W., KRUSE, F. A. & GREEN, R. O. 1995. Mapping target signatures via partial unmixing of AVIRIS data. *In*: GREEN, R. O. (ed.) *Summaries of the Fifth Annual JPL Airborne Earth Science Workshop, Vol. 1, Washington, DC, 23–26 January 1995*, JPL Publications, **95-1**, 23–26.

EVANS, K. G., SAYNOR, M. J., WILLGOOSE, G. R. & RILEY, S. J. 2000. Post-mining landform evolution modelling: 1. Derivation of sediment transport model and rainfall–runoff model parameters. *Earth Surface Processes and Landforms*, **25**, 743–763.

FERRIER, G. 1997. Enhanced estimation of apparent surface reflectance. *International Journal of Remote Sensing*, **17**, 2881–2889.

FERRIS, F. G., TAZAKI, K. & FYFE, W. S. 1989. Iron oxides in iron mine drainage environments and their association with bacteria. *Chemical Geology*, **74**, 321–330.

GREEN, A. A., BERMAN, M., SWITZER, P. & CRAIG, M. D. 1988. A transformation for ordering multispectral data in terms of image quality with implications for noise removal. *IEEE Transactions on Geoscience and Remote Sensing*, **26**, 65–74.

GROVE, C. I., HOOK, S. J. & PAYLOR, E. D. 1992. Laboratory reflectance spectrum of 160 minerals, 0.4–2.5 micron. JPL Publications, **92-2**.

HARSANYI, J. C. & CHANG, C. I. 1994. Hyperspectral image classification and dimensionality reduction: an orthogonal subspace projection approach. *IEEE Transactions on Geoscience and Remote Sensing*, **32**, 779–785.

HUDSON-EDWARDS, K. A., MACKLIN, M. G., JAMIESON, H. E., BREWER, P. A., COULTHARD, T. J., HOWARD, A. J. & TURNER, J. N. 2003. The impact of tailings dam spills and clean-up operations on sediment and water quality in river systems: the Rios Agrio-Guadiamar, Aznalcollar, Spain. *Applied Geochemistry*, **18**, 221–239.

KRUSE, F. A., LEFKOFF, A. B., BOARDMANN, J. W., HEIDEBRECHT, K. B., SHAPIRO, A. T., BARLOON, J. P. & GOETZ, A. F. H. 1993. The spectral image processing system (SIPS)—interactive visualization and analysis of imaging spectrometer data. *Remote Sensing of Environment*, **44**, 145–163.

MEEUSSEN, J. C. L., RIEMSDIJK, W. H. & VAN DER ZEE, S. 1995. Transport of complexed cyanide in soil. *Geoderma*, **67**, 73–85.

MONTERO, I. C., BRIMHALL, G. H., ALPERS, C. N. & SWAYZE, G. A. 2005. Characterisation of waste rock associated with acid drainage at the Penn Mine, California, by ground-based visible to short-wave infrared reflectance spectroscopy assisted by digital mapping. *Chemical Geology*, **215**, 453–472.

RYTUBA, J. J., ARRIBAS, A., JR, CUNNINGHAM, C. G. ET AL. 1990. Mineralised and unmineralised calderas in Spain; Part II, evolution of the Rodaquilar caldera complex and associated gold–alunite deposits. *Mineralium Deposita*, **25**, S29–S35.

SCHWERTMANN, U. & TAYLOR, R. M. 1977. Iron oxides. *In*: DIXON, J. B. & WEED, S. B. (eds) *Minerals in Soil Environments*. Soil Science Society of America, Madison, WI, 145–180.

SETTLE, J. J. & DRAKE, N. A. 1993. Linear mixing and the estimation of ground cover proportions.

International Journal of Remote Sensing, **14**, 1159–1177.

SWAYZE, G. A., SMITH, K. S., CLARK, R. N. ET AL. 2000. Using imaging spectroscopy to map acidic mine waste. *Environmental Science and Technology*, **34**, 47–54.

TOY, J. T. & HADLEY, R. F. 1987. *Geomorphology and Reclamation of Disturbed Lands*. Academic Press, London.

VAN DER MEER, F. & BAKKER, W. 1997. Cross correlogram spectrum matching: application to surface mineralogical mapping by using AVIRIS data from Cuprite, Nevada. *Remote Sensing of Environment*, **61**, 371–382.

VAN OORSCHOTS, I. H. M., GRYGAR, T. & DEKKERS, M. J. 2001. Detection of low concentrations of fine-grained iron oxides by voltammetry of microparticles. *Earth and Planetary Science Letters*, **193**, 631–642.

WRAY, D. S. 1998. The impact of unconfined mine tailings and anthropogenic pollution on a semi-arid environment—an initial study of the Rodaquilar mining district. *Environmental Geochemistry and Health*, **20**, 29–38.

Detection and mapping of shrink–swell clays in SW France, using ASTER imagery

A. BOURGUIGNON[1], G. DELPONT[1], S. CHEVREL[1] & S. CHABRILLAT[2]

[1]Bureau de Recherches Géologiques et Minières (BRGM), Orléans, France
(e-mail: a.bourguignon@brgm.fr)

[2]GeoForschungsZentrum (GFZ), Potsdam, Germany

Abstract: Current mapping methods for shrink–swell clays in France are based on the use of existing 1:50 000 geological maps. However, stratigraphy is the primary basis of current published geological mapping, which is detrimental to the accurate mapping of clay minerals, argillaceous lithologies and clayey soils. In the study region, Pyrenean forefront molasse deposits have been mapped in a stratigraphy-dominated way, grouped into a single mapping unit, even though they are made up of eight sequences of continental sands, gravels, clays and lacustrine limestones. Mapping shrink–swell clay hazardous terrain can be improved by spectral methods, allowing rapid delineation of argillaceous units and the identification of their mineralogy, especially with regard to illite, chlorite and kaolinite. The Advanced Spaceborne Thermal Emission and Reflection Radiometer (ASTER) allows a new approach to regional clay mineral mapping, via the six spectral bands in its short-wave infrared domain, particularly wavelengths between 2.145 and 2.43 μm.

Shrink–swell clays can cause severe damage to buildings and infrastructure during drought periods. Associated indemnities represent the main source of compensation costs for French insurance companies. Mapping shrink–swell clays is thus of major importance in land-use planning (Chabrillat & Goetz 1999; Chabrillat et al. 1999, 2002). The French Ministry for Land-Use Planning and Environment has entrusted the Bureau de Recherches Géologiques et Minières (BRGM) with a national programme for regional mapping of shrink–swell clays.

The study region is located in the northern forefront of the Pyrenean chain (Fig. 1), and corresponds to the Gers administrative department. Mapped as a single geological unit, clay-rich molasse forefront deposits cover 80% of the region and are responsible for substantial deformation in building structures, particularly in recent residential areas.

Methods

Image pre-processing

At the time of the study (early 2000), most of the Advanced Spaceborne Thermal Emission and Reflection Radiometer (ASTER) data were available only as level 1A and hence presented band-to-band registration problems that had to be empirically corrected. Atmospheric corrections were performed using ENVI's Internal Average Relative Reflectance algorithm: an average spectrum is calculated from the entire image scene and is used as the reference spectrum, which is then divided into the spectrum of each pixel in the image.

The early spring image shows large areas covered by bare (ploughed) soils, with high contrasts in soil moisture content after the winter rainy season. The different lithological units can be easily distinguished in Figure 2, a false-colour IR composite image (342 RGB). ASTER band 4 (the first shortwave IR band, SWIR1), clearly enhances areas of high moisture content in bare soils: examples are circled in Figure 3.

Image masking and clay soil extension mapping

Applying a Normalized Difference Vegetation Index (NDVI) vegetation index on visible–near-IR (VNIR) bands allowed the production of a bare-soil mask image. The specific area of interest (AOI) of each of the major lithological units was then generated to mask terraces and alluvial deposits, keeping only the clay-rich molasse deposits as the area of interest for further processing. A digital classification of clayey soil was then performed only on the 'bare-soil image' corresponding to argillaceous molasse geological units.

A supervised classification procedure was applied to the vegetation-masked image. This allowed the mapping of different moisture levels in the clayey soils of the molasse deposits, using the preferential absorption feature in the SWIR1 range (1.6–1.7 μm). This result was validated in the field and confirms the effective use of ASTER data for mapping varying moisture levels in clay-rich zones, something that is not shown on the standard published geological maps (Fig. 4).

From: TEEUW, R. M. (ed.) *Mapping Hazardous Terrain using Remote Sensing*. Geological Society, London, Special Publications, **283**, 117–124.
DOI: 10.1144/SP283.10   0305-8719/07/$15.00   © The Geological Society 2007.

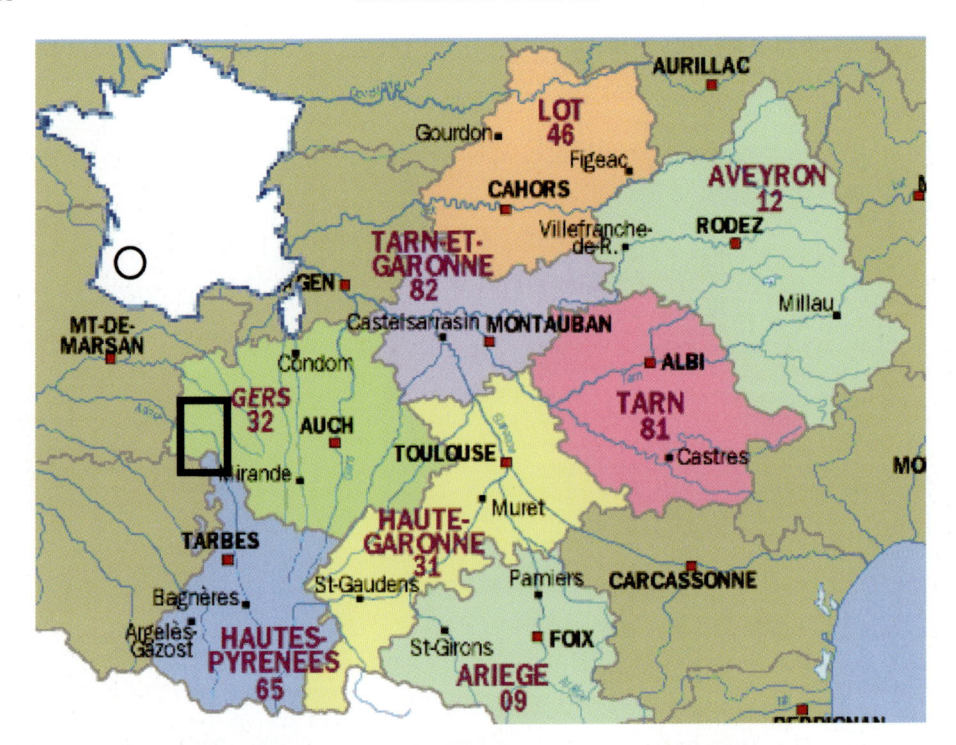

Fig. 1. Location of the study region: Gers, SW France. Produced from NASA SRTM DEM data.

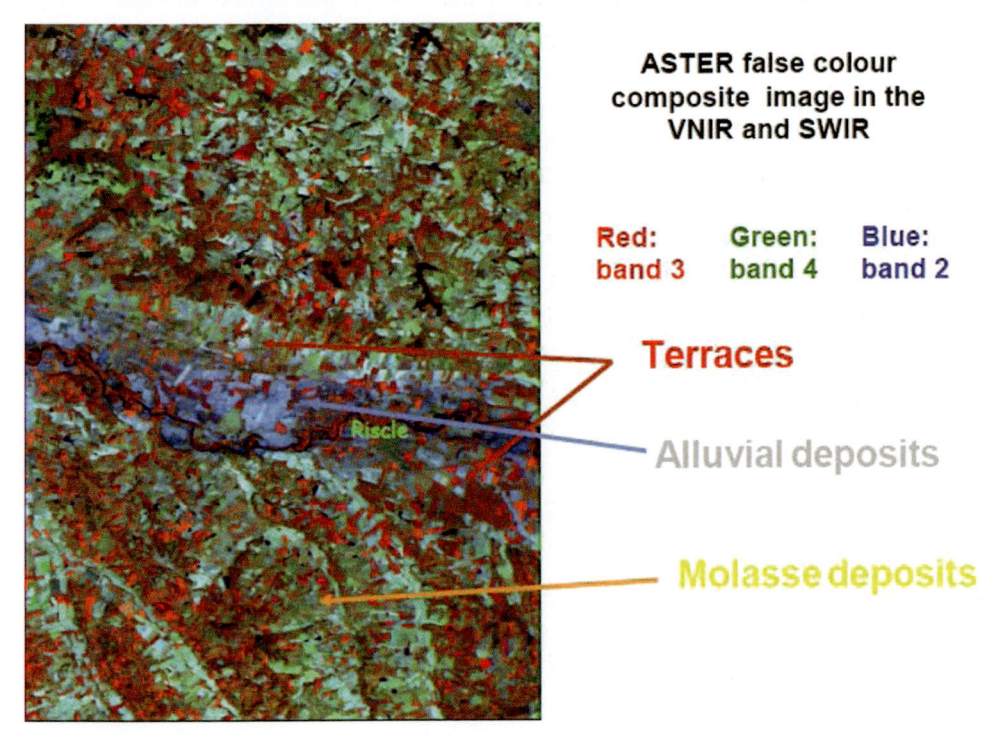

Fig. 2. False-colour IR ASTER image of the study area, centred on the town of Riscle, showing the main geomorphological features of the study region.

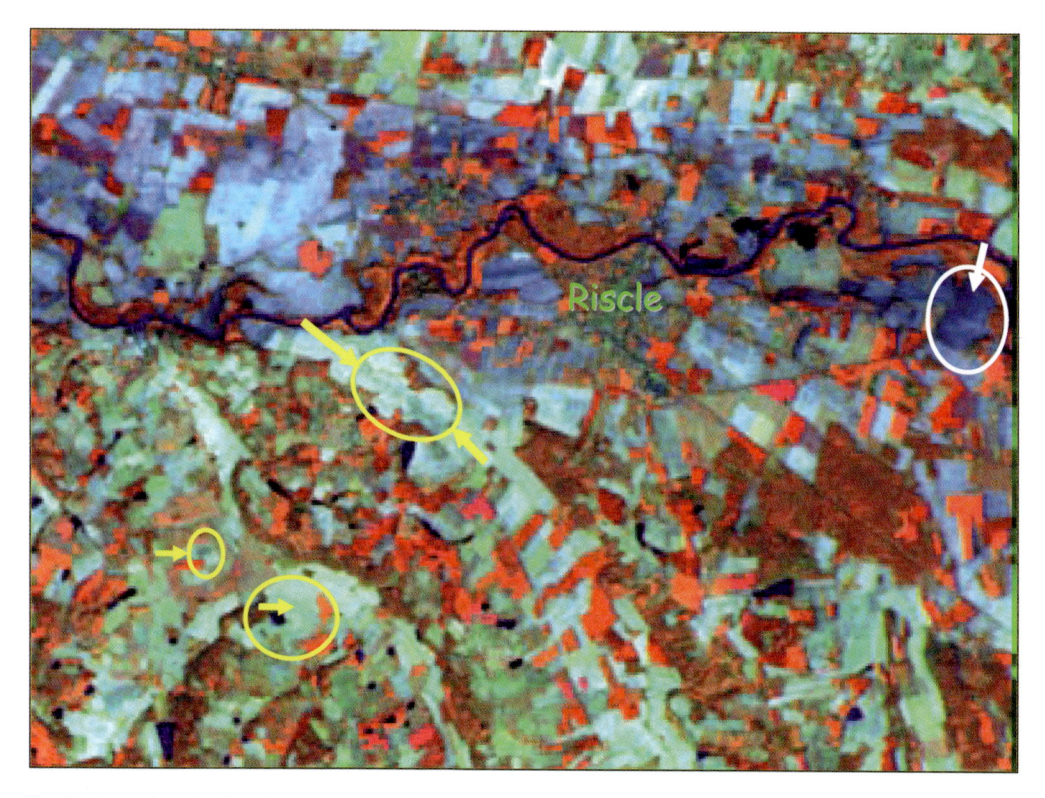

Fig. 3. Examples of soil moisture corresponding to clayey soils (circled) near the town of Riscle. The study area corresponds to the box outline shown in Figure 2.

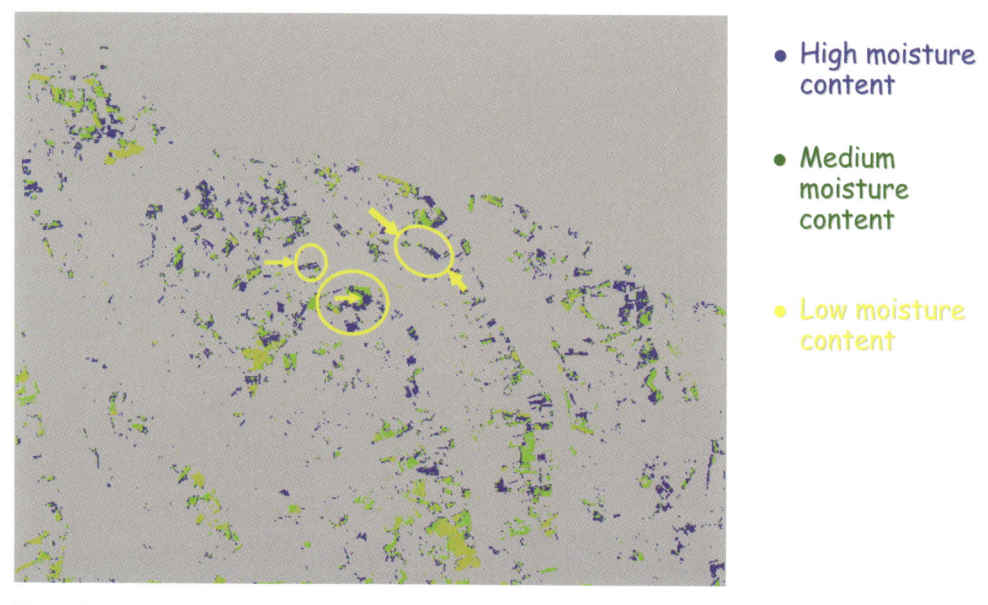

Fig. 4. Supervised classification of moist areas in the molassie deposits. The yellow circles are the same areas as shown in Fig. 3.

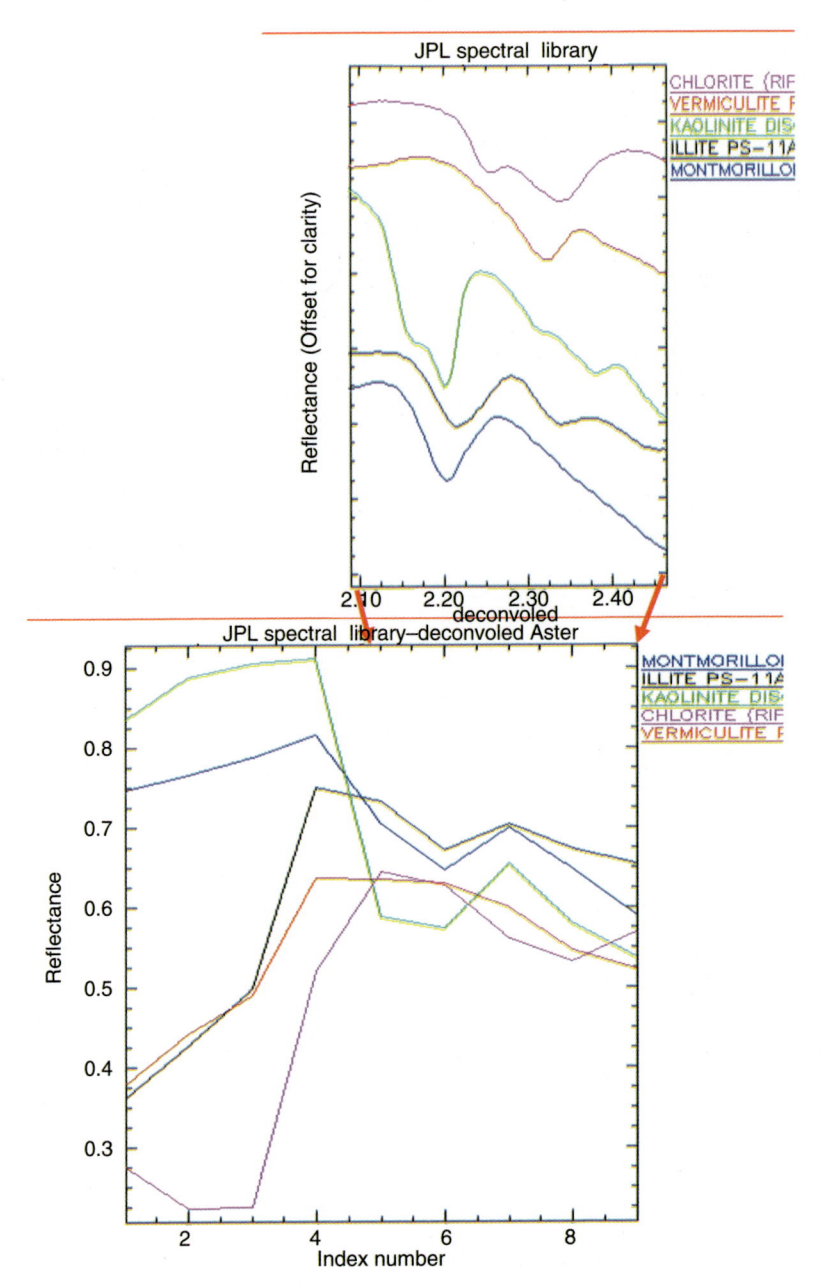

Fig. 5. Top: various clay spectra in the SWIR2 range (wavelength shown on abscissa); bottom: the corresponding ASTER-deconvolved spectra (band number shown on abscissa). Reproduced from the ASTER Spectral Library, courtesy of the Jet Propulsion Laboratory, California Institute of Technology, Pasadena, California. Copyright ©1999, California Institute of Technology.

Clay spectral identification

The US Jet Propulsion Laboratory spectral library (http://speclib.jpl.nasa.gov), deconvolved to ASTER bands in the SWIR2 range, highlights several spectral characteristics of relevance in the classification of argillaceous material, with the following characteristic features (Fig. 5): illite presents an absorption

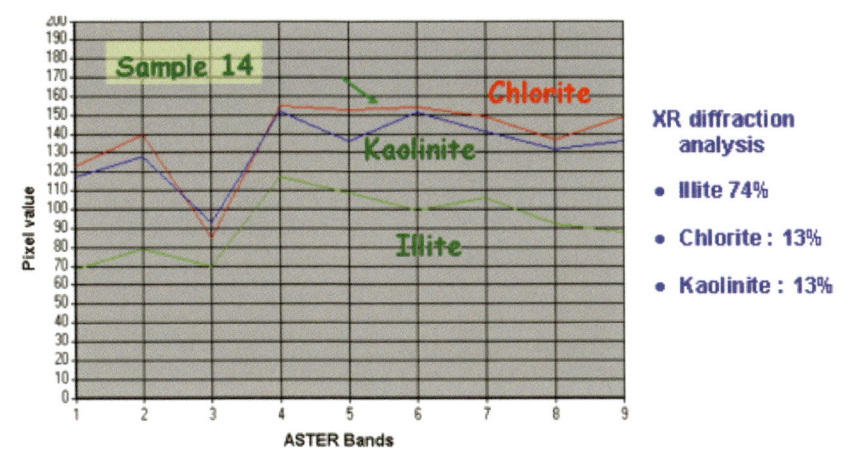

Fig. 6. ASTER spectral profiles at sample location and relevant XRD analysis.

feature in band 6 and a slight inflection ASTER band 8; montmorillonite presents a single characteristic absorption feature in band 8; kaolinite presents a wide absorption feature in bands 5 and 6 and a slight inflection in band 8; chlorite presents a wide absorption feature centred on band 8. These spectral features are sufficiently discriminating to allow the identification of mineral composition from image spectra (Fig. 6).

Clay classification

To allow clay classification, a dedicated spectral library was built with six spectra from ASTER imagery, corresponding to four illite and two chlorite samples, validated with XRD analyses. Two different mapping methods were then used to classify the clayey soils from these spectra; namely, the Spectral Angle Mapper (SAM) and the

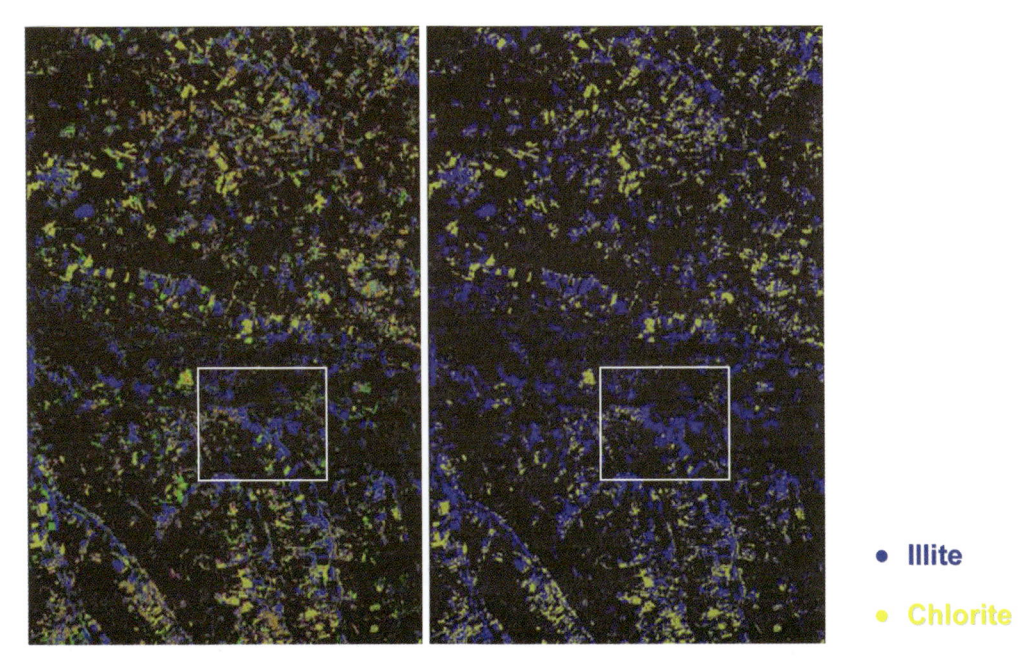

Fig. 7. SAM classification results on the whole ASTER image. Left: results obtained from four illite and two chlorite spectra. Right: illite and chlorite gathered into two classes.

Table 1. *Validation of SAM classification results on field samples*

Sample	X-ray analysis (%)	Spectral profile	SAM
12	Illite (64) + kaolinite (15)	Kaolinite	Illite
13	Illite (78) + montmorillonite (16) + kaolinite (6)	Illite	Illite
14	Illite (74) + kaolinite (13) + chlorite (13)	Chlorite	Chlorite
15	Illite (69) + kaolinite (10) + vermiculite (21)	Chlorite	Chlorite
16	Illite (69) + chlorite (31)	±Chlorite	Illite
17	Illite (75) + kaolinite (25)	Illite	Illite
18	Illite (75) + chlorite (20) + vermiculite (5)	Chlorite	Unclassified
19	Illite (81) + chlorite (19)	Chlorite	Unclassified
20	Illite (88) + kaolinite (12)	Illite	Chlorite

Mixture Tuned Matched Filtering (MTMF) algorithms. Both methods compare reference spectra from the spectral library with each image pixel spectrum, but use different algorithms to compare and measure spectral similarity. Both methods are available through the ENVI software package and are validated extensively in the scientific literature (e.g. Kruse *et al.* 1993; Boardman *et al.* 1995).

The Spectral Angle Mapper algorithm determines the similarity between a reference spectrum and image spectrum by calculating the 'spectral angle' between them. An angular threshold value is then chosen, so that only pixels with smaller angles to the reference spectra will be mapped. The smallest values for the angle indicate the highest similarity between pixel spectrum and reference spectra. The results of SAM classification are presented in Fig. 7 and Table 1.

Mixture Tuned Matched Filtering (MTMF) is a method that estimates the relative degree of match of a given pixel to a reference spectrum and approximates the sub-pixel abundance. The MTMF is then carried out using end-members for the requested materials (Fig. 8). End-members are spectral signatures of 'pure' materials that occur in the area. The outputs from the MTMF are two bands for each end-member; one is an estimate of the abundance of the spectrum in the pixel (MF-score) and the other is an estimate to reduce the 'false positives' (unfeasibility). Correctly classified pixels will have a high MF-score and a low unfeasibility. The higher the MF-score, the higher is the abundance of the material in the pixel. This is done using a subjective threshold on a 2D scatterplot of the MF-score and unfeasibility, based on the related spectra. For illite, only areas

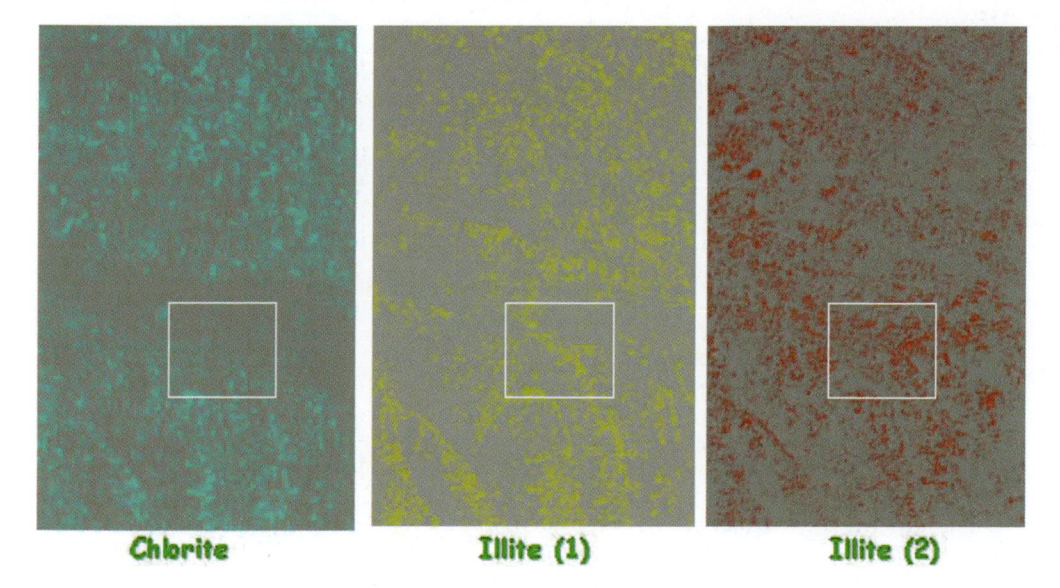

Fig. 8. Results of the MTMF classification on the whole ASTER image.

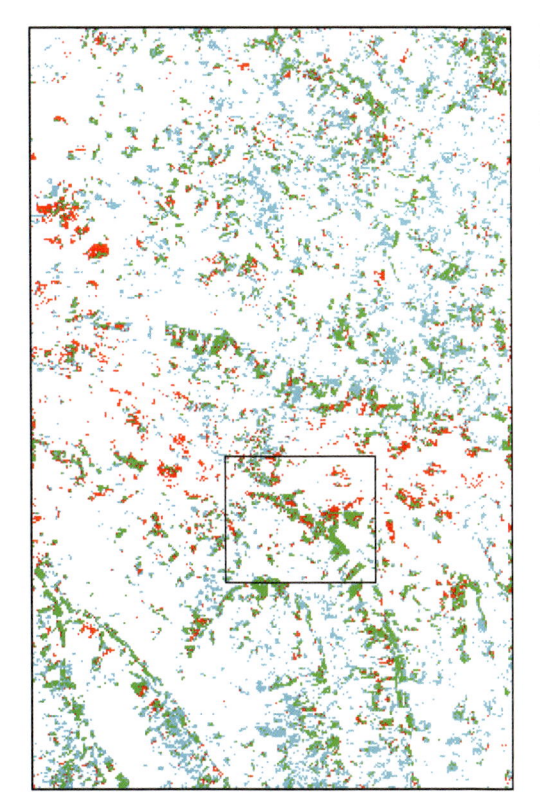

Fig. 9. Comparison of SAM and MTMF classification methods on the whole ASTER image. Green pixels are those that were identically classified with both the SAM and MTMF algorithms.

with MF-score above 0.3 and unfeasibility below five were chosen.

Comparison of classification results

The green pixels in Figure 9 are those identically classified by SAM and MTMF algorithms. A first estimation shows a 68% convergence between the two methods. A better parameter selection would have improved the classification accuracy; for instance, the use of SAM angle threshold, MTMF MF-score and unfeasibility values.

Discussion

Shrink–swell clays can cause substantial damage to buildings and infrastructure, especially during drought periods. Accurate identification and mapping of the lithologies and mineralogies associated with shrink–swell clays is therefore important in land-use planning. Current mapping methods for shrink–swell clays are based on existing 1:50 000 geological maps and are not satisfactory, being based on stratigraphic criteria, rather than mineralogical criteria.

The use of ASTER data for the detection and mapping of shrink–swell clays has been demonstrated in the Pyrenean forefront molasse region of SW France. The potential and limitations of ASTER imagery have been investigated in two directions: soil moisture detection and clay mineral discrimination. Using supervised classification applied to the vegetation-masked image, we were able to map clay-rich zones, which are undifferentiated on the published geological maps. This potential of ASTER imagery for soil moisture detection was also demonstrated, using the SWIR1 spectral range (1.60–1.70 μm, band 4). Application of this simple operational processing result should improve the rapid mapping of argillaceous zones.

The analysis of ASTER images, using two mapping methods (SAM and MTMF algorithms) confirms the potential of the SWIR2 range (bands 5–9, 2.14–2.43 μm) to identify clay minerals; in particular, illite, chlorite and locally kaolinite. A first estimation shows a 68% convergence between the two methods: this could be improved by using better selection parameters. Following field sample validation analysis, it must be noted that the clay mineral identification using ASTER is only possible for a minimal mineral content, which varies depending on the clay mineral: illite: c. 80% minimum; kaolinite–chlorite: 15–30% minimum; montmorillonite: not identified below 16% (one sample only).

Systematic mapping using this method requires ASTER data acquisition during periods with minimal vegetation cover. In this case study, winter ASTER scenes are best, although this has inherent problems of increased cloudiness and possible snow cover. For a better efficiency, ASTER-based clay mineral mapping should use a PIMA portable spectrometer, which would provide reliable field spectral and chemical analyses at a reasonable cost and improve the image classification results.

Conclusion

This study was a first assessment of the possibilities and limitations of ASTER imagery for the detection and mapping of shrink–swell clays, in the context of a rapidly growing urban area in a humid temperate region, affected by severe shrink–swell damage to buildings and infrastructure, especially during drought periods.

ASTER band 4 (SWIR1) has been used to identify damp soils associated with argillaceous bedrock, with a good degree of confidence. ASTER bands 5–9 (SWIR2) allowed discrimination between clay minerals, particularly illite, chlorite and kaolinite. Mineral identification has been validated in the field and through X-ray analyses. In humid temperate regions, systematic mapping using this method requires ASTER data acquisition during winter, the season with the least vegetation cover.

References

BOARDMAN, J. W., KRUSE, F. A. & GREEN, R. O. 1995. Mapping target signatures via partial unmixing of AVIRIS data. *In*: *Summaries, Fifth JPL Airborne Earth Science Workshop*. JPL Publications, **95-1**, 23–26.

CHABRILLAT, S. & GOETZ, A. F. H. 1999. The search for swelling clays along the Colorado Front Range: the role of AVIRIS resolution in detection. *In*: *Summaries of the 8th JPL Airborne Earth Science Workshop, Jet Propulsion Laboratory, Pasadena, CA*. JPL Publications, **99-17**, 69–78.

CHABRILLAT, S., GOETZ, A. F. H., OLSEN, H. W., KROSLEY, L. & NOE, D. C., 1999. Use of AVIRIS hyperspectral data to identify and map expansive clay soils in the Front Range Urban Corridor in Colorado. *In*: *Proceedings of the 13th International Conference on Applied Geologic Remote Sensing, Vancouver, Canada, 1–3 March 1999, Vol. 1*. ERIM, Ann Arbor, MI, 390–397.

CHABRILLAT, S., GOETZ, A. F. H., KROSLEY, L. & OLSEN, H. W. 2002. Use of hyperspectral images in the identification and mapping of expansive clay soils and the role of spatial resolution. *Remote Sensing of Environment*, **82**, 431–445.

KRUSE, F. A., LEFKOFF, A. B., BOARDMAN, J. B., HEIDEBRECHT, K. B., SHAPIRO, A. T., BARLOON, P. J. & GOETZ, A. F. H. 1993. The Spectral Image Processing System (SIPS)–Interactive visualization and analysis of imaging spectrometer data. *Remote Sensing of Environment*, **44**, 145–163.

Remote sensing of onshore hydrocarbon seepage: problems and solutions

H. M. A. VAN DER WERFF, M. F. NOOMEN, M. VAN DER MEIJDE & F. D. VAN DER MEER

ITC, Department of Earth Systems Analysis, Hengelosestraat 99, 7500 AA, Enschede, Netherlands (e-mail: vdwerff@itc.nl)

Abstract: Optical remote sensing has in the last two decades been extensively tested for the detection of hydrocarbons at the Earth's surface. The spectral absorption features of seepage-related hydrocarbons can easily be confused with those of man-made bituminous surfaces such as tarred roads. The characteristic low albedo of bituminous surfaces can, at the same time, easily be confused with other dark surfaces such as shade. This paper presents the results of two pixel-based classifications that have been carried out on hyperspectral imagery acquired over seepage areas. The first classification algorithm is a 'minimum distance to class means' (MDC), which is sensitive to spectral absorption features as well as albedo differences. The second algorithm is a 'spectral angle mapper' (SAM), which is not sensitive to albedo differences. Both algorithms are applied for the detection of crude oil resulting from macroseepage and an anomalous halo of bare soil resulting from microseepage. The results show that, at best, only 48% and 29% of the pixels that respectively contain crude oil and seepage-related bare soil could be detected, with the inclusion of many false anomalies. Confusion mainly results from the physical characteristics of the anomalies, as these are not unique to seepages. It is concluded that remote sensing of natural hydrocarbon seepages can be improved by image processing algorithms that make use of spatial information.

Hydrocarbon leakage into the environment is a major problem with large economic and environmental impacts. Hydrocarbon pollution can be man-induced, through leaking pipelines or storage tanks. If a pipeline leak is large or undiscovered for a long time, substantial volumes of explosive gases in the soil can result in the development of dangerous situations involving costly remediation works. The United States National Transportation Safety Board (NTSB) has reported millions of dollars in losses and several casualties as a result of gas pipeline leaks in recent years (NTSB 2001, 2003).

Leaking of hydrocarbons into the environment can also result from natural processes. Pressure differences in the Earth's subsurface forces hydrocarbons to migrate from subsurface reservoirs to shallower levels and eventually to the surface (Schumacher & Abrams 1996). Natural seepage of hydrocarbons has negative consequences for the environment and society. Upwelling tar and oil, or 'heavy hydrocarbons', cause local pollution of soil and water. The presence of hydrocarbons in the root zone of vegetation is thought to cause stress to crops and to be a cause of chlorosis (Crawford 1986). Gas seepages can have high radon concentrations (locally exceeding $100\ kBq\ m^{-1}$), as reported from Hungary by Tóth and Boros (1994). Upwelling gases, or 'light hydrocarbons', consist mainly of the greenhouse gases CO_2 and CH_4. Gas seepages not only cause local pollution but also contribute to the effect of global warming, although the flux of seeping hydrocarbons at a global scale is still unknown (Brown 2000).

This research will focus on locating hydrocarbons in the environment as a result of natural processes. Natural seepage is not only a potential source of hazards, but is also of interest for hydrocarbon exploration. In historical times, oil from seepages has been used for lubrication, sealing, medical purposes and, of course, as fuel (Faber 1947). Present-day geochemical analysis of seeping hydrocarbons can reveal information on the type of hydrocarbons that are present at depth (Jones & Drozd 1983; Prinzhofer et al. 2000), as well as information on the prospecting viability of a reservoir (Saunders et al. 1999).

Traditional methods for investigating seepage and pollution, such as drilling, are time consuming, destructive and expensive. Remote sensing has proved to be a tool that offers a non-destructive investigation method and has a significant added value compared with traditional methods. Thermal remote sensing has been tested with varying success for detection of thermal anomalies resulting from chemical reactions of seeping hydrocabrons with reactive lithology such as evaporites (Williams 2000). Optical remote sensing has been extensively tested for exploration of onshore hydrocarbon reservoirs (e.g. Lang et al. 1984a, b) and detection of hydrocarbons at the Earth's surface (e.g. Ellis et al. 2001; Hörig et al. 2001; Kühn et al. 2004; Li et al. 2005). Theoretically, remote sensing is a suitable tool for direct and indirect detection of the presence of hydrocarbon seepages (e.g. Abrams et al. 1984; Tedesco 1995; Yang et al. 1998; Schumacher 2001; Noomen et al. 2005). In practice, however, the spectral characteristics can easily be confused with those of other bituminous surfaces such as asphalt. At the same time, most of the observed geochemical and botanical anomalies that result from seeping hydrocarbons are subtle and not unique to seepages.

To demonstrate these problems, this paper presents the results of two pixel-based classifications that are carried out on hyperspectral imagery acquired over seepage areas. The algorithms are applied for the identification of both macro- and microseepages in grassland areas. The applicability of remote sensing techniques for detection of hydrocarbon seepages is assessed. Techniques that are needed to improve the obtained results are presented, with implications for future studies on remote detection of hydrocarbon seepage.

Geochemical anomalies resulting from seepages

Link (1952) was the first to separate macroseepages, which consist of seeping liquids (crude oil) and gases that are visible to the human eye, from microseepages, which can be detected only by geochemical means. The migration of heavy hydrocarbons, in the process of forming a macroseepage, requires considerable space in the Earth's subsurface. Possible migration paths are unconformities, tectonic structures that breach reservoirs and seals, reservoir rocks acting as a carrier bed, and surface expressions of intrusions such as mud volcanoes and salt domes.

In microseepages, trace quantities of light hydrocarbons, such as methane, ethane, propane, butane and pentane, migrate rapidly through a microfracture network, also called a chimney (Tedesco 1995). A review of possible migration mechanisms for hydrocarbon seepage has been given by Brown (2000). In general, migration can vary from near-vertical to lateral movement over long distances (Schumacher & Abrams 1996). An understanding of the interaction between geochemistry, migration paths and spatial distribution at the surface is possible in areas with a simple geological setting. These interactions become more difficult to understand when the geology is complex.

At the surface, long-term leakage of hydrocarbons can lead to their bacterial oxidation. This can establish locally anomalous redox zones that favour the development of a diverse array of chemical changes that influence the surrounding environment (Hoeks 1983; Schumacher & Abrams 1996). An overview of seepage-induced alterations and consequences for remote sensing has been given by Schumacher (1996), Yang et al. (1998) and Saunders et al. (1999). A schematic diagram of possible mineralogical, botanical and geomorphological alterations is shown in Figure 1.

Bacterial oxidation of methane (CH_4) and the presence of carbon dioxide (CO_2) deplete oxygen from the soil (Smith et al. 2004). The soil can even become anaerobic, eventually causing roots of plants to die (Hoeks 1983; Pysek & Pysek 1989). Observed mineralogical alterations in soils include the formation of pyrite, calcite, uraninite, elemental sulphur, specific magnetic oxides and iron sulphides (Schumacher 1996). However, the chemical processes involved in seepages and resulting surface expressions are still not fully understood. Most alterations are not unique to the redox environment of seeping hydrocarbons, which makes it difficult to separate them from alterations caused by other natural soil processes (Schumacher 1996).

Figure 2 shows the area containing macroseepage in Upper Ojai Valley, California. According to local observers, an earthquake in the 1980s caused these seepages to appear in an orchard that

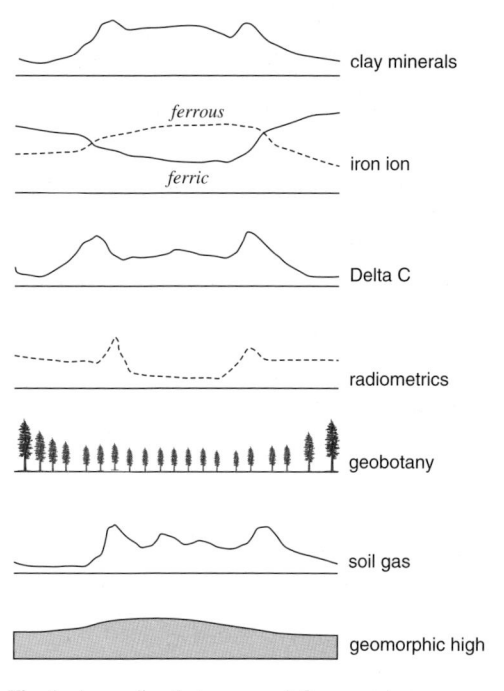

Fig. 1. Anomalies that may result from seeping hydrocarbons, modified after Yang (1999).

(a) aerial view of macroseepages

(b) aerial view of microseepages

(c) a macroseepage vent with oil

(d) an anomalous halo with less grass

Fig. 2. The macroseepages in Upper Ojai Valley, California, and microseepages in Parádfürdo, Hungary, studied in this research. (**a**) A color aerial photo with the locations of macroseepage vents shown as dashed circles (subvertical seepages) and an oval (subhorizontal flow out of a 2 m high ridge that probably results from a fault). The dashed line indicates the orientation of a likely subsurface structure that acts as a migration path. The area covered by the aerial photo is approximately 1200 m × 1200 m. (**b**) Locations of microseepage vents and anomalous halos in Hungary, shown by dashed circles. The dashed line indicates the orientation of a likely subsurface structure that acts as a migration path. The area covered by the image is c. 270 m × 270 m. (**c**) A vent of crude oil surrounded by some bare soil, which results from oxygen shortage owing to upwelling gases. (**d**) In the foreground is shown a part of an anomalous halo resulting from seeping carbon dioxide and methane. The unaltered grassland with flowering weeds can be seen in the background.

has since been abandoned. The seepages consist of central vents of crude oil surrounded by a halo of bare soil, probably as a result of oxygen shortage. Figure 2 also shows an area containing microseepage in the Hungarian spa Parádfürdo, located in the Mátra mountains. This seepage is caused by a high geothermal gradient in this area (Tóth & Boros 1994). The seepages became active after groundwater pumping activities in this mining area ceased two decades previously (T. Zelenka, Pers. Comm.). They consist of central vents with c. 90% CO_2 and 10% CH_4 (Nagy & Turtegin 1992) and are surrounded by a halo of decreased vegetation height and density resulting from the upwelling gases (Fig. 2d). There is a slight colour difference between the soil inside and outside the halo that is mainly a result of the presence of some organic matter and rust mottles.

Remote sensing of hydrocarbon seepages

In the electromagnetic spectrum, oil has absorption features at 1.7 μm and between 2.3 and 2.6 μm, as well as at shorter wavelengths (Cloutis 1989). Hörig et al. (2001) and Kühn et al. (2004) concluded that the shortwave infrared (SWIR) can be used to detect bituminous surfaces and the visible & near infrared (VNIR) for distinction between different hydrocarbons. Together with Li et al. (2005), they showed that in HyMap airborne hyperspectral imagery (Cocks et al. 1998) the presence and the type of bitumen can be seen as subtle intensity differences in the 1.7 μm absorption bands. These results, however, were obtained by measuring artificial bituminous targets against a rather homogeneous background. When mapping hydrocarbons in an image that covers a natural area, both the spectral detection and the distinction between different bituminous surfaces becomes problematic (Li et al. 2005; Salem et al. 2005).

Two gases that are commonly found in hydrocarbon seepages, CH_4 and CO_2, have absorption features in the reflective (0.4–2.5 μm) part of the spectrum (Pieters & Englert 1993). Unfortunately, direct detection of gases that originate from seepages is problematic, as CO_2 is already present in the atmosphere and the absorption feature of CH_4 is too narrow to allow detection by present-day airborne or spaceborne sensors. The flux of escaping hydrocarbons also changes over time (Jones & Burtell 1996) and seepages are not the only natural source of CH_4 and CO_2 (Schumacher 1996). The subtle spectral signals of gases generally do not show up against a non-homogeneous background (de Jong 1998).

Optical remote sensing can be used to detect indirect evidence of hydrocarbon seepages, such as botanical and mineralogical alterations that develop as a result of the presence of hydrocarbons in the soil (see above). Almeida-Filho (2002) detected microseepage by mapping bleached red-beds in a c. 6 km area, using several ratios of Landsat TM bands 2, 3 and 4. These results appeared to be consistent with soil gas anomalies that had been measured at the same location. The question remains as to whether this detection would have been possible for smaller targets or without prior knowledge.

Detection methods and results

Two pixel-based classifications have been carried out on hyperspectral imagery. The first algorithm is a 'minimum distance to class means' (MDC) classification (Richards & Jia 1999), which is widely used and uses both the albedo and spectral absorption features. The second classification is the widely used 'spectral angle mapper' (SAM) algorithm (Kruse et al. 1993) that uses only the spectral absorption features and ignores the albedo (Bakker & Schmidt 2002).

Detection of macroseepages

The two pixel-based classifications were first carried out on a Probe-1 (HyMap) hyperspectral image that covers an area with macroseepages in California (see above). Pixels that are known from the field to contain mainly oil are selected as reference for the classification. The mean spectrum of these 69 pixels has been used as a classification endmember. The threshold used to obtain a hard classification from the grey-scale image is set in such way that the classification results optimally cover the known seepage pixels while keeping a minimum number of false anomalies.

The results of the MDC classification are given in Table 1 and Figure 3d. Of the seepage pixels in Figure 3b, 33 are correctly identified whereas 36 cannot be detected. Furthermore, the classification gives 380 false anomalies, which occur mainly as dark pixels that contain shade from trees and shrubs, and at pixels that have a similar spectral signal, such as roads. The results of the SAM classification are given in Table 1 and Figure 3d. Of the seepage pixels in Figure 3d, 32 are correctly identified, 37 cannot be detected and 1103 are false anomalies. The number of false anomalies at dark pixels decreases in comparison with the MDC classification. However, the number of false anomalies at bituminous surfaces, especially asphalt roads, increases substantially.

In summary, the MDC and SAM classifications correctly identify 48% and 46%, respectively, of the reference seepage pixels. Of the total seepages identified by the respective algorithms, 92% and 97% were false anomalies.

Detection of microseepages

The two pixel-based classifications were carried out on a Digital Airborne Imaging Spectrometer 7915 (DAIS7915) hyperspectral image (DLR German

Table 1. *Confusion matrices of the macroseepage detection results*

Class	Ground truth		
	Unclassified	Seeps	Total
MDC			
Unclassified	24832	36	24868
Seeps	380	33	413
Total	25212	69	25281
SAM			
Unclassified	24109	37	24146
Seeps	1103	32	1135
Total	25212	69	25281

(a) soft MDC classification (DN)

(b) hard MDC classification (0.1 DN)

(c) soft SAM classification (rad)

(d) hard SAM classification (0.1 rad)

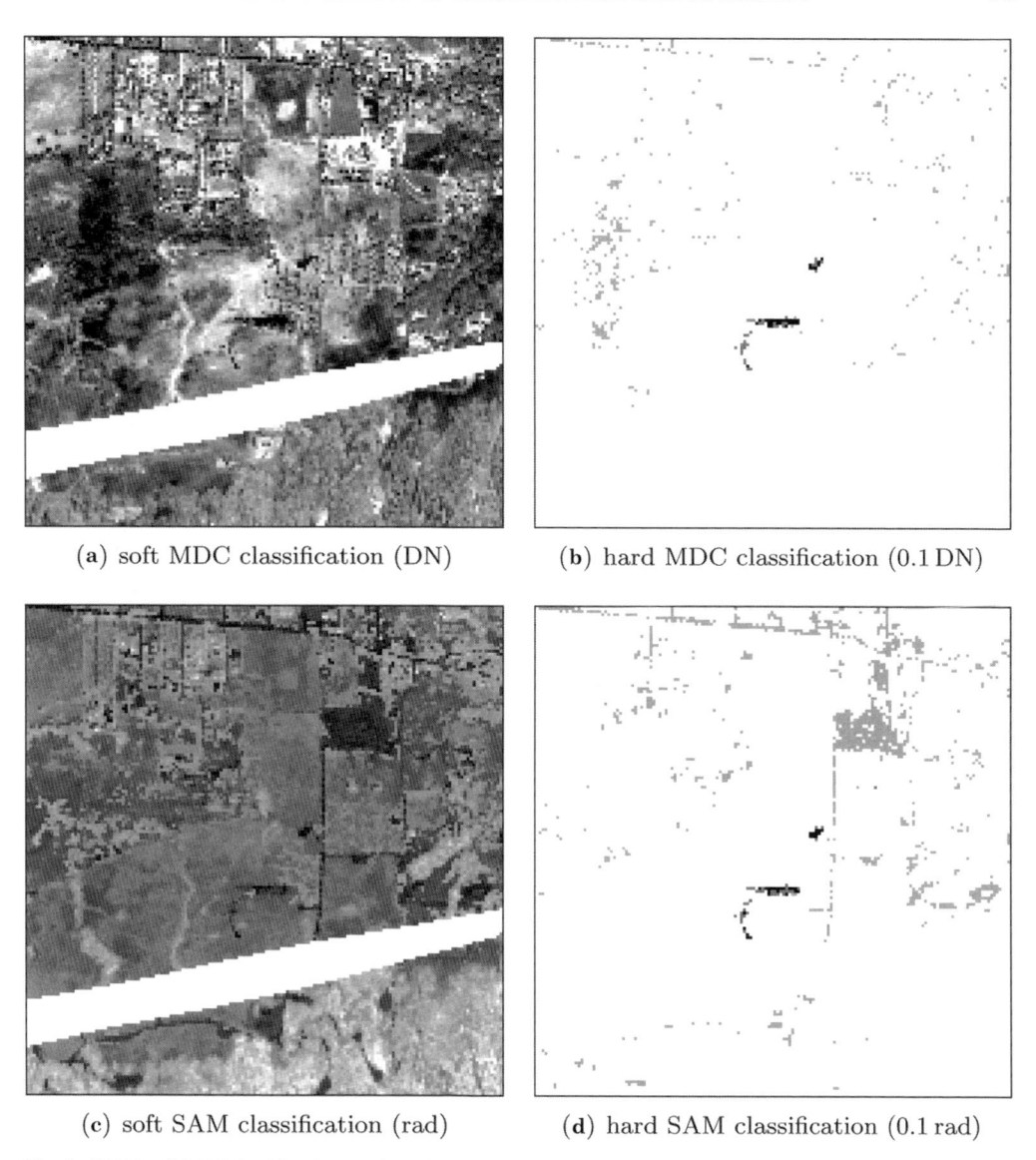

Fig. 3. MDC and SAM classification resulting from a 1200 m × 1200 m macroseepage area in Upper Ojai Valley, California, obtained from Probe-1 imagery. (**a, c**) Soft classification results in a grey-scale image: dark tones represent a relatively good fit with the reference spectrum whereas light tones represent a poorer fit. (**b, d**) A hard classification obtained by thresholding the grey-scale image at 0.1 DN and 0.1 radians, respectively. The grey-scale values indicate whether a pixel is correctly identified (black), not recognized as seepage (dark grey), or is a false anomaly (light grey). The black and dark grey pixels are known to contain mainly oil and were selected as reference for this classification. A confusion matrix of this classification result is shown in Table 1.

Aerospace Center 2005) that covers a microseepage area in Hungary (see above). The classification setup is identical to that for macroseepage detection described above. The spectral reference in this classification is that of a bare soil seepage halo. The pixels that are known from the field to fall within a seepage halo are selected as reference pixels for the classifications.

The results of the MDC and SAM classifications are displayed in Table 2 and Figure 4. The results of the MDC classification are disappointing: of the seepage pixels in Figure 4a, only nine are correctly identified and 22 cannot be detected. Furthermore, the classification gives 71 false anomalies, which are mainly roof tiles, shade resulting from trees, or grassland. Of the SAM classification results,

Table 2. *Confusion matrices of the microseepage detection results*

Class	Ground Truth		
	Unclassified	Seeps	Total
MDC			
Unclassified	2814	22	2836
Seeps	71	9	80
Total	2885	31	2916
SAM			
Unclassified	2859	24	2883
Seeps	26	7	33
Total	2885	31	2916

only seven of the seepage pixels in Figure 4c are correctly identified, 24 cannot be detected and there are 26 false anomalies. Unlike the MDC algorithm results, these do not occur at dark pixels related to shade, but primarily at pixels indicating bare soil created by other natural or human causes.

In summary, the MDC and SAM classifications correctly identify 29% and 23%, respectively, of the reference seepage pixels. Of the total seepages identified by the respective algorithms, 89% and 79% were false anomalies.

Discussion

Both hard classifications obtained in this research are a direct result of the choice of threshold set from the soft classification images, and the threshold is set according to *a priori* knowledge of the spatial extent of the seepages. A classification that would be made without this prior knowledge would probably perform far more poorly. When setting a wider threshold, the number of correct identifications would increase but the number of false anomalies also increases, and *vice versa*. The results of detecting macroseepages in the Californian study area show that oil is spectrally easily confused with other dark or bituminous objects, supporting the findings of Salem *et al.* (2005). However, this is not the main problem in the remote detection of macroseepages. This research has identified the major problem to be that no distinction could be made between hydrocarbons that are present as a result of seepage activity and those that are present as a result of human activity, such as asphalt roads. The only distinction that can be observed in the field is the presence of a chemically anomalous halo around an actual hydrocarbon seepage vent. This halo results from light hydrocarbons (gases) that migrate upwards along with the heavier hydrocarbons. Consequently, the detection of the macroseepage vents would benefit from the detection of microseepage-induced anomalies.

The results of detecting microseepage-induced anomalies in the Hungarian study area show that such anomalies cannot reliably be detected by using standard image processing methods. The confusion in detection of pixels with microseepage-induced alteration differs from the spectral confusion that was seen in the detection of macroseepages. The problem in the detection of microseepage-induced alteration is that the observed alterations are not unique to the seepages. An anomalous halo that is purely defined by bare soil cannot be distinguished from other bare soil areas by any pixel-based classification algorithm.

The human eye is capable of recognizing the seepages in the field from a distance. With *in situ* observation of the natural seepages in California and Hungary, and by simulating gas leakage in laboratories in Wageningen University (Hoeks 1983) and Nottingham University (Noomen *et al.* 2005), one can easily identify seepage vents that produce halos of *c.* 4–14 m width (natural) or *c.* 1–4 m width (simulated). These halos are defined by the composition of different surface covers: a background (grassland in the presented seepages) and a halo of bare soil with possibly tar or oil in the centre of the halo. Aerial observation (dashed lines in Fig. 2a and b) shows that the circular halos line up along geological structures that are present in the substrate, as was also found by Hernandéz *et al.* (2000). This complex spatial pattern, circles that are on a line, can be described by simple mathematical shapes (van der Werff *et al.* 2006).

The combination of the non-unique spectral characteristics and the observed spatial patterns provides a unique 'fingerprint' of these seepages. Table 3 gives the results of classifications made with a pixel-based classifier and an algorithm based on Hough transforms (Ballard 1981). This table shows the distance between seepage locations observed in the field and image pixels that were found to have a high seepage probability. The algorithm based on Hough transforms finds the relatively high probabilities close to the actual seepage locations, whereas the pixel-based classifier shows high probabilities up to 100 pixels away from the actual seepages. Hence, the number of false anomalies that appear with the pixel-based classifier is greatly reduced (van der Werff *et al.* 2006). The detection of seepage-induced anomalies and their distinction from other objects by optical remote sensing needs a contextual image processing technique that incorporates the spatial information on seepages (van der Werff & Lucieer 2004). A contextual approach also has applications in the monitoring of pipelines. From a spatial point of view,

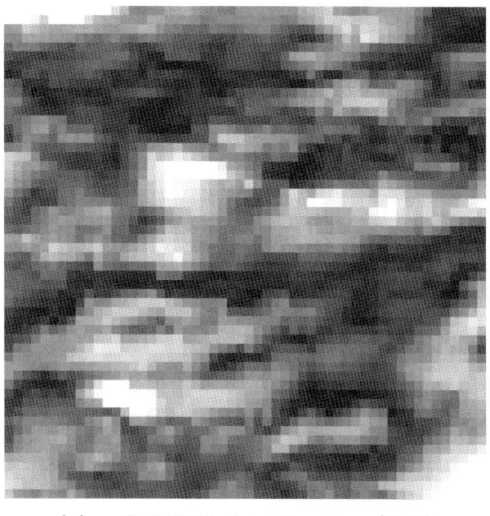

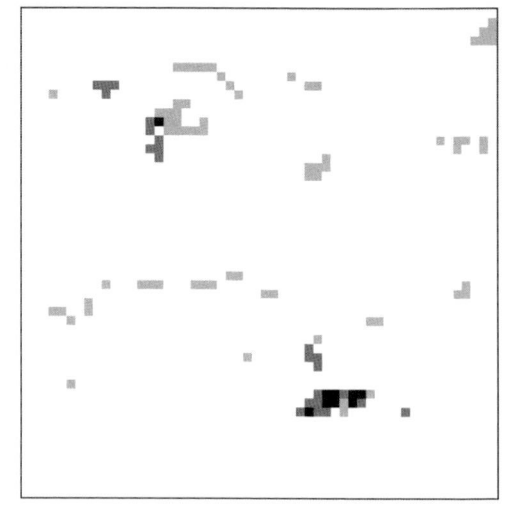

(a) soft MDC classification (DN)

(b) hard MDC classification (80 DN)

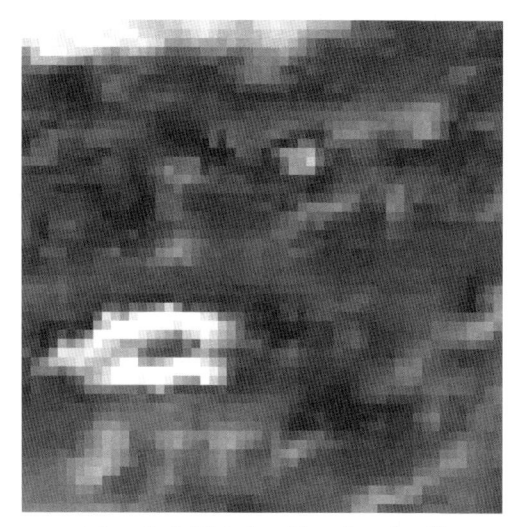

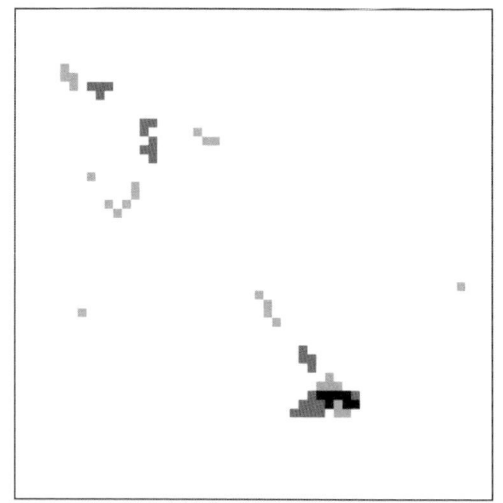

(c) soft SAM classification (rad)

(d) hard SAM classification (0.035 rad)

Fig. 4. MDC and SAM classification results of a field containing microseepages in Parádfürdo, Hungary, obtained from DAIS7915 imagery. (**a, c**) Soft classification results in a grey-scale image (DN and rad, respectively). Dark tones represent a relatively good fit with the endmember spectrum whereas light tones represent a poorer fit. (**b, d**) Hard classifications obtained by thresholding the grey-scale image at 80 DN and 0.035 rad, respectively. The grey values indicate whether a pixel is correctly identified (black), not recognized as seepage (dark grey), or is a false anomaly (light grey). The black and dark grey pixels are known to belong to an anomalous halo resulting from seeping CO_2 and CH_4. A confusion matrix of the classification result is shown in Table 2.

pipeline leakages can be considered as point sources with a radial flow pattern. Using the spatial pattern of a pipeline and the radial flow pattern resulting from leaks, which result in halos similar to natural seepage, might lead to improvements in pipeline leakage detection and monitoring (van der Meijde *et al.* 2005).

Conclusions

This paper showed the results of two pixel-based classifications on the detection of oil resulting from a macroseepage and bare soil resulting from a microseepage. The detection of oil related to macroseepages had a maximum score of 48%,

Table 3. *Comparison of a pixel-based Euclidean distance (ED) classifier and an algorithm based on the Hough transform (HT)*

Distance	ED score	HT score
0	0.75	0.80
10	0.71	0.42
20	0.77	0.27
30	0.83	0.00
40	0.77	0.00
50	0.46	0.00
60	0.86	0.00
70	0.76	0.09
80	0.79	0.05
90	0.60	0.21
100	0.90	0.18
110	0.91	0.00

Column 1 shows the distance between seepages observed in the field and pixels with a high seepage probability, indicated by either algorithm. Columns 2 and 3 show the seepage probability indicated by the ED and HT classifier, respectively. To allow a comparison of these results, output values are relatively scaled between 0.00 and 1.00 DN for low and high fit, respectively. After van der Werff *et al.* (2006).

whereas the detection of bare soil related to microseepage had a maximum score of 29%. The low score for the detection of seepage-induced oil and geochemical anomalies is only partly due to spectral confusion. The confusion is mainly the result of the physical characteristics of the anomalies, as these are not unique to seepages. The spatial pattern of seepages that have been observed in the field consists of a series of circles that tend to line up along a subsurface geological structure. It is concluded that the remote sensing of hydrocarbon seepages has to be carried out by image processing algorithms that include the use of pattern recognition.

The authors wish to thank T. Zelenka of the Hungarian Geological Institute (MÁFI) for his input and guidance in the field. R. Teeuw is acknowledged for his input and, together with N. Knox, thanked for correcting an earlier version of this paper.

References

ABRAMS, M., CONEL, J., LANG, H. & PALEY, H. (eds) 1984. *The joint NASA/Geosat Test Case Project*. Vol. 1. American Association of Petroleum Geologists, Tulsa, OK.

ALMEIDA-FILHO, R. 2002. Remote detection of hydrocarbon microseepage-induced soil alteration. *International Journal of Remote Sensing*, **23**, 3523–3524.

BAKKER, W. & SCHMIDT, K. 2002. Hyperspectral edge filtering for measuring homogeneity of surface cover types. *ISPRS Journal of Photogrammetry and Remote Sensing*, **56**, 246–256.

BALLARD, D. 1981. Generalizing the Hough transform to detect arbitrary shapes. *Pattern Recognition*, **13**, 111–122.

BROWN, A. 2000. Evaluation of possible gas microseepage mechanisms. *AAPG Bulletin*, **84**, 1775–1789.

CLOUTIS, E. 1989. Spectral reflectance properties of hydrocarbons: remote-sensing implications. *Science*, **245**, 165–168.

COCKS, T., JENSSEN, R., STEWART, A., WILSON, I. & SHIELDS, T. 1998. The HyMap airborne hyperspectral sensor: the system, calibration and performance. *In*: SCHAEPMAN, M., SCHLÄPFER, D. & ITTEN, K. (eds) *Proceedings of the 1st EARSeL Workshop on Imaging Spectroscopy, 6–8 October 1998, Zürich*. Remote Sensing Laboratories, University of Zürich, 37–42.

CRAWFORD, M. 1986. Preliminary evaluation of remote sensing data for detection of vegetation stress related to hydrocarbon microseepage: mist gas field, Oregon. *In: Proceedings of the 5 Thematic Conference: Remote Sensing for Exploration Geology*. Reno, NV, Environmental Research Institute of Michigan (ERIM), Ann Harbour, 161–177.

DE JONG, S. 1998. Imaging spectrometry for monitoring tree damage caused by volcanic activity in the Long Valley caldera, California. *ITC Journal*, **1**, 1–10.

DLR 2005. The Digital Airborne Imaging Spectrometer DAIS 7915. World Wide Web address: http://www.dlr.de.

ELLIS, J., DAVIS, H. & ZAMUDIO, J. 2001. Exploring for onshore oil seeps with hyperspectral imaging. *Oil and Gas Journal*, **99**, 49–58.

FABER, M. 1947. *Petroleum zoeken en ontdekken*, 2nd edn. W. J. Thieme, Zutphen, Netherlands.

HERNANDÉZ, P., PÉREZ, N., SALAZAR, J., SATO, M., NOTSU, K. & WAKITA, H. 2000. Soil gas CO, CH, and H distribution in and around Las Cañadas caldera, Tenerife, Canary Islands, Spain. *Journal of Volcanology and Geothermal Research*, **103**, 425–438.

HOEKS, J. 1983. *Gastransport in de bodem*. Technical Report 1470, Instituut voor Cultuurtechniek en Waterhuishouding, Wageningen.

HÖRIG, B., KÜHN, F., OSCHÜTZ, F. & LEHMANN, F. 2001. Hymap hyperspectral remote sensing to detect hydrocarbons. *International Journal of Remote Sensing*, **22**, 1413–1422.

JONES, V. & DROZD, R. 1983. Predictions of oil or gas potential by near-surface geochemistry. *AAPG Bulletin*, **67**, 932–952.

JONES, V. III & BURTELL, S. 1996. Hydrocarbon flux variations in natural and anthropogenic seeps. *In*: SCHUMACHER, D. & ABRAMS, M. (eds) *Hydrocarbon Migration and its Near-surface Expression*. American Association of Petroleum Geologists, Memoirs, **66**, 203–221.

KRUSE, F., LETKOFF, A., BOARDMANN, J., HEIDEBRECHT, K., SHAPIRO, A., BARLOON, P. & GOETZ, A. 1993. The spectral image processing system (SIPS)—interactive visualization and analysis of imaging spectrometer data. *Remote Sensing of Environment*, **44**, 145–163.

KÜHN, F., OPPERMANN, K. & HÖRIG, B. 2004. Hydrocarbon index—an algorithm for hyperspectral detection of hydrocarbons. *International Journal of Remote Sensing*, **25**, 2467–2473.

LANG, H., ALDEMAN, W. & SABINS, F., JR. 1984a. Patrick Draw, Wyoming—petroleum test case report. *In*: ABRAMS, M., CONEL, J., LANG, H. & PALEY, H. (eds) *The Joint NASA/Geosat Test Case Project, Vol. 1.* American Association of Petroleum Geologists, Tulsa, OK, 11-1–11-28.

LANG, H., CURTIS, J. & KOVACS, J. 1984b. Lost River, West Virginia—petroleum test site report. *In*: ABRAMS, M., CONEL, J., LANG, H. & PALEY, H. (eds) *The Joint NASA/Geosat Test Case Project, Vol. 1.* American Association of Petroleum Geologists, Tulsa, OK, 12-1–12-96.

LI, L., USTIN, S. & LAY, M. 2005. Application of AVIRIS data in detection of oil-induced vegetation stress and cover change at Jornada, New Mexico. *Remote Sensing of Environment*, **94**, 1–16.

LINK, W. 1952. Significance of oil and gas seeps in world oil exploration. *AAPG Bulletin*, **36**, 1505–1540.

NAGY, G. & TURTEGIN, E. 1992. A Mátraderecskei gázsívárgás környezetvédelmi-földtani vizsgálata, Environmental–geological investigation of the gas-seepage at Mátraderecske. Hungarian Geological Institute, Budapest, Technical Report T15764.

NOOMEN, M., VAN DER MEER, F. & SKIDMORE, A. 2005. Hyperspectral remote sensing for detecting the effects of three hydrocarbon gases on maize reflectance. *In: Proceedings of the 31st International Symposium on Remote Sensing of Environment: Global Monitoring for Sustainability and Security.* International Society for Photogrammetry and Remote Sensing (ISPRS), St. Petersburg State University. St. Petersburg, Russia.

NTSB 2001. *Natural gas exploitation and fire in South Riding, Virginia, July 7, 1998.* National Transportation Safety Board, Washington, DC, Technical Report.

NTSB 2003. *Natural gas pipeline rupture and fire near Carlsbad, New Mexico.* National Transportation Safety Board, Washington, DC, Technical Report.

PIETERS, C. & ENGLERT, P. 1993. *Remote Geochemical Analysis: Elemental and Mineralogical Composition.* Cambridge University Press, Cambridge.

PRINZHOFER, A., ROCHA MELLO, M. & TAKAKI, T. 2000. Geochemical characterization of natural gas: a physical multivariable approach and its applications in maturity and migration estimates. *AAPG Bulletin*, **84**, 1152–1172.

PYSEK, P. & PYSEK, A. 1989. Veränderungen der Vegetation durch experimentelle Erdgasbehandlung. *Weed Research*, **29**, 193–204.

RICHARDS, J. & JIA, X. 1999. *Remote Sensing Digital Image Analysis, an Introduction*, 3rd edn. Springer, Heidelberg.

SALEM, F., KAFATOS, M., EL-GHAZAWI, T., GOMEZ, R. & YANG, R. 2005. Hyperspectral image assessment of oil-contaminated wetland. *International Journal of Remote Sensing*, **26**, 811–821.

SAUNDERS, D., BURSON, K. & THOMPSON, C. 1999. Model for hydrocarbon microseepage and related near-surface alterations. *AAPG Bulletin*, **83**, 170–185.

SCHUMACHER, D. 1996. Hydrocarbon-induced alteration of soils and sediments. *In*: SCHUMACHER, D. & ABRAMS, M. (eds) *Hydrocarbon Migration and its Near-surface Expression.* American Association of Petroleum Geologists, Memoirs, **66**, 71–89.

SCHUMACHER, D. 2001. Petroleum exploration in environmentally sensitive areas: opportunities for non-invasive geochemical and remote sensing methods. Annual Convention of the ASPG (American Society of Professionl Geographers), 012-1–012-5.

SCHUMACHER, D. & ABRAMS, M. (eds) 1996. *Hydrocarbon Migration and its Near-surface Expression.* American Association of Petroleum Geologists, Memoirs, **66**.

SMITH, K., STEVEN, M. & COLLS, J. 2004. Use of hyperspectral derivative ratios in the red-edge region to identify plant stress responses to gas leaks. *Remote Sensing of Environment*, **92**, 207–217.

TEDESCO, S. 1995. *Surface Geochemistry in Petroleum Exploration.* Chapman & Hall, New York.

TÓTH, E. & BOROS, D. 1994. High radon activity in Northeast Hungary. *Physica Scripta*, **50**, 726–730.

VAN DER MEIJDE, M., NOOMEN, M., VAN DER WERFF, H., KOOISTRA, J. & VAN DER MEER, F. 2005. Pipeline leakage revealed by vegetation anomalies measured with reflectance spectroscopy (abstract). *In: Proceedings of the GRSG Annual Meeting, London, Geological Remote Sensing Group* (http: www.grsg.org/abs_vol_GRSG05_final.doc, page 7).

VAN DER WERFF, H. & LUCIEER, A. 2004. A contextual algorithm for detection of mineral alteration halos with hyperspectral remote sensing. *In*: DE JONG, S. & VAN DER MEER, F. (eds) *Remote Sensing Image Analysis: Including the Spatial Domain. Remote Sensing and Digital Image Processing, Vol. 5.* Kluwer, Dordrecht, 201–210.

VAN DER WERFF, H., BAKKER, W., SIDERIUS, W. & VAN DER MEER, F. 2006. Combining spectral signals and spatial patterns using multiple Hough transforms: an application for detection of natural gas seepages. *Computers & Geosciences*, **32**, 1334–1343.

WILLIAMS, A. 2000. The role of satellite exploration in the search for new petroleum reserves in South Asia. SPE–PAPG Annual Conference, Islamabad.

YANG, H. 1999. *Imaging spectrometry for hydrocarbon microseepage.* PhD thesis, Delft University of Technology, Delft.

YANG, H., ZHANG, J., VAN DER MEER, F. & KROONENBERG, S. 1998. Geochemistry and field spectrometry for detecting hydrocarbon microseepage. *Terra Nova*, **10**, 231–235.

Remote sensing for terrain analysis of linear infrastructure projects

J. MANNING

Arup, 13 Fitzroy Street, London W1T 4BQ, UK
(e-mail: jason.manning@arup.com)

Abstract: This paper examines recent developments in remote sensing applied to civil engineering projects, particularly those involving terrain analysis of linear infrastructure features. Applications using high-resolution satellite data (e.g. IKONOS and Quickbird), laser altimetry and digital elevation models (DEMs) are discussed. Guidelines are given for the stages of a project in which different remote sensing techniques could be employed. The case is also made for greater use of archive remote sensing datasets, which can be a valuable source of information about landscape dynamics.

Linear infrastructure projects (e.g. highways, railways, pipelines, power lines, canals) often extend across large geographical areas and may pass through a variety of terrain types. Remote sensing can greatly assist linear project mapping, particularly where linear construction projects pass through areas of challenging and inaccessible terrain, such as mountains, deserts or wetlands, and will often provide more detail than published mapping or other sources of information.

Terrain analysis allows for characterization of the various terrain types along, and adjacent to, the infrastructure alignment. This allows the identification and mapping of potential hazards that may affect construction and/or operation of the asset. Types of hazard fall into two key groups: natural (geohazards and ground-related hazards) and man-made hazards, as detailed in Table 1. Identification of the hazard prior to construction allows engineers to take into consideration the hazard and to either design to deal with the hazard, or consider potential realignments of the proposed infrastructure. Early identification of the hazard reduces the risk of costly delays to construction.

Remote sensing provides many useful tools for terrain analysis. Historically, aerial photography is the most widely used technique. However, satellite remote sensing is better suited to projects covering large geographical areas and has seen a progressive increase in use since the mid-1970s. Both airborne and satellite imagery provide useful sources of image layers and elevation data for terrain analysis. Table 2 summarizes key remote sensing systems for terrain analysis applications.

Airborne systems

Aerial survey photography

Aerial survey photography is the earliest method of remote sensing, used widely since the 1940s, primarily for production of topographic maps. Aerial survey photography uses overlapping contact prints (typically 230 mm × 230 mm, 9 inches × 9 inches) that allow for stereoscopic interpretation. Small-format photography is of limited analytical use. The advantages of aerial photographic interpretation, for both geological and geomorphological mapping, are well documented (e.g. Brunsden *et al.* 1975; Van Zuidam 1985; Drury 1987). The resolution and stereo capabilities of aerial photography exceed those from satellite imagery. Image scale, resolution, accuracy and format can all be varied to suit particular project requirements. Infrared films can be used in some applications (vegetation distinction, vegetation stress and soil moisture content). Photogrammetric software can be used to generate contour models, digital elevation models (DEMs), topographic base maps and orthophotograph datasets, all of which can be used within a geographical information system (GIS) for terrain analysis. Reviews of the use of aerial photography, GIS and photogrammetrical methods for assessing slope instability are given by Ho *et al.* (2006) and Walstra *et al.* (2007).

Airborne laser scanning (LiDAR)

LiDAR (light detection and ranging) is a relatively new remote sensing technique using laser technology. Laser-scanned data can be captured from fixed-wing aircraft or helicopter and provides 3D points (*xyz*) at densities ranging from 1 to 25 m^2. Swath widths typically range from 50 m to over 1 km. Helicopter LiDAR survey can acquire about 100 line km a day. Lower-resolution aeroplane LiDAR can collect up to 150 line km per hour, dependent on terrain type. Outputs can include 'point cloud' data, digital terrain model (DTM) grids and contour models. Some LiDAR systems allow for the simultaneous capture of high-resolution digital photographs from which orthophotographs can be generated. The accuracy of the laser data

From: TEEUW, R. M. (ed.) *Mapping Hazardous Terrain using Remote Sensing*. Geological Society, London, Special Publications, **283**, 135–142.
DOI: 10.1144/SP283.12 0305-8719/07/$15.00 © The Geological Society 2007.

Table 1. *Principle hazard types affecting linear infrastructure*

Natural hazards (geo-hazards and ground-related hazards)	Man-made hazards
Soft ground (compressible soils, high water table) Rock outcrop and shallow rockhead Dissolution features (karst) Slope instability (landslides, mudflows, etc.) Soil erosion Wind-blown sands Expansive clays Saline soils River channel migration and erosion Coastal erosion Volcanic hazards Seismicity (ground rupture, soil liquefaction) Floods	Contaminated land and groundwater Made ground (spoil and backfill) Excavated ground (quarries and pits) Mining (shafts, subsidence, contamination) Groundwater abstraction (wells, subsidence) Not ordnance (UXO, military firing ranges, World War Two bombs) Existing structures and infrastructure Third-party interference (e.g. to pipelines)

varies depending on the system and other variables such as flying height. With low-altitude helicopter systems survey accuracies of as good as ±50 mm (vertical root mean square error (RMSE)) are achievable. As with other types of survey and investigation work, the appropriate specification for data capture, processing and output is critical to the successful use of LiDAR.

LiDAR offers many advantages over conventional survey methods, including fast and accurate

Table 2. *Appropriate remote sensing data for linear infrastructure projects*

SYSTEM	TECHNICAL DETAILS	PROJECT STAGE (Site Selection / Feasibility / Concept Design / Detailed Design / inspection)	ARCHIVE RESOURCES
AIRBORNE			
Aerial photography	Assorted scales (typically 1:5000 to 1:25 000) Black & white, colour & Infra-red films; Stereo (3D); Excellent detail; Contact prints or orthophotos (0.1 m to 2 m)		Extensive historic archives, pre-1940s to present
LiDAR	Laserscanning (helicopter or fixed wing platform) Detailed DTM generation & 3D visualisation Ability to 'see' through tree canopy		Limited archive
Hyperspectral	Assorted sensors available Large number of multispectral bands enable detailed mineralogical & vegetational mapping		Limited archive
Thermal scanning	Assorted sensors available		Limited archive
SATELLITE			
Mid-resolution satellites:			
ASTER	14 multispectral wavebands (VNIR, SWIR, TIR) 15 m, 30 m & 90 m resolution Stereo capability at 15 m resolution VNIR		1999 to present
Landsat	Landsat 7: 8 multispectral wavebands (VNIR, SWIR, TIR) 15 m, 30 m & 60 m resolution		Considerable historic archive: Landsat 1 1972 Landsat 7 1999 to present
SPOT	Spot 5: 2 Panchromatic bands (2 x 5 m resolution, 2.5 m if merged) 4 multispectral bands (3 x 10 m resolution, 1 x 20 m)		Considerable historic archive: SPOT 1 1986 SPOT 5 2002 to present
High-resolution satellites:			
Ikonos-2	1 Pan band (1 m resolution) 4 multispectral bands (4 m resolution, VNIR)		2000 to present
QuickBird-2	1 Pan band (0.6 m resolution) 4 multispectral bands (2.4 m resolution, VNIR)		2001 to present
OrbView-3	1 Pan band (1 m resolution) 4 multispectral bands (4 m resolution, VNIR)		2003 to present
Satellite photography:			
Corona	1.8 m (6 ft) to 140 m (460 ft) spatial resolution Some stereo images collected		1959–1972
KVR-1000	2–5 m resolution		1984 to present
Satellite RADAR			
InSAR	Interferometric Synthetic Aperture Radar Analysis of multiple radar images to detect ground movement		From early 1990s Requires detailed processing
SRTM	Shuttle Radar Topography Mission (DTM dataset) Absolute horizontal accuracy = 20 m Absolute vertical accuracy = 16 m		2000 only

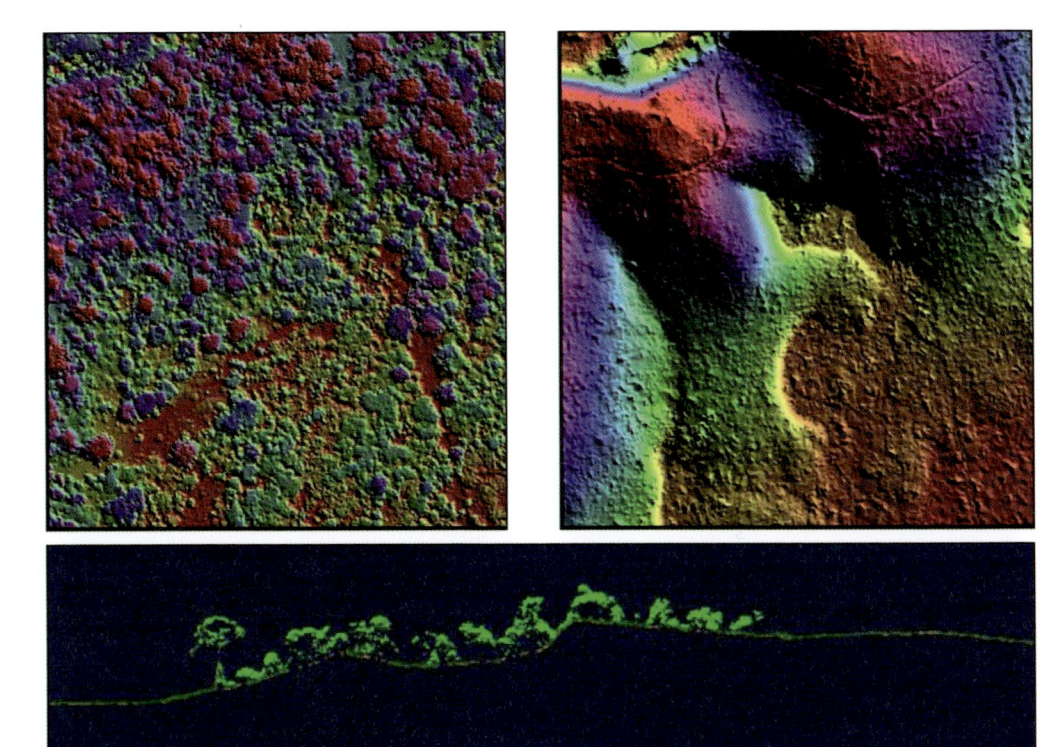

Fig. 1. Airborne LiDAR elevation survey over tropical rainforest. LiDAR records levels of both the canopy (a digital surface model: top left) and the ground surface (a digital terrain model: top right). Detailed profiles of both vegetation cover and the ground surface can be readily generated (bottom).

mapping of large areas at very dense height point densities, and has the unique capability to provide terrain data below dense vegetation canopy (Fig. 1). LiDAR is particularly useful for terrain analysis in forested areas. Profiles can be readily generated within image processing software, thereby allowing detailed vertical alignments of proposed routes to be prepared. Shaded relief models can be used to map terrain types and identify potential hazards, such as floodplains and unstable slopes.

Hyperspectral and thermal scanning

Airborne hyperspectral and thermal scanners have proven useful for identifying vegetation stress and potential leakages from existing pipelines (see, e.g. the study by van der Werff *et al.* 2007). These techniques show some promise for use in monitoring existing pipelines, useful for both maintenance and planning for decommissioning of existing pipeline infrastructure.

Digital elevation data

In the UK there are a number of 'off-the-shelf' DEM datasets available that can be used effectively with other remote sensing data. The DEM data can be derived from aerial photography (photogrammetrically), LiDAR or airborne radar. Digital surface models (DSMs) show the elevation of all features in a survey area, including non-terrain features, such as woodland canopies and roof tops; DSMs are therefore are limited in their ability to accurately portray terrain variations. On the other hand, digital terrain models (DTMs) show the land surface and its morphology, making them ideal for terrain analysis. Of the airborne sensors, only LiDAR can detect through tree cover, and the resulting DTMs can highlight ground conditions that would otherwise be undetected by aerial photography or radar.

Satellite systems

Earth observation from space started in the late 1950s, initially via military spy-satellites. Commercial Earth observation satellites started with the Landsat programme in 1970, followed by the SPOT programme in the 1980s. Further developments continued through the 1990s with increasing spatial and spectral resolutions (number of wavebands), with the first sub-metre commercial satellite data being made available in 2000.

Satellite imagery is particularly useful in remote or restricted areas where there is a lack of existing mapping or other data. Some sensors can provide stereo data allowing the production of DEMs, for combination with imagery and other digital datasets (geological maps, topographic maps, etc.) within GIS. One of the key advantages of satellite systems for long linear infrastructure projects is that they collect data over large geographical areas, hence one dataset can often be used for the study or analysis of both the whole alignment and also the regional setting. Satellite imagery is available in a wide range of spatial resolutions and spectral properties. Those considered of benefit for engineering projects mostly fall in the range from 90 m to <1 m spatial resolution. Satellites with higher spectral resolution allow greater distinction of geology, vegetation and land use.

Mid-resolution satellites

Mid-resolution (5–90 m per pixel) satellites (ASTER, Landsat and SPOT) are most suited to be used at the early stage of projects. ASTER (Advanced Spaceborne Thermal Emission and Reflection Radiometer) has a particularly useful combination of moderate spatial resolution (15 m, 30 m and 90 m pixels) with high spectral resolution (14 wavebands, from visible to thermal IR wavelengths). ASTER has the considerable advantage of low-cost stereo capability, with the VNIR (visible and near-IR) band 3 having both a nadir view and a backward-looking view. Using image processing software it is possible to view these data in stereo at 15 m resolution. Other satellite systems generally require the purchase of two different image scenes to acquire stereo capability.

SRTM data

The Shuttle Radar Topography Mission (SRTM) data provide a particularly useful DTM dataset that can be combined with satellite imagery. These terrain radar data were collected from the NASA Space Shuttle in 2000 and provide a DTM database covering the majority of the Earth's surface (between latitudes 60°N and 54°S), at 3 arc second spacing (c. 90 m grid posting), with absolute horizontal accuracy of 20 m and absolute vertical accuracy of 16 m. Data for the US landmass are available at greater precision.

High-resolution satellites

Currently there are three key commercial high-resolution satellite systems (IKONOS, launched 1999; QuickBird 2000; OrbView 2003). These have spatial resolutions of between 0.6 and 1.0 m resolution in the panchromatic waveband, and between 2.4 and 4 m resolution in the multispectral VNIR range. This combination allows the generation of either true colour or false colour images providing exceptional detail of the ground, which prove particularly useful for generation of large-scale base maps. These data can be used effectively for producing base-mapping at scales between 1:25 000 and 1:2500. A 0.41 m resolution satellite is planned for launch in 2007 (GeoEye 1), and a 0.46 m resolution satellite planned for 2008 (WorldView 2). Within the next 5–10 years new satellite sensors with 0.25 m resolution are anticipated.

Data selection

Given the wealth of remote sensing data available, careful consideration of the most appropriate data to select and use is required. Data types include both imagery and DEMs. Choice of data will depend on individual project requirements, including type and stage of engineering project, geographical location, and type of environment and terrain. Choice of individual dataset should also consider temporal influences (year, season, time of day of the data). For any particular sensor type selected it may be appropriate to obtain more than one date of data. The key technical parameters influencing choice of data are spatial resolution, spectral resolution and stereo capability. Choice of data is often controlled significantly by programme and budget constraints. Survey aerial photography remains a key dataset, primarily because of its stereoscopic viewing capabilities, which can greatly assist geomorphological interpretation and terrain analysis. However, for some countries security concerns can make acquiring photography difficult, hence in some instances the only available image data will be from satellites.

Terrain analysis for linear infrastructure projects is most commonly undertaken at route planning stages, although it can be applied to all stages of project lifespan. Key stages of project lifespan include route selection, design, construction, maintenance (operation of asset) and decommissioning. Table 2 lists the remote sensing datasets that are considered most appropriate for use on linear infrastructure engineering projects. An indication is also given of which project stage(s) most benefits from the use of the data.

ASTER data provide a useful low-cost tool (from $80 per 60 km × 60 km scene) for terrain analysts, as they allow more detailed geomorphological (landform) assessments than can usually be made by combining satellite imagery with conventional DEM data. Radar satellite imagery has the ability to 'see through clouds', which can be of benefit in some tropical areas where persistent cloud cover can make it difficult to acquire other data. High-resolution satellite data add value to many projects by providing an up-to-date and accurate base map. However, the limited multispectral properties of the high-resolution satellites, such as IKONOS or Quickbird, mean that the distinction of land use, soils and rock types is more limited than can be achieved with mid-resolution satellites that have greater spectral resolution (e.g. Landsat). Therefore, in many cases it is beneficial to obtain a combination of both mid-resolution and high-resolution data.

Archive datasets

There is a considerable archive of both aerial photography and satellite data that can provide considerable benefit to projects. Archive data extend back more than six decades with aerial photography and four decades with satellite imagery (Table 2). World War One vertical aerial photography exists covering the battlefield areas of northern France. Reconnaissance photography from World War Two and the post-war period provides a valuable historical dataset for many areas of the globe. Luftwaffe, RAF and USAF coverage exists for many parts of Europe and Asia.

Archive satellite imagery tends to be more easily searched and accessed than archive aerial photography, because of the significant number of on-line satellite data archives. Recent declassification of 'Cold War' spy-satellite imagery has provided a useful source of archive imagery dating back to 1959. Figure 2 is a 1960s CORONA satellite image of the Russian Far East. The CORONA satellite acquired stereo photography at variable resolutions, which allows stereoscopic (3D) interpretation and is of considerable benefit to terrain analysts and geomorphologists.

Comparing historical imagery with current imagery provides a useful record of change, and thus gives valuable information for routing studies. Past land uses of potential hazard can be identified, ranging from past industrial use (contaminated land and obstructions) to areas of backfilled pits, landfill and past mining activity (Fig. 3). Natural geohazards can be more readily identified and geomorphological processes can be assessed (landslides, coastal erosion, river migration and wind-blown sands). Many natural processes show considerable change over the lifespan of engineering projects. Examination of historical image datasets for changes to the landscape over the last 50 years helps to indicate

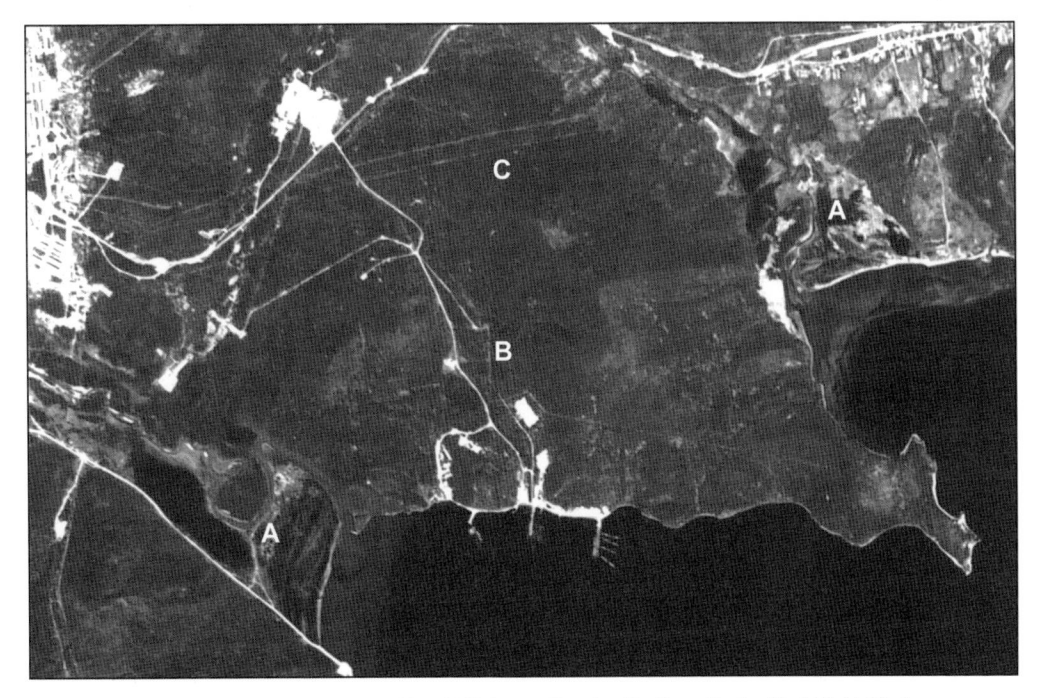

Fig. 2. US CORONA spy satellite photography, 1968 image, Russian Far East. Declassified 'Cold War' imagery provides record of site history prior to subsequent development. Coastal terrain with wetlands. Compressible soils and high water table (A) are visible together with existing infrastructure including pipelines (B) and power lines (C). Image size 7 km × 4 km.

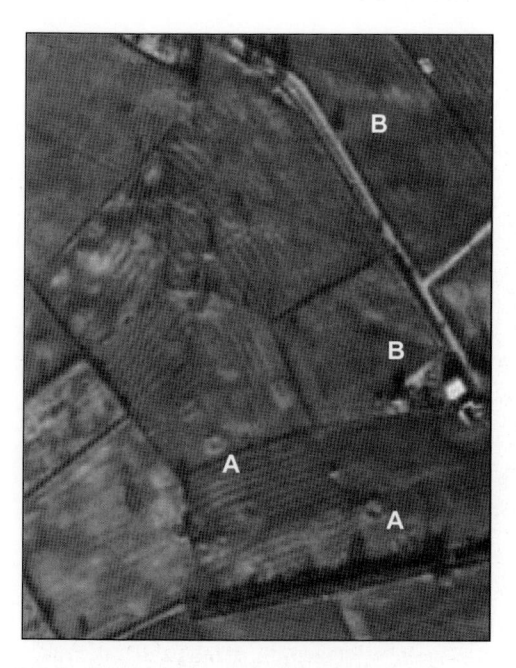

Fig. 3. A 1950 archive aerial photography showing evidence of mining. 'Bell-pits' are identified by ring-shaped spoil mounds around former shaft locations (A). Ground displacement along faults is also evident (B).

what future changes may occur during the lifespan of the project. Some potential geohazards are clearly defined and in some instances can be identified on the historical imagery but not current imagery. Historical imagery is uniquely useful for change detection studies applicable to geomorphological processes (particularly coastal and fluvial erosion or deposition and slope instability) and 'brown-field' sites (past mining, quarrying, land fill, industry). Historical survey photography allows the viewing of earlier states of the landscape in great detail in three dimensions over the past 50–60 years.

Terrain analysis

Terrain analysis should preferably be undertaken by experienced terrain analysts with geomorphological training, who have experience of 'reading the landscape' for evidence of geohazards and man-made hazards. Terrain analysis should be carried out in combination with consulting other appropriate datasets such as published geological and topographic mapping. An understanding of study area landscape evolution is essential, including aspects such as past changes in bioclimatic conditions, weathering intensity and sea-level changes. Where possible, terrain analysis should include walk-over surveys to determine 'ground truth'. Image interpretation is typically undertaken using a combination of conventional stereoscopic examination, image processing software and GIS. When examining images (air photo, satellite and DEMs) image interpreters consider the following elements to read the terrain and identify hazards: form (3D, shape of landforms); tone and colour; texture; pattern; time (date, season, time of day, temporal changes (over years)); association. Remote sensing data can be incorporated into GIS to allow 2D and 3D analysis, mapping and visualization. Civil engineering CAD drawings (dwg, dxf) can be superimposed on imagery to assess the impact of terrain-related features to the project. GIS tools also allow the efficient planning of project tasks, such as determining ground investigation locations, construction access and borrow pit locations.

Although linear infrastructure assets occupy only narrow strips of land, any appropriate terrain analysis work needs to also consider areas beyond the corridor. Areas beyond the corridor, the geomorphological setting, might contain hazards that could affect the construction area, such as landslides upslope or downslope of the corridor. The setting may provide clues on the nature of features within the corridor that are masked by sediments, such as faults or karst landforms. For these reasons it is essential that the data be selected appropriately, such that both regional setting and the corridor detail can be analysed in sufficient detail. This typically means selecting a combination of both mid- and high-resolution satellite data for analysis.

The multispectral properties of satellites and airborne hyperspectral sensors allow the distinction of geology, vegetation and land use. Figure 4 shows a false-colour Landsat TM image of a delta environment. This shows clearly how even at mid-resolution (30 m pixels) landforms can be readily identified, soil types distinguished and geological models developed. This information can be of great asset for routing studies of linear infrastructure, such as pipelines. High-resolution satellite imagery is capable of identifying smaller-scale, more localized, geohazard features, such as sinkholes and landslides (Fig. 5). False-colour near-IR images can be used to identify areas of vegetation distress and changes in soil moisture content. Even subtle shallow-seated solifluction slope instability can be identified.

Conventional interpretation using stereo aerial survey photographs still provides the highest level of detail for landform and feature identification during geomorphological terrain mapping, surmounting that which may be obtained by computer visualization using DEM datasets. The possible exception to this is high-resolution LiDAR data, which allow detailed 3D visualizations to be undertaken. The draping of orthophotography or satellite

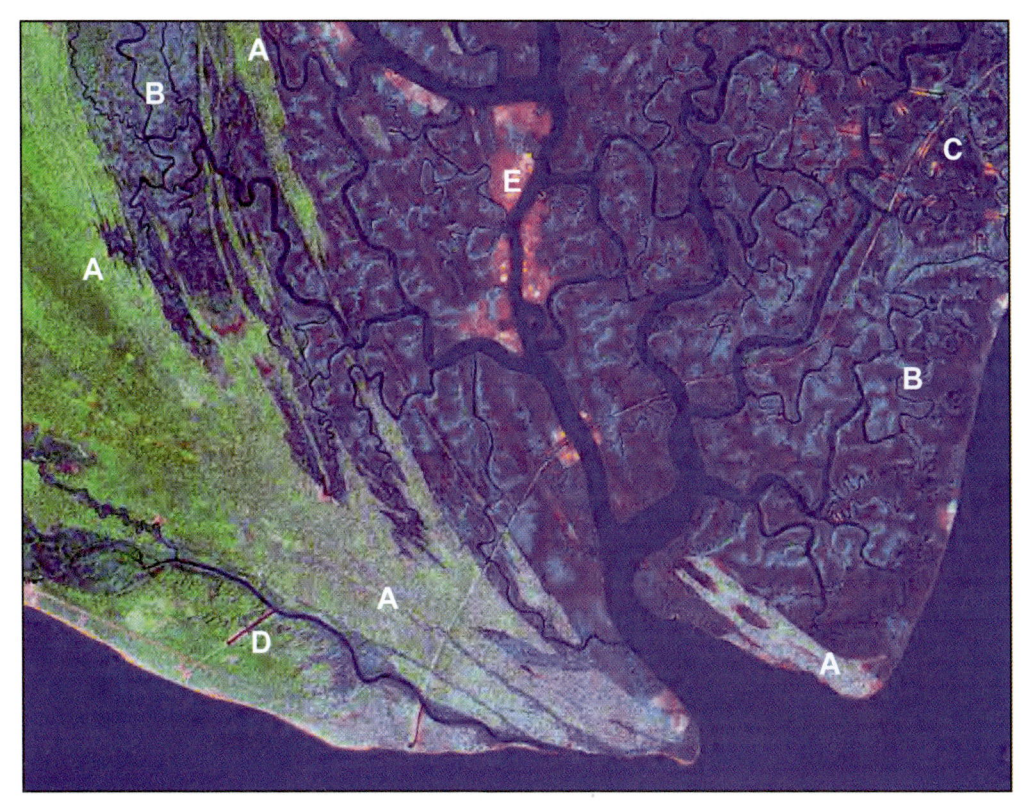

Fig. 4. Landsat 5 1986 false colour image (bands 247 RGB). Tropical wetlands–delta environment. Parallel ridges of barrier island beach sands (A) indicate sandy soils. Intertidal zone of mangroves (purple shades) indicates highly compressible clay and silt soils. Pink colours denote disturbed ground of pipeline slots (C), dredged channels (D), developed areas (E) and settlements. Image size: 24 km × 18 km.

Fig. 5. IKONOS high-resolution satellite image (false-colour IR: bands 421 RGB), temperate zone, low to moderate relief terrain with clay soils. Both deep-seated instability (left image) and shallow solifluction instability (right image) can be seen on high-resolution satellite imagery. Each image is 600 m × 600 m.

Fig. 6. A 3D visualization using satellite imagery and DTM. QuickBird high-resolution satellite image (false-colour IR, bands 421) draped over DTM. Village in foreground; volcano with lava field at horizon on right side of image.

imagery over DEM datasets for 3D visualizations provides a powerful tool that can be used effectively both by engineers and also for public consultation exercises (Fig. 6).

Summary

Remote sensing provides key data for terrain analysis and for the identification of hazards, both natural and man-made. Recent developments of high-resolution satellite data, LiDAR technologies and widely available DEMs have proven especially useful for linear infrastructure projects. The integration of remote sensing datasets with civil engineering design data within GIS packages allows efficient project working, with the data being usable at many different stages of a project.

Given the relatively low cost of many of the satellite remote sensing datasets and the high value that image interpretation can give to linear engineering projects, there can be little justification for not using this type of data. Commissioning bespoke aerial survey or LiDAR surveys generally involves higher costs, but these datasets bring considerable value to projects by providing very detailed geomorphological information, with LiDAR having the advantage of being able to show the land surface under tree cover.

In addition to acquiring current data, the case has been made for greater use of historical archive datasets, some of which date back to the 1940s. An understanding of how a landscape has changed over the last 50 years can provide a good indication of how the landscape may change in the next 50 years, which can frequently be the key lifespan of many engineered structures.

References

BRUNSDEN, D., DOORNKAMP, J. C., FOOKES, P. G., JONES, D. K. C. & KELLY, J. M. H. 1975. Large scale geomorphological mapping and highway engineering design. *Quarterly Journal of Engineering Geology*, **8**, 227–530.

DRURY, S. A. 1987. *Image Interpretation in Geology*. Allen & Unwin, London.

HO, H. Y., KING, J. P. & WALLACE, M. I. 2007. *A Basic Guide to Air Photo Interpretation in Hong Kong*. Applied Geoscience Centre, University of Hong Kong.

VAN DER WERFF, H. M. A., NOOMEN, N. F., VAN DER MEIJDE, M. & VAN DER MEER, F. D. 2007. Remote sensing of onshore hydrocarbon seepage: problems and solutions. *In*: TEEUW, R. M. (ed.) *Mapping Hazardous Terrain using Remote Sensing*. Geological Society, London, Special Publications, **283**, 125–133.

VAN ZUIDAM, R. A. 1985. *Aerial Photo-Interpretation in Terrain Analysis and Geomorphological Mapping*. Smits, The Hague.

WALSTRA, J., CHANDLER, J. H., DIXON, N. & DIJKSTRA, T. A. 2007. Aerial photography and digital photogrammetry for landslide monitoring. *In*: TEEUW, R. M. (ed.) *Mapping Hazardous Terrain using Remote Sensing*. Geological Society, London, Special Publications, **283**, 55–65.

Mapping remote areas using SRTM and ASTER digital elevation model data: a solution to orientation problems

I. DOWMAN & P. BALAN

Department of Geomatic Engineering, University College London, London WC1E 6BT, UK (e-mail: idowman@ge.ucl.ac.uk)

Abstract: Digital elevation models (DEMs) are used in many applications and some of them, such as geomorphological or geological applications, need DEMs to be generated over a very large area, with high resolution. As high-resolution sensors cover smaller areas, DEMs could be generated only for those areas that could be mosaiced later to cover the total area. If no ground control points are accessible, individual DEM tiles pose orientation problems during mosaicing, as the stereo models are not adjusted to each other. These orientation problems are addressed in this paper. Differences between the adjacent DEM tiles include tilt and vertical offset. A solution is proposed to remove such tilt and vertical offset and/or to register the relative DEMs to a reference DEM, if available. Results are presented of a case study over part of the Zagros Mountains, Iran, covering 72 065 km^2. The software tool developed in this study was applied to mosaic 14 ASTER DEMs: no ground control points were used for any of those DEM tiles or for the mosaic. An SRTM DEM of 90 m grid spacing was used to register the mosaic to achieve the absolute orientation.

Topography contains a wealth of information that allows an interpreter to identify faulting and other features that have shaped terrain into what it is today, and also holds the clue to how it could be evolving tomorrow. Digital elevation models (DEMs) that cover a large area give an overall understanding of the fault activity in that region, and high-resolution DEMs provide us with better information. A 90 m Shuttle Radar Topography Mission (SRTM) DEM, available for a large part of the Earth's land surface, provides a homogeneous elevation model but does not give sufficient information on faulting. However, DEMs from Advanced Spaceborne Thermal Emission and Reflection Radiometer (ASTER) data, generated at 30 m resolution over an area of 242 km × 297 km, can provide the required detail (Hirano et al. 2003; Toutin 2002). A significant problem with DEM generation is the lack of suitable ground control in remote areas. Although relative DEMs can be generated, if these are not oriented, and a mosaic is required, unacceptable gaps and discontinuities will occur. These and other difficulties are discussed in this paper.

Identifying faults and analysing past tectonic activities that have resulted in changes in the geomorphology of the terrain is one of the major objectives of the UK Centre for Observing and Modelling Earthquakes and Tectonics (COMET). One of COMET's study areas is the Zagros mountain chain in Iran. This study reports on the experience of generating high-resolution DEMs using ASTER scenes and mosaicing them with the help of SRTM data.

Digital elevation models from optical images

Digital elevation models can be formed from two or more overlapping images taken from different positions in the same way as from aerial photographs. To obtain suitable geometry and a good base to height ratio it is necessary for the sensors to point sideways, or forwards and backwards. A number of satellite sensors are designed to do this and data are available at various pixel sizes. Table 1 lists some of the satellites that can be used for DEM generation.

Shuttle Radar Topographic Mission (SRTM)

Following the success of the experimental Shuttle Imaging Radar (SIR) missions A, B and C, as well as successful demonstration of interferometric techniques in deriving 3D measurement of terrain, the Shuttle Radar Topographic Mission carried an interferometric synthetic aperture radar (SAR) that acquired data during its 11 day mission in February 2000, from which a near-global DEM was generated. SRTM DEMs are currently available for the USA at 30 m (1 arc sec) grid spacing and at 90 m grid spacing (3 arc sec) for the rest of the world. SRTM DEM data of 90 m grid spacing were downloaded from JPL NASA's site, and were used in this study as the reference DEM.

From: TEEUW, R. M. (ed.) *Mapping Hazardous Terrain using Remote Sensing*. Geological Society, London, Special Publications, **283**, 143–147.
DOI: 10.1144/SP283.13   0305-8719/07/$15.00   © The Geological Society 2007.

Table 1. *Stereoscopic satellite imagery available for topographic mapping*

Platform	Sensor	Launch date	Type	Pixel size (m)	Swath width (km)
SPOT 1–4	HRV	1986–	Cross track, push broom	10, 20	60
SPOT 5	HRS	2002	Along track, push broom	10	60
JERS-1	OPS	1992	2-sensor along-track stereo	20	75
IRS-1C	Pan	1995	Cross track-push broom	5.8	70
Priroda	MOMS	1996	Fore–nadir–aft stereo	6	80
ASTER	TERRA	1999	Nadir–aft stereo	15	60
IKONOS		1999	Cross track, fore and aft pointing	1	12
Quickbird		2002	Cross track, fore and aft pointing	0.6	17

Orientation of stereo data

Exterior orientation is normally carried out using ground control points and/or orbital information that gives the co-ordinates of the exposure stations of the stereo pair. This ensures that correct orientation is achieved and subsequent DEMs will fit together, but accuracy is variable. Exterior orientation can be split into relative orientation and absolute orientation. Relative orientation will form a good stereo model, but in arbitrary orientation. Absolute orientation requires control points: a minimum of three for a single model, and additional points for a block, which can be adjusted by aerial triangulation. In some software packages such as PCI Orthoengine, orbital information and approximate elevation of the area are used in finding tie points automatically, which are used in relative orientation of the model. Both ASTER and SPOT data come with orbital information in their header record. This information is extracted and used in Orthoengine while setting up the project. Using this orbital information and approximate elevation, PCI can generate tie points automatically and relative orientation could be achieved to acceptable levels of accuracy. However, when the variation in elevation is too different from the single value that defines the approximate elevation of the area, the automatic tie point generation process often results in poor identification of the points. Inputting few ground control points gives a better approximation of the terrain; this helps to identify a few more tie points automatically and hence the stereo pair could be successfully relatively oriented.

Satellite data can be acquired in strips and when this is the case, can be treated as an extended model, and no mosaicing is required. When the images are not in single strip, tie points are generated for all overlapping scenes, and epipolar images have to be generated for every possible pair that will be later used in the DEM generation process. It should noted that relative orientation of individual pairs provides models that are independent of each other and do not fit together. Although the pairs are independently oriented, the epipolar images that cross-connected the pairs provide approximate orientation between the tiles. This could be used only if there were just a few pairs to handle. When there are too many tiles, this procedure becomes very time consuming. This method could not be applied to ASTER even for a few pairs, as the stereo pair is formed by backward looking and nadir images.

Proposed solution

A program that estimates tilt and offset between the individual DEMs was developed. Tiles of relative DEMs showing tilt and offset between the tiles could be mosaiced by removing the tilt and offset in the adjacent DEMs using the program developed. As this program could be used to remove tilt and offset between any two DEMs, it could be used to register a relative DEM with a reference DEM, which absolutely orients the relative DEM. Hence individual DEMs or a mosaic could be absolutely oriented using this program. Tiles of relative DEMs could be adjusted relatively to each other first, and could then be absolutely oriented. Individual DEMs could be absolutely oriented with the reference DEMs and the tiles could be mosaiced. In the latter method there will still be some amount of tilt and/or datum offset between the tiles because of difference in accuracy of absolute orientations of the individual DEMs.

Data used

Thirty-four ASTER scenes were acquired that covered the Zagros study area. Not all of them provided good DEMs, for several reasons, such as snow and cloud coverage. Fourteen DEM tiles that were reasonably good and covered the area of interest were selected. No ground control points were available and hence all of them were relative

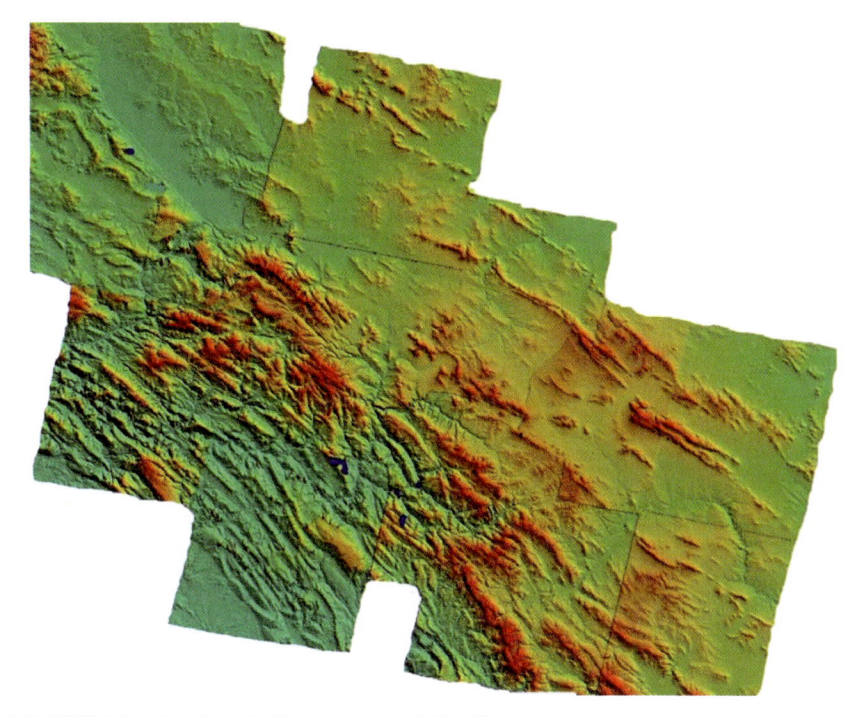

Fig. 1. Initial DEM showing tilt and offset between the DEM tiles.

DEMs and not adjusted to each other. The initial mosaic showed tilt and vertical offset between the DEMs (Fig. 1).

A new method for orientation and mosaicing

The SRTM DEM of 90 m grid spacing was used as the reference DEM to which all the DEMs were registered. Registering DEMs at 90 m spacing resulted in underestimation of the tilt because of the resolution being low. Hence the SRTM DEM was sub-sampled to 30 m pixels, which is the spacing of the ASTER DEM, to estimate the tilt and offset. Sub-sampling of SRTM was done by repeating the pixel value over the 90 m area; for example, pixel value is repeated for 3 × 3 pixels in the output image, provided the pixel corners of the input and output pixels are same. When that differs, some of the output 30 m pixels contained some percentage of area from more than one 90 m input pixels, and in such cases, area-weighted average was taken to the output pixel. A difference image between SRTM DEM and ASTER DEM was generated. Tilt and offset were estimated from the difference image. As the difference image is created by subtracting elevation values pixel-by-pixel, care was taken to keep the geo-referencing of the pixel corners the same for both SRTM and ASTER DEMs. This was achieved by resampling the SRTM DEM from 90 m to 30 m, as explained above. Figure 2 shows a flow diagram of the program. Tilt was estimated on the difference image as its x component and y component separately. The greater of the two components was removed first from the ASTER DEM during the first iteration. In the second iteration, tilt was estimated again for both of the components, and the greater magnitude of the two was removed in the second iteration, and so on, until either the greater magnitude of the components is insignificant or the program has carried out maximum iterations, as specified by the user. Then the vertical offset was estimated from the difference image and removed from the tilt-free ASTER DEM. Two adjacent single DEM tiles were registered to the SRTM DEM as described above. Also, the tiles were checked for any discrepancy in tilt and offset between them, and this was removed if present. Now these two tiles could be mosaiced before moving on to a third one.

Pairs or groups of tiles showing good overlap with the adjacent tile were selected first. Care was also taken to preserve the features of interest, and tiles showing less noise along these features were selected in one group. The following procedure was followed: groups of ASTER DEM tiles were mosaiced into separate groups, where good overlap occurred between tiles. These tile groups were registered to the SRTM DEM and between each other, and the whole set was then mosaiced. Figure 3a shows Group 1, a mosaic of four tiles that has a fault running across it from NW to SE (which otherwise could have been compromised). Group 2 is a mosaic of another four single DEM tiles and is shown in Figure 3d. In this group (Group 2), the mosaic had to be prepared two at a time (two subgroups of two single tiles each), as the noise in one of the tiles did not yield a seamless mosaic in any of the combinations. The subgroups were mosaiced later. In Group 3, two single tiles were mosaiced (Fig. 3c). Another two single tiles were added to form a mosaic tile of Group 4, and this is shown in Figure 3b. The mosaic tile of Group 3 and another single DEM tile were added to group 1, forming a larger mosaic tile. Group 4 was added to Group 2. These two mosaic tiles were registered and added to form a mosaic tile that contained 13 single tiles. Adding a fourteenth tile to this large mosaic tile covered the area of interest.

PCI Orthoengine could adjust minor tilt and datum offset between the tiles using the 'colour balance' (histogram matching) option. This has to be used very carefully, as it was developed to 'balance' the pixel values of ortho images at the joints of the mosaics, and hence would change the elevation values when used for DEM mosaicing.

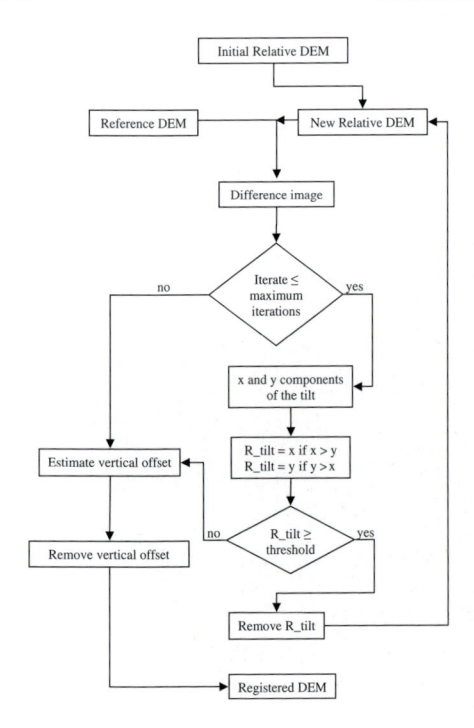

Fig. 2. Scheme showing the program flow.

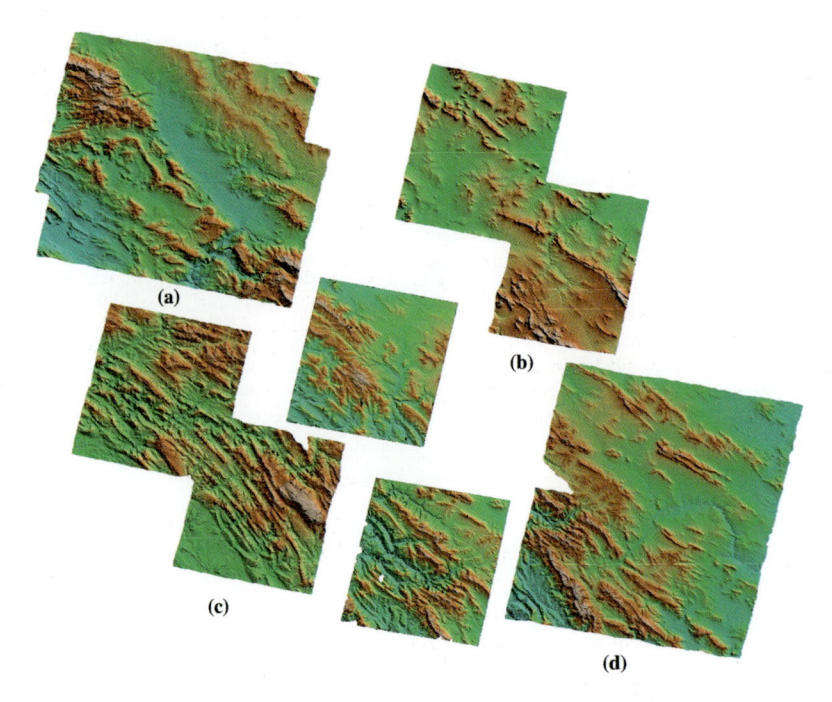

Fig. 3. Mosaic tiles after initial grouping.

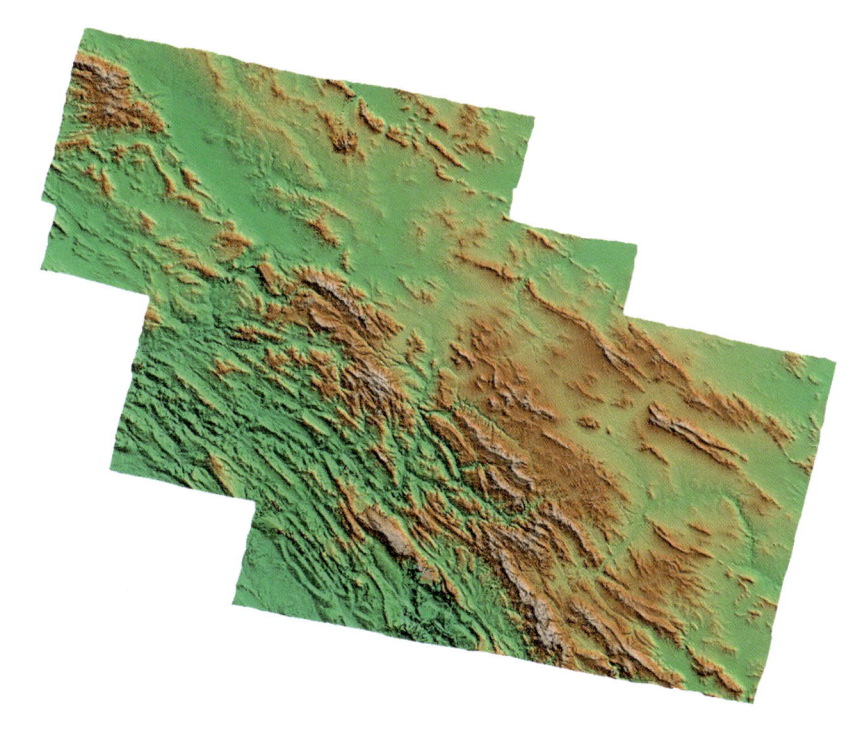

Fig. 4. Mosaic of ASTER DEM tiles with holes filled using SRTM DEM values.

This option could not be used when the magnitude of the tilt/and or offset between the DEMs is large.

Discussion of results

The errors in ASTER DEM data can pose huge problems, as they lead to incorrect estimation of tilt and offset. Removing all small and large isolated errors was not practically possible and hence the problem could not be eliminated altogether. It was minimized by removing large areas of significant errors. After the mosaic was prepared, the holes in ASTER DEMs were replaced by SRTM DEM and vice versa. Figure 4 shows the final product of this study. The method could be improved by the addition of matching salient topographical features on the terrain, to ensure better registration of the DEMs.

Conclusions

A new method to orient DEMs, which have been produced without the use of ground control, to a reference DEM is described. This involves comparing adjacent edges and removing tilt and offset between them. The program we developed is used to mosaic 17 DEMs derived from ASTER stereo pairs over a part of Zagros Mountains in Iran. SRTM data were used as the reference DEM; this proved to be very suitable for this purpose by providing a reference frame that covered a wide area with homogeneous accuracy. Holes in ASTER DEM were replaced by SRTM values. The method is designed to give a seamless mosaic, which can be used for geomorphological interpretation using data of a higher resolution than the SRTM DEM.

The authors acknowledge the UK Natural Environment Research Council and COMET for sponsoring this work. Acknowledgements are also due to JPL NASA for providing ASTER data. SRTM DEM data used in this study were downloaded from JPL's website.

References

HIRANO, A., WELCH, R. & LANG, H. 2003. Mapping from ASTER stereo image data: DEM validation and accuracy assessment. *Photogrammetry and Remote Sensing*, **57**, 356–370.

TOUTIN, T. 2002. Three-dimensional topographic mapping with ASTER stereo data in rugged topography. *IEEE Transactions on Geoscience and Remote Sensing*, **40**(10), 2241–2247.

Space technology for disaster management: data access and its place in the community

M. E. ANDREWS DELLER

*Earth Sciences, Open University, Milton Keynes MK7 6AA,
UK (e-mail: m.e.andrews@open.ac.uk)*

Abstract: The ways in which remotely sensed data can be used to manage and alleviate the consequences of natural disasters have never been fully exploited. If prediction of impending disaster is to be useful, those affected by catastrophe and those who come to their aid must work together. A global strategy based on a vision for disaster management will fail if those affected by hazards are not involved in their own disaster preparation, relief efforts and rehabilitation. Local people are experts in ground knowledge; all that is needed is for those with expertise in remote sensing to pass on their skills, knowledge and data in a way that can be understood and valued. To do this, those threatened by catastrophe must understand how disaster relates to their lives, how satellite data can be used with confidence to prepare for local hazards and how to apply information that can help lessen the effects of catastrophe. This can be achieved by making available remotely sensed images with simple explanations that enhance vegetation, topography and geology. With timely, reliable information, preventive measures can be taken; surface features identified on satellite images can guide refugee placement and hazards can be anticipated and contained.

Although the mapping of hazardous terrain using remote sensing is well documented (King 1984, 1994; Massonnet 1995; Bjorgo 2000; Drury & Andrews Deller 2002; Huang & Fu 2002; Hunt 2002; Wright *et al.* 2002; Kerle *et al.* 2003; Oštir *et al.* 2003; Marsh 2004; Ramsay & Flynn 2004; UNOOSA 2004), its application in the hands of people who live and work in these areas is limited. With minimal training and easily understood satellite images, however, local inhabitants and relief workers can quickly become proficient in simple image interpretation and competent at producing their own informative maps and information. There are many examples of the general public's fascination with remotely sensed images, for instance:

- 'When the first satellite image posters of Southern California and Los Angeles went on sale several years ago they were extremely popular. From what I saw, the public was quick to see them as a new type of map, with new information, in addition to the aesthetic aspects of the imagery. People would gather around the displays at the shopping mall and be amazed at what they saw.' (R. E. Crippen, NASA/JPL, pers. comm. 2006);
- 'In 1998 I did a compilation of satellite images covering Eritrea and had it printed at A1 size so the Eritrean Water Resources Department could sell it to expatriates to raise funds for water exploration but copies ended up in schools, public offices and tiny fuel stations throughout the country.' (S. A. Drury, Open University, pers. comm. 2001; see also Drury 1997), and

- 'We've been developing a project called Terra-Look (formerly the Protected Area Archive) to make it easy for "ordinary" people to get images of sites they have an interest in. Originally it was geared towards Ecosystem and Park managers but now it is really for anybody.' (M. Abrams & G. N. Geller, NASA/JPL, pers. comm. 2006; see Table 2).

Specific reports of people's capacity to use satellite data are not hard to find, for example:

- 'As to people's ability to use images ... our experience has been that people with ground knowledge can very quickly orient themselves to an image and start using it—in fact, they much enjoy doing that. The image is generally treated as an extension of their ground knowledge—for example, they may have known that there was agricultural encroachment into their park but did not know how much. An image can tell them that. In fact, that ground knowledge usually seems to act as pre-existing ground truth, greatly aiding interpretation.' (M. Abrams & G. N. Geller, NASA/JPL, pers. comm. 2006; see also Abrams & Geller 2005);
- 'Several years ago some of us at JPL were doing aerial reconnaissance over the Mojave Desert from a NASA airplane, looking at earthquake faults. There was some difficulty telling the pilots exactly where we wanted to go. I had Landsat images of the area for our use, but quickly realised that the pilots would better understand the geography if I supplied them with images too. The pilots' maps (contour lines, etc.) looked not much like the view out the window but the satellite images showed them what to seek.'

From: TEEUW, R. M. (ed.) *Mapping Hazardous Terrain using Remote Sensing.* Geological Society, London, Special Publications, **283**, 149–164.
DOI: 10.1144/SP283.14 0305-8719/07/$15.00 © The Geological Society 2007.

(R. E. Crippen, NASA/JPL, pers. comm. 2006), and
- 'I found Tanzanian farmers could well read satellite images (Landsat 3 RBV) and understand their value better than some government officials.' (R. B. King, pers. comm. 2006, regarding a project funded by the British Overseas Development Authority, the International Developmental Research Centre and the US Agency for International Development (USAID); see also King 1981).

However, the ways in which remote sensing data can be used to manage and alleviate the consequences of disasters have never been fully exploited. The problem is concentrated in areas like the Horn, one of Africa's most disaster-prone regions with poor map coverage, insufficient roads or funds, endemic drought, famine and disease, unreliable rainfall and inhospitable terrain (Andrews Deller 2005). In such territory, the importance of multispectral remote sensing in geological and environmental applications, many of which relate directly to natural hazards, is unsurpassed (Vincent 1997). Given cheap, efficient means of acquiring appropriate data (Abrams 2000; Andrews Deller 2000; Goossens 2002) and the potential uses of satellite technologies for disaster relief (J. McCluskey, pers. comm.; see Table 1), remotely sensed data provide a way forward and can be used to aid local communities. As such its socio-economic value is considerable, as much of the rapidly growing archive of data (Drury 1998) required to enable local people to manage the hazards that affect them is freely available (Andrews Deller 2006).

Data and materials

Several operational satellite systems provide image data that can be used with simple interpretation to manage the effects of catastrophe. The most notable are the Enhanced Thematic Mapper Plus (ETM+) deployed on Landsat-7 and the Advanced Spaceborne Thermal Emission and Reflection Radiometer (ASTER) (Abrams 2000). Complementing ASTER and ETM+, the Shuttle Radar Topography Mission (SRTM), with near-global coverage, provides free, moderate- to high-resolution (90 and 30 m) topographic elevation data useful for detailed terrain mapping (Jenson & Domingue 1988; Crippen 2001).

From the wealth of data on offer, images selected as appropriate for raising public awareness of natural hazards should be clear and simple and ought to include the following:

- perspective views based on SRTM digital elevation models (DEM) with images draped over them; these provide natural-looking renditions of familiar landscapes or specific localities (Jenson & Domingue 1988; Goossens 2002), and
- stereo anaglyphs using co-registered left and right images of stereopairs (such as those from ASTER) as red and cyan components of red, green, blue (RGB) images, viewed using simple colour filters (Crippen & Blom 2000); these simulate 3D maps of natural landscape (Abdallah et al. 2005), and
- Landsat ETM+ 742 and ASTER 631 bands combined as RGB images; these display vegetation in natural-looking colours and provide a good basis for general geological interpretation.

The above datasets are free or low cost and can be easily appreciated by people of all cultures. Accessed from sites listed in Table 2, images can be printed by individuals or personnel from government departments, local surveys, educational establishments, non-governmental organizations (NGOs) or United Nations (UN) agencies and used for training or distribution in local communities via schools, places of worship, relief agencies and administrative buildings.

Although cheap Landsat TM, ASTER and SRTM data for the Earth's land surface are readily available (Table 2), the cost of image processing software, adequate computers and peripherals is still too high for most people affected by disasters. Websites such as Google Earth, which utilize satellite imagery and vector maps, and are 'well accepted as a new way to understand places' (R. E. Crippen, NASA/JPL, pers. comm. 2006) are very useful for obtaining information about a given region quickly, but few people in the Third World have access to such electronic digital data. Relief workers may have computers in the field, together with some skills and software for image display and interpretation, but the local population will probably not. In this case colour lithographic printing offers a solution. It is cheap and an excellent source of educational aid (Andrews Deller et al. 2004, 2005). A good all-in-one printer–photocopier–scanner costs no more than £65 (US$121 or €94) at 2006 rates, a price that continues to fall. A good quality colour-printed A4 page costs just over one penny or roughly 2 US cents. In durable form (e.g. with plastic laminated protection) images and explanations of image features that indicate hazards of different kinds can be delivered to at-risk communities as warnings and sources of valuable information for educational purposes. Lower-cost copies (e.g. colour newsprint), suitable for individual families, might also be considered.

Geoscientists who use hard-copy satellite images have noticed the interest shown by local people from many countries in such data and their ability to understand the material about terrain displayed (King 1981; Drury 1998; Setzer 2004;

Table 1. Potential uses for satellite technologies for disaster relief

Natural disaster	Maps	Migration	Site planning	Environmental degradation	Communications/infrastructure	Vector control	Food security/agriculture
Flood	Flood prediction: flood-prone areas; flood progress monitoring; areas affected	Where populations have moved to, to facilitate assistance	Mapping of potential flood shelter locations: out of flood risk	Effect of flood on drinking water supply and installations	Effect on roads, rail and river networks; alternative access	Monitoring duration of flood - sustained environmental/ecological changes	Destruction of crops; recovery of crops
Drought	Progress of drought	If populations and livestock have migrated out of area	Locating new water resources	Mapping of soil or land degradation; change in water bodies: reduction and quality; change in groundwater levels		Potential for increase in vector breeding sites i.e. pools of water instead of flowing water in rivers	Effect on crop production; change in cultivated area; change in crops; availability of pasture for animals
Earthquake and Tsunami	Extent of quake for assessment purposes; map of tsunami potential effect	Where populations have moved to, to facilitate assistance	Mapping of seismically stable areas: locations for camps and unsuitable locations	Mapping of areas at risk from aftershocks, rockfalls, landslides, etc.; effect of saline intrusions due to tsunami on water supply	Effect on roads, rail and bridges; alternative access; extent of building damage		Effect of saline intrusions due to tsunami on agriculture
Cyclone, hurricane, tropical depression	± Real-time tracking of weather event	Where populations have moved to, to facilitate assistance	Mapping of probable affected and safe areas	Mapping of areas at risk from mudslides; effect of saline intrusions on water supply	Effect on roads, rail and bridges; alternative access; extent of building damage		Effect of saline intrusions on agriculture

Source: Modified from a table provided by J. McCluskey (unpubl. data).

Table 2. *Free and low-cost remotely sensed data*

Web site	Brief description of available data
landsat.usgs.gov	Landsat 7 ETM+ to purchase or browse, and ordering facilities
glcfapp.umiacs.umd.edu: 8080/esdi/index.jsp	Free Landsat TM and ETM+ 742 compressed mosaics; free Landsat TM and ETM+ scenes for selected dates (often post-wet and dry seasons)
zulu.ssc.nasa.gov/mrsid/	Free Landsat TM and ETM+ 742 compressed mosaics
edcimswww.cr.usgs.gov/pub/imswelcome/	Low-cost or free ASTER, MODIS and AVHRR data; other data; browse and ordering facilities
ftp://e0mss21u.ecs.nasa.gov/srtm/	Free SRTM digital elevation data
www.maproom.psu.edu/dcw/ or http://www.libraries.psu.edu/crsweb/maps/	Free on-line digital chart of the world
www.ngdc.noaa.gov/mgg/image/2minrelief.html	Digital elevation data
edcdaac.usgs.gov/gtopo30/gtopo30.html	Free *c.* 1 km resolution digital elevation data
www.oosa.unvienna.org/SAP/stdm	Links to web sites on space technology and disaster management, including the International Charter 'Space and Major Disasters' (set up specifically for disaster response), maps, satellite images and other data assembled by D. Stevens of the UN Office for Outer Space Affairs
http://www.geodata.gov/gos	US government's Geospatial One-Stop E-Gov Initiative: free high-resolution maps and images for search, rescue disaster areas (e.g. Indian Ocean tsunamis Disaster link)
ftp://ftp.glcf.umiacs.umd.edu/glcf/SRTM/	Free *c.* 90 m resolution SRTM digital elevation data
http://earth.google.com/	Google Earth: Free on-line, global Landsat TM natural colour, perspective views of topography overlain on SRTM elevation data; high-resolution data available in some areas
http://www.nasa.gov/home/	Directory and links to many NASA sites
http://edcimswww.cr.usgs.gov/pub/imswelcome/	NASA Earth Observing System Data Gateway (EROS Data Center): low-cost or free ASTER, MODIS and AVHRR data; other data; browse and ordering facilities
http://asterweb.jpl.nasa.gov/paa	TerraLook makes it easy for non-specialists to obtain images
www.nws.noaa.gov	NOAA National Oceanic and Atmospheric Administration sites: weather, tropical storms, climate etc.
http://www.ncdc.noaa.gov/oa/climate/onlineprod/drought/xmgr.html	CLIMVIS: online access to drought, precipitation, and temperature data
http://TerraWeb.wr.usgs.gov/projects/eolian/	USGS TerraWeb: use of remotely sensed images for aeolian and rainfall mapping, detection of active dust storms and/or detection and mapping of areas vulnerable to aeolian erosion

Abrams & Geller 2005). Before global positioning system (GPS) receivers became available, Drury and Berhe reported getting lost topographically and geologically when using Landsat TM images to map the geology of remote western Eritrea while working with Oxfam in 1990. Their guide, a semi-nomadic herdsman, intervened, pointing out on the image the ridge on which they stood, the same rock type occurring on two further ridges to the west, landscape features, their destination and a nearby gold-bearing stream. (S. A. Drury & S. M. Berhe, Open University, pers. comm. 1998). Across the globe, 'a company named Solid Terrain Modelling ... in Fillmore, California, created a new process to drape satellite images onto high-density foam blocks that are carved into the shape of a DEM'. Amongst their customers, 'lawyers who need to get the jury to understand the geography of a court case ... found that imagery, especially in 3D, is well understood by the general public' (R. E. Crippen, NASA/JPL, pers. comm. 2006).

User-friendly images

People affected by disaster are generally unaware of the potential for space technology and the usefulness of remotely sensed images, so starting with the familiar can instil confidence. This can be achieved by making available remotely sensed images that clearly enhance features of the local environment (vegetation, water, roads, topography and geology), as well as those features that typify hazards (volcanoes, landslides, active faults, flash floods and areas desiccated by drought) (Andrews Deller 2004, 2005). Indigenous peoples readily appreciate images in printed form and the basic information about the features depicted, a fact confirmed by the author's experience on several occasions. For example, Alamou, a guide, driver and navigator in Ethiopia (2002), proved an excellent image interpreter, anticipating and guiding geologists to clay- and iron-rich localities noted on ASTER 631 images of the country.

A good starting point and an image that mirrors many people's general perception of their surroundings, is an ASTER 631 image combined as RGB. Equivalent to Landsat TM 742 (Andrews Deller 2006), it simulates natural-looking vegetation and highlights rock differences (Fig. 1). A clear informative image, it is easily understood by relief workers and local people (Andrews Deller et al. 2004, 2005). Linking the familiar with what can be seen on the image, vegetation is green (not an aesthetically disturbing red tone (Drury 1993) as on standard false-colour ASTER 321 images) and iron-rich rocks and soils are red, as they are with human vision, so there is no confusion. This type of image bears out experience and encourages trust in Earth observation for assessing resources or potential hazards.

Equally useful, ASTER 631 combined with SRTM DEM data (Fig. 2) produces a natural-looking perspective view that can be used for location and realistic graphic representation of landscape. Such 3D renditions of Earth observation data are more easily understood than 2D prints. The positive as well as the dangerous can be highlighted on this image. The combination of bands clearly distinguishes outcrop (high-ground basalt, clay and multicoloured basement) from green vegetation. Abundant vegetation indicates sources of water and land fit for agriculture, whereas the distinct blue clay horizons warn of potential land slippage and areas best avoided for buildings, roads or grain storage. Early signs of trouble can be noted and once features observed on the image are understood in the context of hazardous geological processes, potential disasters can be avoided.

An ASTER 631 image and stereo anaglyph of part of the Ogaden, Ethiopia (Fig. 3) makes the point. Failure of the clay zones (blue in Fig. 3) produces landslides that range in size from a few square kilometres to thousands of square kilometres (Temesgen et al. 2001; Andrews Deller 2004; Singhroy & Molch 2004; Abdallah et al. 2005). Images like this not only identify areas that are potentially hazardous, but also provide visually stimulating ways of teaching entertainingly. Using a 20 pence (c. 40 cents) anaglyph viewer, adults and children are fascinated by the dramatic features of landslides that take on 3D form before their eyes. Furthermore, a serious point is made that such hazards, highlighted by remote sensing, can be mitigated by local government planning and by individuals taking precautions.

Sadly, this was not the case on 26 December 2004, when one of the biggest natural disasters in recorded history took place (see McCluskey & Choudhury 2005). Tragically, no one was prepared for the tsunamis of the northern Indian Ocean (see Drury 2005). Small signs, such as the receding coastal waters, that might have alerted many and saved lives were not understood. Fascinated by the withdrawal of water, people walked towards the sea rather than run to higher ground. The need for any foreknowledge of both disaster-affected districts and those areas where people have found refuge in the past has been noted by many relief teams (e.g. Sartori et al. 2000; HAIC Secretariat 2001, Sanyal & Lu 2006), particularly in regions affected by severe flooding. Even moderate-resolution DEMs from SRTM data (Fig. 4) provide useful indicators of safe and unsafe localities, and such information derived from stereoscopic, high-resolution satellite image data (e.g. Ikonos and Quickbird) far surpasses the topographic maps of most Third World countries.

Fig. 1. ASTER 631 combined as RGB simulates natural-looking vegetation and enhances rock differences in the Ethiopian Escarpment. Key shows colours of various elements of the terrain. CFB, continental flood basalts.

SPACE TECHNOLOGY FOR DISASTER MANAGEMENT 155

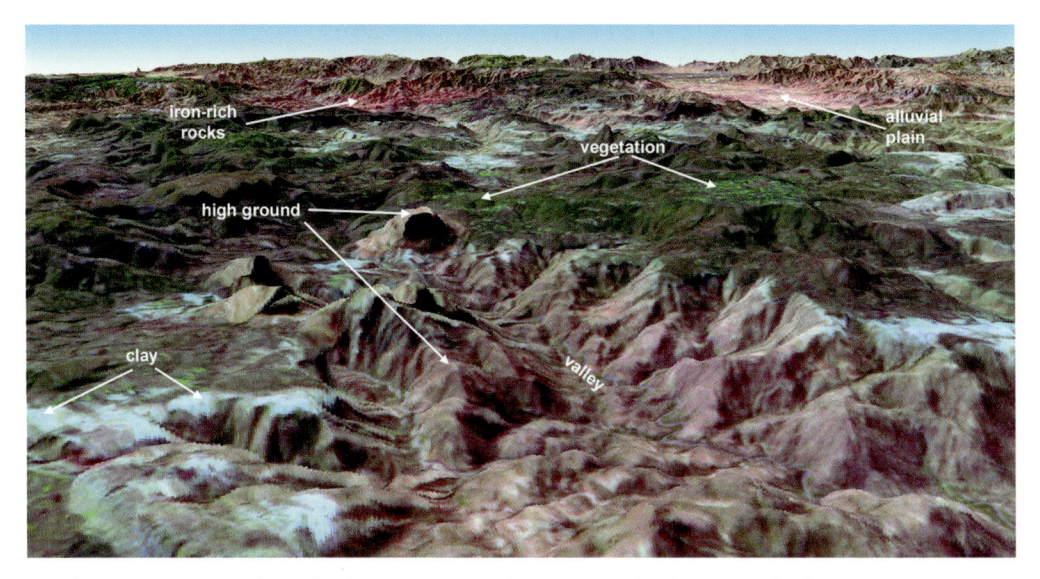

Fig. 2. ASTER 631, draped over SRTM DEM data, produces a natural-looking perspective landscape of the Areza area, Eritrea, used for terrain visualization, main features are indicated.

Fig. 3. ASTER 631 (left) and stereo anaglyph (right) showing landslides in the Ogaden, Ethiopia.

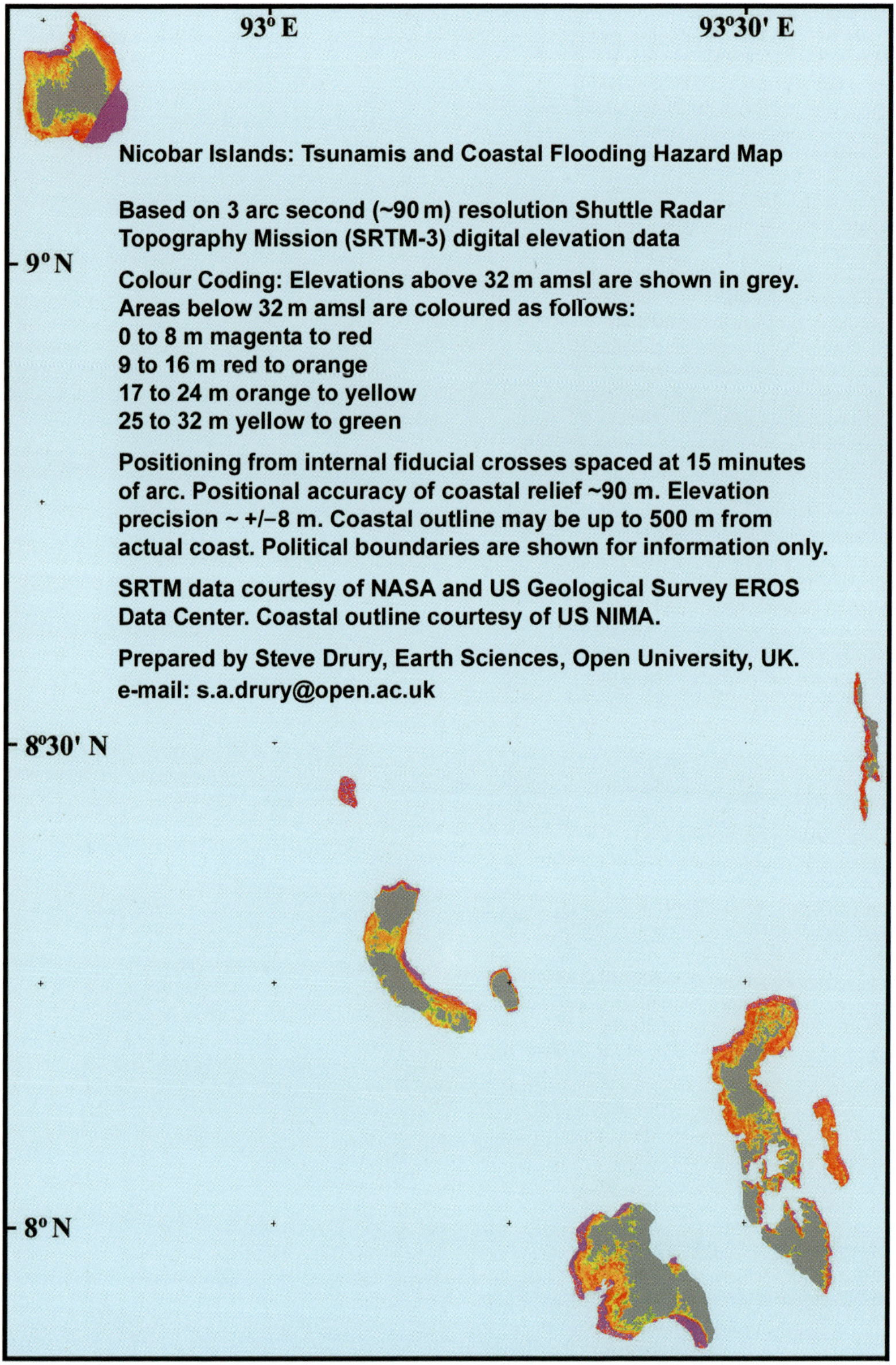

Fig. 4. Coastal flooding hazard map of the Nicobar Islands, Bay of Bengal, based on SRTM DEM data.

Figure 4 displays coastal areas of the Nicobar Islands below 15 m elevation in magenta, red and orange (Drury 2005; see also Indian Ocean Disaster link, at http://www.geodata.gov/gos). These areas, clearly identifiable on the image, correlate with the tsunamis' worst effects. Remotely sensed data of this kind, which can be understood by anyone, are instructive and alert local people to possible hazards. They indicate areas to avoid when warnings of imminent danger are made, highlight areas of safety when disaster strikes, and mark out areas in which to concentrate relief efforts should a tsunami strike again. Understandably, people are attracted to areas with fertile soils or other useful attributes on which their livelihoods depend, even when fully aware of the associated risks from local natural hazards. Nevertheless, educational measures outlined above increase their chances of surviving disasters where legislation for safe siting of dwellings is lacking.

At their most useful, remotely sensed images indicate what geologists know or believe to have happened and what might happen again. Large-scale topographic maps derived from DEM data are ideal for highlighting land surface features related to active faults and volcanoes, whereas multispectral remote sensing data distinguish vegetation and rock type clearly.

Examples of areas that ought to be highlighted on images include:

- steep slopes with thick soil or debris, prone to landslides and debris flows (Jordan *et al.* 2000*b*);
- low-lying inland or coastal areas liable to flood (Drury 2005; Sanyal & Lu 2006);
- highly vegetated areas prone to forest fire when dry (e.g. Setzer 2004);
- sparsely vegetated or poor kaolinitic soils prone to drought (Andrews Deller 2006);
- areas where geology dictates that groundwater will endanger life (e.g. McArthur *et al.* 2001);
- recent lava flows and volcanic gas emissions (Wright 1999; Alvarez *et al.* 2000); and
- active fault lines (Fu *et al.* 2004; Kaya *et al.* 2004).

Armed with appropriate data, hazardous situations are easily avoided, as Figure 5, an ASTER 631 image, illustrates. In 2000, an already disastrous situation in the drought-ridden Gode area of Ethiopia was exacerbated when new wells were drilled in iron-rich sediments. These produced red-stained water, which the indigenous population refused to use (E. Abate, Oxfam, pers. comm. 2001). The image clearly shows areas that should be avoided as water sources because of the presence of iron-rich rocks. These are highlighted in red because of the spectral properties of iron(III) oxy-hydroxides (Andrews Deller 2006). Such knowledge could have facilitated strategic placing of more suitable wells for people already desperate for clean water (Andrews Deller 2004).

However, given that disaster relief situations are often already desperate, as in the continuing humanitarian crisis in Darfur, Sudan (Buchanan-Smith 1997), remote sensing can be used to alleviate steadily worsening conditions and help relocate relief camps near clean water. Compressed Landsat ETM+ 742 mosaics (e.g. Fig. 6), freely available from the University of Maryland Global Land Cover Facility, are useful for planning access and provision of clean drinking water. A case in point is Figure 6, showing part of Western Darfur, downloaded from glcfapp.umiacs.umd.edu:8080/esdi/index.jsp and prepared in 1 hour. Roads can be added to the image in 30 minutes, using the free, on-line Digital Chart of the World and VMAP as rough guides, until more accurate road mapping data can be accessed from UN humanitarian information centres (Table 2). Natural features enhanced on the image can be used to locate sources of water. In the NE quadrant of Figure 6, for example, fossil sand dunes, which form excellent aquifers and could provide new sources of water within walking distance of overcrowded established camps, show up clearly as pale thin linear features with vegetation following them. As a result of only half a day's analysis, ASTER-derived geological maps of all water sources in areas around camps can be compiled to help refugees, aid workers and local people cope better with limited water supplies (Drury & Andrews Deller 2005).

Disaster mitigation and preparedness: infrastructure, water and education

'Many areas of the world are poorly mapped by conventional topographic mapping so satellite images are a big advantage when working in such terrains and are absolutely essential for determining location, geological features of interest such as faults and contacts, as well as the zones of rock alteration that may be the first indicators of economic concentrations of metals.' (D. Taranik, Anglo American, pers. comm. 2006).

Dealing with disasters is important, but improving people's normal living conditions is more so. A person's ability to survive disasters is enhanced if he or she has good food, water, financial security and a safe environment. The usefulness of remote-sensing technologies in long-term planning and economic management should not be underestimated. With basic training, local people can use satellite images to plan infrastructure: the best

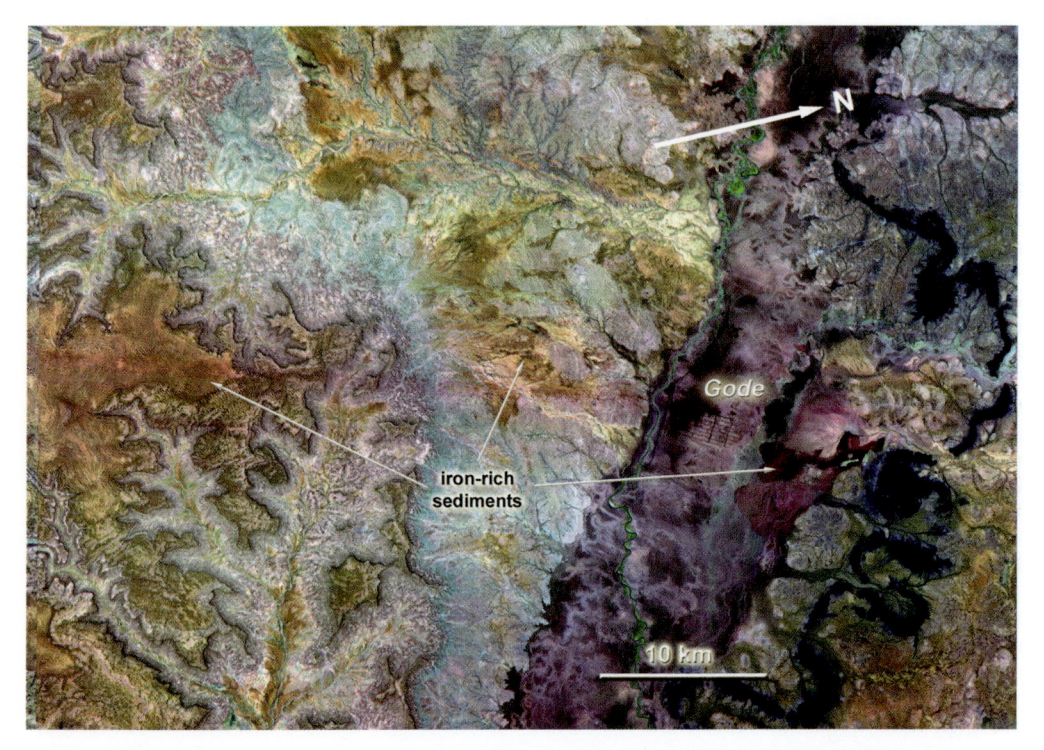

Fig. 5. ASTER 631 of Gode, Ethiopia: a region affected by the 1998–2001 drought: red indicates iron-rich sediments, the areas to avoid for groundwater boreholes.

Fig. 6. Landsat 742 mosaic of part of Western Darfur, Sudan, where excellent aquifers in fossil dunes (linear features in NE) could help in the siting of relief camps.

routes for roads; the whereabouts of building materials; agricultural possibilities; the management of mineral wealth to finance projects; areas best avoided because of risk from natural hazards, and safe water supplies. Once the geology of an area is mapped, its resource potential is clear. Figure 7, a geological map of laterite facies in the Asmara area, Eritrea, based on image data for the region (Andrews Deller 2000, 2006), provides a simple example. The uppermost ferricrete provides aggregate for roads and the mottled zone building material for houses. The clay zone with its infertile, poorly draining, kaolinitic soils susceptible to land slippage, is best avoided for agriculture. The weathered basement, easily pinpointed in Figure 7 as sub-laterite regolith where the process of lateritization has rotted primary minerals in the underlying bedrock, together with lateritic ferricretes and the mottled zone are potential aquifers whose waters can be beneficial (Andrews Deller 2004, 2006). Strategic siting of boreholes is crucial, however, as ground water in some lateritized terrain may have high levels of arsenic, concentrated fluoride that leads to dental fluorosis (Zerai 1996), excessive levels of magnesium sulphate and sodium chloride, which cause diarrhoea, and high ferric hydroxide content that results in pipe blockages. With very little input, specialist and non-specialist can use remotely sensed data for positive ends within the community.

Adequate planning to manage disasters that come with little or no warning is often difficult; earthquakes (Ellsworth 2006) and volcanic eruptions are cases in point. But drought and famine are entirely foreseeable, even though the new millennium began with nearly 20% of the world drought-stricken (Kogan 2005). Drought, desertification and crop yield monitoring data (Chavez *et al.* 2002) are available, as are comprehensive Drought Early Warning (DEW) and Famine Early Warning (FEW) systems with alerts and appropriate data collections that can be used to provide timely warning (Davies *et al.* 1991; Buchanan-Smith 1997; Table 2). Discernible deterioration of vegetation captured on satellite images over time coupled with local people's expertise in ground information can facilitate strategic planning for disaster mitigation months before drought takes hold.

However, central to any disaster that disrupts infrastructure and displaces people from their homes is the need for rapid provision of clean drinking water, an even greater priority than food and shelter because of the ever-present threat of epidemic disease. Remotely sensed data together with pictorial representation

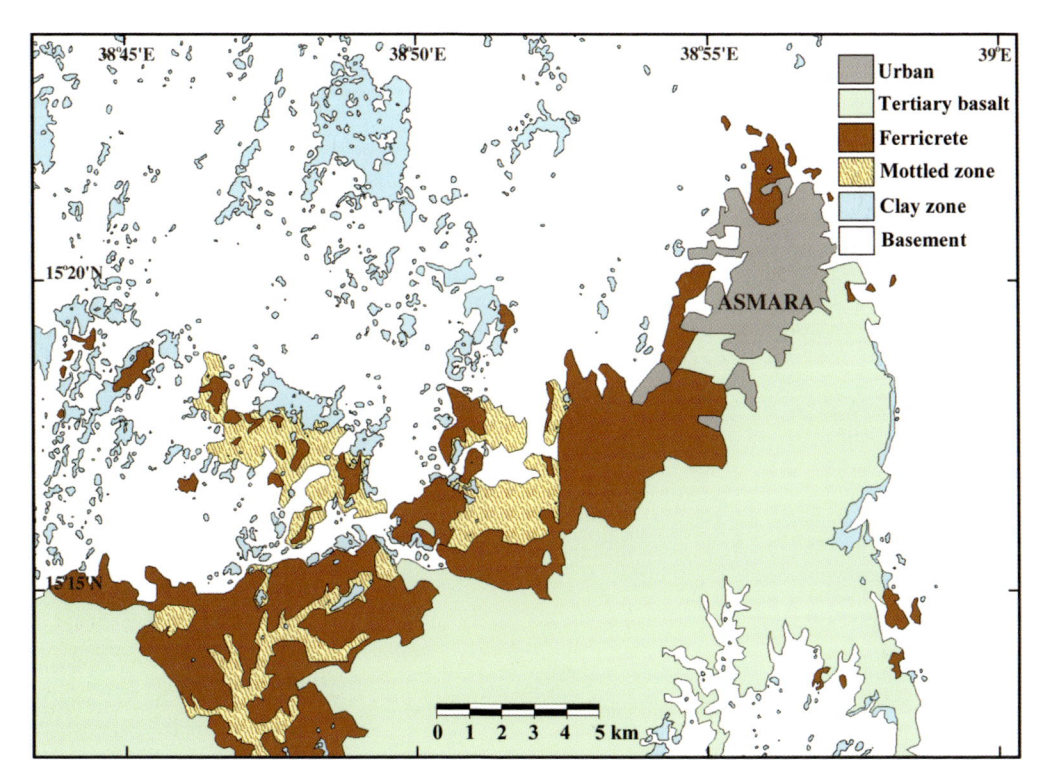

Fig. 7. Geological map of Eritrean laterite facies near Asmara, Eritrea, interpreted from Landsat TM and ASTER data (Andrews Deller 2006).

of digital elevation data of topography and drainage (see Jenson & Domingue 1988) matched to the skills of local hydrogeologists, provide a rapid means of assessing and prioritizing new well sites; an efficient means of detailed hydrological reconnaissance, and increase the prospects of success. Both Landsat and ASTER provide image data that are useful for efficient water exploration. These data, designed to highlight vegetation and rock-related information, together with indigenous people's knowledge of their surroundings, vital if remotely sensed data are to be validated, can form the basis for educational training programmes. 'Other special purpose imagery such as radar and thermal help extend human vision beyond the traditional optical wavelengths and reveal a whole new landscape of physical and chemical terrain features' (D. Taranik, Anglo American, pers. comm. 2006).

Using reflected and thermal data, local people, relief workers and scientists can assess potential water resources and work out the best sites for extraction, storage and distribution. ASTER 631 and TM 742 images illustrate the use of remote sensing for groundwater exploration simply. On these images green vegetation can be traced along spring lines, an elementary observation that allows for the best placing of wells. Although TM data allow some discrimination of potential aquifers, ASTER data have the spectral capabilities to identify sediments likely to have high yields (Drury & Andrews Deller 2002). Major aquifers are rocks and superficial deposits rich in quartz and feldspar, and/or carbonates. These minerals produce porous and permeable conditions in sediments, but also in crystalline rocks that have been pervasively fractured. Quartz and feldspar have unique red to magenta signatures (Drury 1998) in the standard multispectral thermal combinations of ASTER bands 14, 12 and 10 in RGB images and can be identified easily because of that coloration. Quartz- and feldspar-rich alluvial sediments therefore are often the most easily found aquifers and readily pinpointed where the ground is not obscured by thick vegetation. True thermal images are darkened by vegetation, but the predicted hues still show unless vegetation cover is dense.

Carbonates show in yellow hues in ASTER bands 5, 6 and 8 in RGB images, but so too do epidote- and chlorite-bearing rocks, which have to be checked according to their spectral 'signatures' in other band combinations. ASTER 631 images are ideal for such screening, as carbonates show as pale grey to white because of their flat visible and near infrared (VNIR) to short-wave infrared (SWIR) spectra, whereas chlorite and epidote show up in shades of red to magenta (Drury & Andrews Deller 2002, 2004). Use of these two ASTER images therefore identifies carbonates effectively. ASTER 631 images can also be used to screen quartz- and feldspar-rich sediments for low permeability caused by pore clogging of clays and white micas, which show up as cyan or sky blue, and by detrital chlorite, which gives a magenta tint (Drury & Andrews Deller 2005).

Moreover, both TM and ASTER instruments have sufficient resolution to delineate major joint systems and large fracture zones (Drury et al. 2001). Fracture zone mapping is most easily achieved using stereoscopic ASTER bands 3 N and 3B in anaglyph form. For example, International Committee of the Red Cross (ICRC) relief workers in Eritrea, 'with diverse backgrounds and no remote sensing experience', quickly grasped the essentials of image interpretation and proficiently targeted 'locations for sinking wells on simple 3-band images and ASTER anaglyphs... An Eritrean ICRC drilling team produced even quicker responses using images in the field' (S. A. Drury, Open University, pers. comm. 2006).

To determine water table depth, field follow-up with geophysical methods is necessary (Drury et al. 2001), but rapid remote sensing narrows the areas of search significantly and increases the prospects of finding water dramatically. Observation borne out by successful experience instils confidence and none of the observations made above, essential for water exploration, are beyond local people. In rural areas, where individuals specialize in locating near-surface water supplies, their expertise coupled with a remote sensing approach can benefit all.

Perhaps a key to successful community-based disaster management is education (Andrews Deller et al. 2004, 2005). Long-term and emergency water supplies, important at all times and crucial during periods of disaster and post-crisis recovery, might be a topic used to raise public awareness of the usefulness of space technology for disaster management and a starting point for training adults and children. Students and school children assimilate new ideas quickly, and older people, who are experts in ground truth, have a great respect for education in less-developed and rural areas. Sowden et al. (1997), Plester et al. (2002) and Blaut et al. (2003), report on pre-school children's innate ability to recognize terrain features and navigate using aerial photos. This is borne out by the author's experience in rural Eritrea (2003) with young children who mastered aerial photos and ASTER images within an hour and selected appropriate samples correctly all over the locality; a pre-school child even located a house made out of the correct material. Nothing can replace field observation as a basis for building up knowledge in depth (Marsh 2000). Local people of all ages at any standard of education, given motivation, are capable of collecting and locating data (Delson

2006; see also Ferreira 1792; Dehaene et al. 2006) that can add to hazard early warning by reporting instances of:

- unusual changes in the level of water in wells, a common precursor of seismic events in earthquake-prone areas;
- ground fracturing on and near unstable slopes, an indication of landslides;
- recent fumarole activity and ground changes in the vicinity of volcanoes, common precursors to volcanic eruption. Moreover, in areas prone to flooding, tsunamis, seismic activity, landslides or volcanism, people with ready access to graphic remotely sensed images that indicate potential refuge can prepare themselves with greater confidence to escape future hazards.

In the long term, data distribution on a global level is being addressed by the UN Office for Outer Space Affairs (UNOOSA), NASA, NOAA, the US Geological Survey (USGS) EROS Data Centre (Table 2) and the Integrated Global Observing Strategy (IGOS) (Marsh 2004). Training opportunities in remote sensing are available through organizations such as the UN Development Programme, the World Bank, USAID, ICRC and the British Department for International Development (Drury 1998). However, space technology for disaster management is often centred round specific problem areas or hazards, and few people in disaster-prone areas currently benefit from accurate regional information, data, training or funding (HAIC Secretariat 2001; Rochon et al. 2005).

Conclusion

'Developing countries still do not have wide access to space-based technologies used for early warning and emergency response, not only during the disaster response phase, but also during the more important preparedness phase of the disaster cycle, which invariably results in unnecessary loss of life and property when disaster strikes' (Stevens 2005). The place for remotely sensed data and access to it should be within the local community and at the disposal of the indigenous population. Yet the fact that multispectral remote sensing data remain little understood by local people, emergency relief workers and civil protection agency personnel is indicative of the lack of communication between specialists in space technology and the people who need their help. The problem is not small but can be tackled. Remote sensing applications provide an opportunity for progress. Country-specific information and guidance on the use of appropriate cost-effective remote sensing data (Jordan et al. 2000a) are available. Given the chance, local people and those involved in humanitarian relief quickly grasp the essentials of image interpretation in the context of their lives or duties. Once dialogue is established between locals, relief workers and specialists, those who

Table 3. *Useful types of satellite imagery for assessing geohazards*

Geoscientific hazard	Useful remotely sensed data for unmapped terrains
Volcanic	ASTER stereo (bands 3N and 3B in anaglyph form: landform overview), MODIS (thermal: near real-time thermal mapping), AVHRR (thermal: near real-time thermal mapping), MODIS and AVHRR (visible: near real-time emission mapping), perspective views (ASTER 631 or ETM+ 742 draped over SRTM DEM)
Seismic	ASTER stereo (bands 3N and 3B in anaglyph form: landform overview, active fault mapping), ASTER or Landsat ETM+ (631 or 742: clay-rich strata prone to liquefaction), perspective views (ASTER 631 or ETM+ 742 draped over SRTM DEM)
Landslide	ASTER stereo (bands 3N and 3B: landform overview), SRTM (DEM: slope mapping), ASTER or Landsat ETM+ (631 or 742: clay-rich lubricant strata), perspective views (ASTER 631 or ETM+ 742 draped over SRTM DEM)
Flood and tsunami	SRTM (DEM: mapping flood-prone areas, slope and drainage assessment for flash floods)
Drought (water provision)	ASTER (631: geological and vegetation mapping; 568 and 14 12 10: discrimination of quartz- and carbonate-rich aquifers; bands 3N & 3B in anaglyph form: fracture zone mapping); SRTM (DEM: fracture zone mapping)

have a common interest in safety, improvement in living conditions and disaster mitigation can work together. Easily understood images in printed form can be used to assess geohazards in mapped and unmapped terrain. Free or low-cost ASTER, MODIS, Landsat ETM+, SRTM-DEM and AVHRR data can be used to warn, educate and enable local people to deal with the dangers and hazards they live with (Table 3). As a result, local people can take charge of their own lives. Environmental factors affecting human health and wellbeing can be catered for, hazards anticipated and remotely sensed data used by local people to address the problems they face.

Special thanks go to W. S. Deller and S. A. Drury, who helped and encouraged me with this work. My thanks also go to D. Stevens and A. Teklegiorgis for their support, to R. M. Teeuw for asking me to contribute my work to this publication, and to the Eritrean and Ethiopian children and peoples.

Note

Versions of this paper were presented in the 2004 Munich United Nations International Workshop on the Use of Space Technology for Disaster Management; the 2005 Saint Petersburg 31st International Symposium on Remote Sensing of Environment, and as part of an Open University PhD thesis, 'Lateritic palaeosols of N.E. Africa: a remote sensing study.'

References

ABDALLAH, C., CHOROWICZ, J., BOU KHEIR, R. & KHAWLIE, M. 2005. Detecting major terrain parameters relating to mass movements' occurrence using GIS, remote sensing and statistical correlations, case study Lebanon. *Remote Sensing of Environment*, **99**, 448–461.

ABRAMS, M. 2000. The Advanced Spaceborne Thermal Emission and Reflection Radiometer (ASTER): data products for the high spatial resolution imager on NASA's Terra platform. *International Journal of Remote Sensing*, **21**, 847–859.

ABRAMS, M. & GELLER, G. N. 2005. Increasing access and usability of remote sensing data. The Protected Area Archive Tool applied to UNESCO Heritage Conservation Sites/NASA/Jet Propulsion Lab (JPL). In: *31st International Symposium on Remote Sensing of Environment, Saint Petersburg, Russia, 20–24 June 2005* (http://www.isprs.org/publications/related/ISRSE/html/papers/308.pdf).

ALVAREZ, R., ORTIZ, M. I., GOMEZ, G. & VIDAL, R. 2000. Geographic evaluation of hazardous volcanic emissions of populations: the plumes of Popocatepetl in March, 1996. In: *Proceedings of the 14th International Conference on Applied Geological Remote Sensing, Las Vegas, Nevada, USA, 6–8 November 2000*. Environmental Research Institute of Michigan, Ann Arbor, MI, 255–262.

ANDREWS DELLER, M. E. 2000. Facies discrimination in laterites, using Landsat Thematic Mapper (TM) data: an example from Eritrea (NE Africa). In: *Proceedings of the 14th International Conference on Applied Geological Remote Sensing, Las Vegas, Nevada, USA, 6–8 November 2000*. Environmental Research Institute of Michigan, Ann Arbor, MI, 302.

ANDREWS DELLER, M. E. 2002. Facies discrimination in laterites, using remotely sensed data: Landsat (TM) versus ASTER, ALI and Hyperion data—an example from Eritrea. *Annual General Meeting of the Geological Remote Sensing Group, Geological Society, London, 5–6 December 2002*.

ANDREWS DELLER, M. E. 2004. Space technology for disaster management: data access and its place in the community. In: *United Nations International Workshop on the Use of Space Technology for Disaster Management, Munich, 18–22 October 2004*. World Wide Web Address: www.zki.caf.dlr.de/events/2004/unoosa_workshop/unoosa_programme_en.html.

ANDREWS DELLER, M. E. 2005. Space technology for disaster management: examples from NE Africa. In: *31st International Symposium on Remote Sensing of Environment, Saint Petersburg, Russia, 20–24 June 2005*. World Wide Web Address: www.isprs.org/publications/related/ISRSE/html/papers/592.pdf.

ANDREWS DELLER, M. E. 2006. Facies discrimination in laterites using Landsat Thematic Mapper, ASTER and ALI-EO1 data: examples from Eritrea and Arabia. *International Journal of Remote Sensing*, **27**, 2389–2409.

ANDREWS DELLER, M. E., DELLER, W. S., DRURY, S. A. & TEKLEGIORIS, A. 2004. Use of space technology for disaster management: data access and its place in the community. In: *United Nations International Workshop on the Use of Space Technology for Disaster Management, Munich, 18–22 October 2004*. World Wide Web Address: www.zki.caf.dlr.de/events/2004/unoosa_workshop/unoosa_programme_en.html.

ANDREWS DELLER, M. E, DELLER, W. S., DRURY, S. A. & TEKLEGIORIS, A. 2005. Use of space technology for disaster management: data access and its place in the community. In: *31st International Symposium on Remote Sensing of Environment, Saint Petersburg, Russia, 20–24 June 2005*. World Wide Web Address: www.isprs.org/publications/related/ISRSE/html/papers/665.pdf.

BJORGO, E. 2000. Using very high spatial resolution multispectral satellite sensor imagery to monitor refugee camps. *International Journal of Remote Sensing*, **21**, 611–616.

BLAUT, J. M., STEA, D., SPENCER, C. & BLADES, M. 2003. Mapping as a cultural and cognitive universal. *Annals of the Association of American Geographers*, **93**, 165–185.

BUCHANAN-SMITH, M. 1997. What is a famine early warning system? Can it prevent famine? *Internet Journal for African Studies*, March 1997, No. 2. World Wide Web Address: http://www.bradford.ac.uk/research/ijas/ijasno2/ijasno2.html.

CHAVEZ, P. S., MACKINNON, D. J., REYNOLDS, R. L. & VELASCO, M. G. 2002. Monitoring dust storms and mapping landscape vulnerability to wind erosion using satellite and ground-based digital images. *Aridlands Newsletter*, No. 51, May–June 2002. World Wide Web Address: ag.arizona.edu/OALS/ALN/aln51/chavez.html.

CRIPPEN, R. E. 2001. Global topography at high resolution: shuttle radar topography mission/Jet Propulsion Lab (JPL). *Applications and New Opportunities in Geologic Remote Sensing. Geological Society of America Annual Meeting, 5–8 November 2001.* Session 145 (paper no. 145–0). World Wide Web Address: gsa.confex.com/gsa/2001AM/final program/abstract_17784.htm.

CRIPPEN, R. E. & BLOM, R. G. 2000. Remote sensing in three dimensions: the integration of new global data sets. *In: Proceedings of the 14th International Conference on Applied Geological Remote Sensing, Las Vegas, Nevada, USA, 6–8 November 2000.* Environmental Research Institute of Michigan, Ann Arbor, MI, 121.

DAVIES, S., BUCHANAN-SMITH, M. & LAMBERT, R. 1991. *Early warning in the Sahel and Horn of Africa: The state of the art. A review of the literature. Volume 1.* University of Sussex, Research Report, *20*.

DEHAENE, S., IZARD, V., PICA, P. & SPELKE, E. 2006. Core knowledge of geometry in an Amazonian indigene group. *Science*, **311**, 381–384.

DELSON, R. M. 2006. Examining knowledge of geometry. *Science*, **312**, 1309–1310.

DRURY, S. A. 1993. *Image Interpretation in Geology*, 2nd edn. Chapman and Hall, London.

DRURY, S. A. 1997. The national image mosaic of Eritrea. *International Journal of Remote Sensing*, **18**, 2897–2898.

DRURY, S. A. 1998. *Images of the Earth—a Guide to Remote Sensing*, 2nd edn. Oxford University Press, Oxford.

DRURY, S. A. 2005. *Earth Pages News*, January 2005 issue. Oxford: Blackwell Science. World Wide Web Address: www.earth-pages.com/news.asp.

DRURY, S. A. & ANDREWS DELLER, M. E. 2002. Remote sensing and locating new water sources. *In: United Nations International Workshop for Space Technology and Disaster Management, Addis Ababa, Ethiopia, 1–5 July 2002.* World Wide Web Address: www.unoosa.org/pdf/sap/2002/ethiopia/presentations/12 speaker01_1.pdf.

DRURY, S. A. & ANDREWS DELLER, M. E. 2004. ASTER multispectral thermal data and finding water. *In: United Nations International Workshop on the Use of Space Technology for Disaster Management in Munich, 18–22 October 2004.* World Wide Web Address: www.zki.caf.dlr.de/media/download/unoosa_workshop_presentations/16_pres_session06c_chair-tarabzouni/58_UNOOSA-DLR_Drury_OpenUni v_UK.pdf.

DRURY, S. A. & ANDREWS DELLER, M. E. 2005. Advances in groundwater exploration: the roles of ASTER and SRTM data. *In: 31st International Symposium on Remote Sensing of Environment, Saint Petersburg, Russia, 20–24 June 2005.* World Wide Web Address: www.isprs.org/publications/related/ISRSE/html/papers/661.pdf.

DRURY, S. A., PEART, R. J. & ANDREWS DELLER, M. E. 2001. Hydrogeological potential of major fractures in Eritrea. *Journal of African Earth Sciences*, **32**, 163–177.

ELLSWORTH, W. L. 2006. Halfway through Reid's cycle and still counting. *Science*, **312**, 203–204.

FERREIRA, A. R. 1792. *Viagem Filosofica pelas capitanias do Grão Pará, Rio Negro, Mato Grosso e Cuiabá*, vol. 2, Memorias zoologia e botanica. Conselho Federal de Cultura, Rio de Janeiro (reprinted 1974).

FU, B. H., NINOMIYA, Y., LEI, X. L., TODA, S. & AWATA, Y. 2004 Mapping active fault associated with the M_w 6.6 Bam (SE Iran) earthquake with ASTER 3D images. *Remote Sensing of Environment*, **92**, 153–157.

GOOSSENS, M. 2002. ASTER in mineral exploration: A brief review. *Newsletter of the Geological Remote Sensing Group*, Geological Society of London, London, 33, 16–21.

HAIC SECRETARIAT, 2001. *Establishment of a regional Humanitarian Assistance Information Centre (HAIC).* Report of a UNEP and UNDP (Somalia) Workshop, Nairobi, Feb. 1–2, 2001. World Wide Web Address: www.depha.org/proceedings/HAIC%20Proceedings.DOC.

HUANG, W. & FU, B. 2002. Remote sensing for coastal area management in China. *International Journal of Remote Sensing*, **30**, 271–276.

HUNT, J. C. R. 2002. Floods in a changing climate: a review. *Philosophical Transactions of the Royal Society of London, Series A*, **360**, 1531–1543.

JENSON, S. K. & DOMINGUE, J. O. 1988. Extracting topographic structure from digital elevation data for geographic information system analysis. *Photogrammetric Engineering and Remote Sensing*, **54**, 1593–1600.

JORDAN, C. J., O'CONNOR, A. P., PASSMORE, J., SUCRE, B. & CHULUUM, O. 2000a. Appropriate technology for low-cost geological mapping—the case for publicising geological, mineral and remote sensing information of developing countries to the investor sector via the internet with case studies from Guyana and Mongolia. *In: Proceedings of the 14th International Conference on Applied Geological Remote Sensing, Las Vegas, Nevada, USA, 6–8 November 2000.* Environmental Research Institute of Michigan, Ann Arbor, MI, 303–310.

JORDAN, C. J., O'CONNOR, E. A., MARCHANT, A. P. ET AL. 2000b. Rapid landslide susceptibility mapping using remote sensing and GIS modelling. *In: Proceedings of the 14th International Conference on Applied Geological Remote Sensing, Las Vegas, Nevada, USA, 6–8 November 2000.* Environmental Research Institute of Michigan, Ann Arbor, MI, 113–120.

KAYA, S., MUFTUOGLU, O. & TUYSUZ, O. 2004. Tracing the geometry of an active fault using remote sensing and digital elevation model: Ganos segment, North Anatolian Fault zone, Turkey. *International Journal of Remote Sensing*, **25**, 3843–3855.

KERLE, N., FROGER, J. L., OPPENHEIMER, C. & VAN WYK DE VRIES, B. 2003. Remote sensing of the 1998 mudflow at Casita volcano, Nicaragua. *International Journal of Remote Sensing*, **24**, 4791–4816.

KING, R. B. 1981. An evaluation of Landsat 3 RBV imagery for obtaining environmental information in Tanzania. *In*: ALLAN, J. A. (ed.) *Matching Remote Sensing Technologies and their Applications.*

Proceedings of the 9th Annual Conference of the Remote Sensing Society, London, 16–18 December 1981. Remote Sensing Society, University of Reading, Reading, 85–95.

KING, R. B. 1984. *Remote Sensing Manual of Tanzania,* Surbiton: Land Resources Development Centre, Surbiton, Surrey.

KING, R. B. 1994. The value of ground resolution, spectral range and stereoscopy of satellite imagery for land system and land-use mapping of the humid tropics. *International Journal of Remote Sensing,* **15**, 521–530.

KOGAN, F. 2005. Remote sensing contribution to early drought detection and monitoring. *In*: *31st International Symposium on Remote Sensing of Environment, Saint Petersburg, Russia, 20–24 June 2005.* World Wide Web Address: www.isprs.org/publications/related/ISRSE/html/papers/454. pdf.

MARSH, S. 2000. Remote mapping technologies for temperate, vegetated terrain. *In*: *Proceedings of the 14th International Conference on Applied Geological Remote Sensing, Las Vegas, Nevada, USA, 6–8 November 2000.* Environmental Research Institute of Michigan, Ann Arbor, MI, 11–17.

MARSH, S. 2004. The IGOS Geohazard Report. *In*: *United Nations International Workshop on the Use of Space Technology for Disaster Management in Munich, 18–22 October 2004.* World Wide Web Address: www.zki.caf.dlr.de/media/download/unoosa_workshop_presentations/05_pres_session03_c hairwade/12_UNOOSA-DLR_Marsh_BGS_UK.ppt.

MASSONNET, D. 1995. Application of remote sensing data in earthquake monitoring. *Advances in Space Research,* **15**, 37–44.

MCARTHUR, J. M., RAVENSCROFT, P., SAFIULLAH, S. & THIRLWALL, M. F. 2001. Arsenic in groundwater: testing pollution mechanisms for sedimentary aquifers in Bangladesh. *Water Resources Research,* **37**, 109–117.

MCCLUSKEY, J. & CHOUDHURY, Z. 2005. *Assessment and Scoping Report, Aceh, Indonesia, 4–13 February 2005.* World Wide Web Address: www.hapinternational.org/pdf_word/908-Aceh%20trip%20report%20QMP%20scoping.pdf.

OŠTIR, K., VELJANOVSKI, T., PODOBNIKAR, T. & STANČIČ, Z. 2003. Application of satellite remote sensing in natural hazard management: the Mount Mangart landslide case study. *International Journal of Remote Sensing,* **24**, 3983–4002.

PLESTER, B., RICHARDS, J., BLADES, M. & SPENCER, C. 2002. Young children's ability to use aerial photographs as maps. *Journal of Environmental Psychology,* **22**, 29–47.

RAMSAY, M. S. & FLYNN, L. P. 2004. Strategies, insights, and the recent advances in volcanic monitoring and mapping with data from NASA's Earth Observing System. *Journal of Volcanology and Geothermal Research,* **135**, 1–11.

ROCHON, G. L., QUANSAH, J. E., MOHAMED, M. A. ET AL. 2005. Applicability of near-real-time satellite data acquisition and analysis & distribution of geoinformation in support of African Development. *In*: *United Nations Economic Commission for Africa: 4th Meeting of the Committee for Development Information, Addis Ababa, Ethiopia, 23–28 April 2005.* World Wide Web Address: www.uneca.org/codi/Documents/Word/ECA_CODI_ IV_Paper_Rochon.doc.

SANYAL, J. & LU, X. X. 2006. GIS-based flood hazard mapping at different administrative scales: A case study in Gangetic West Bengal, India. *Singapore Journal of Tropical Geography,* **27**, 207–220.

SARTORI, G., NEMBRINI, P. G., JANSEN, P., SALONE, G. L., WERDMULLER, M. & CORTHÉSY, P. 2000. An appraisal of data collection methodologies: Somalia flood relief operation 1977 with special reference to using a GIS approach. Interagency Meeting, 18–19 December 2000, Oxfam, Oxford.

SETZER, A. 2004. The operational fire alert system of Brazil: a detection and management tool using multiple satellites. *In*: *United Nations International Workshop on the Use of Space Technology for Disaster Management in Munich, 18–22 October 2004.* World Wide Web Address: www.zki.caf.dlr.de/media/download/events/2004/unoosa_workshop_presentations/06_pres_session04a_chair-manikiam/19_UNOOSA-DLR_Setzer_INPE.ppt.

SINGHROY, V. & MOLCH, K. 2004. Characterizing and monitoring rockslides from SAR techniques. *Advances in Space Research,* **33**, 290–295.

SOWDEN, S., STEA, D., BLADES, M., SPENCER, C. & BLAUT, J. M. 1997. Mapping abilities of four-year-old children in York, England. *Journal of Geography,* **95**, 107–111.

STEVENS, D. 2005. Space-based technologies for disaster management—making satellite imagery available for emergency response in developing countries. *In*: *Geoscience and Remote Sensing Symposium, 25–29 July 2005, Seoul, S. Korea, IGARSS '05. Proceedings. 2005 IEEE International* **6**, 4366–4369.

TEMESGEN, B., MOHAMMED, M. U. & KORME, T. 2001. Natural hazard assessment using GIS and remote sensing methods, with particular reference to the landslides in the Wondogenet Area, Ethiopia. *Physics and Chemistry of the Earth, Part C,* **26**, 665–675.

UNOOSA (The United Nations Office for Outer Space Affairs) 2004. *In*: *Proceedings of the United Nations International Workshop on the Use of Space Technology for Disaster Management, Munich, 18–22 October 2004.* World Wide Web Address: www.zki.caf.dlr.de/events/2004/unoosa_workshop/unoosa_programme_en.html.

VINCENT, R. K. 1997. *Image Fundamentals of Geological and Environmental Remote Sensing,* 2nd edn. Prentice Hall, Englewood Chiffs, NJ.

WRIGHT, R. 1999. *Infrared satellite studies of Mount Etna volcano:1991 to 1999.* PhD thesis, The Open University.

WRIGHT, R., FLYNN, L., GARBEIL, H., HARRIS, A. & PILGER, E. 2002. Automated volcanic eruption detection using MODIS. *Remote Sensing of Environment,* **82**, 135–155.

ZERAI, H. 1996. Groundwater and geothermal resources of Eritrea with the emphasis on their chemical quality. *Journal of African Earth Sciences,* **22**, 415–421.

Index

acid mine drainage discharge, 107–116
Advanced Spaceborne Thermal Emission and Reflection Radiometer *see* ASTER
aerial photography, 1, 2
   coastal and riverine environments, 93–94, 97, 100–101, 103
   landslide monitoring, 54–56, 61, 66, 69–73
   landslide susceptibility, 78, 80, 90
   linear infrastructure projects, 135, 136, 138, 139–140
   mine waste dispersion, 110
   see also digital aerial photography
Africa, disaster management, 150–161
airborne digital sensors, 55
airborne digital videography, 101
Airborne Thematic Mapper *see* ATM
Airborne Visible Infrared Imaging Spectroradiometer *see* AVIRIS
AirSAR (Airborne InSAR), 13
Alaska, 13
Aleutians, 13
ALOS, 27, 28, 46
Alps, Swiss, 60
Anianchak volcano, 11
Andes, 8
aquatic weed infestation, 97
Arc Info, 82
ArcGIS, 34, 80, 82–85, 110
archive datasets, 136, 139–140, 142
ArcMap, 19, 34, 36
ARGON, 9
ASAR, 27
asphalt survaces, 126, 130
ASTER, 1, 2, 8, 95, 99, 102, 136, 138, 150, 153, 160, 161
   DEMs, 14, 17–23, 95, 143–147, 150
   shrink-swell clay detection and mapping, 117–124
   volcanic terrain mapping and monitoring, 6, 8–9, 14–23, 27
ATM (Airborne Thematic Mapper)
   acid mine drainage discharge, 109
   coastal and riverine environments, 94, 101, 102
   Daedalus 1268 AZ-16 ATM, 32
   thermal imagery of subglacial volcano, 31–43
atmospheric attenuation, 39, 42
atmospheric effects, and PSI, 49
Autocad, 80
automatic extraction of displaced vectors, 60–61
AVHRR, 152, 161
AVIRIS, 109, 110, 115
AVNIR-2, 27
AZGCORR software, 32

backscatter intensity, 11, 45
bacterial oxidation of hydrocarbons, 126
Banda Aceh (Indonesia), 1

Bárðarbunga volcanic system (Iceland), 32
'bare-soil image', 117
bathymetric mapping, 95, 101
Bayesian probability modelling, 85–90
beach morphology, 101
bituminous surfaces, 126, 128, 130
Black Ven (Dorset, UK), 99
botanical anomalies, hydrocarbon seepage, 126, 128
brightness index, 71
British Geological Survey, 1
Bukavu (R.D. Congo), 65–75
buried structures, 99
   see also pipelines
butane, 126

C-band SAR, 7, 11, 13–14, 46–47
CAD drawings, 140
California (USA), 127, 128, 130, 143
calorimetry, 32, 41
carbon dioxide, 128
carbonate rocks, 160
CartoSAT, 28
CASI (Compact Airborne Spectrographic Images), 94, 97, 102, 103, 109, 110, 115
Catania (Sicily), 14
CATs (Compact Active Transponders), 50–51
Centre for Observing and Modelling Earthquakes and Tectonics (COMET), 143
Chile, 9
chlorite, 121, 123, 160
civil engineering projects, terrain analysis, 3, 135–142
clays
   slope failure, and disaster management, 2, 153
   see also shrink-swell clays
cliffs, coastal, 93, 97–99
CLIMVIS, 152
cloud cover, 95–96, 97, 123, 135, 139, 144
   volcanic terrain mapping, 5, 9, 11, 13, 14, 17, 27
coastal environments, 3, 93–106
Colorado (USA), 60
'colour balance', 146
COMET (UK Centre for Observing and Modelling Earthquakes and Tectonics), 143
Compact Active Transponders (CATs), 50–51
Compact Airborne Spectrographic Imager *see* CASI
cone morphometry, 20–23
Congo, R.D., 65–75
contour models, 135
Corner Reflectors (CRs), 25–26, 50–51
Corniglio landslide (Italy), 60
CORONA spy-satellite data, 2, 6, 7, 9–11, 14, 136, 139
   DEMs, 9–11, 14
cost considerations, 102, 150, 161
cross-corelogram technique, 109
crosstalk problem, 17
cyanide, 107, 108, 110, 113–114

INDEX

DAIS 7195, 128–129
Darfur (Sudan), 157
declassified satellite imagery, 9–11, 139
    DEMs, 9–11
'DEM-of-difference', 59
DEMs, 1, 6, 94, 102, 103, 135, 138, 140–142, 161
    landslide mapping and monitoring, 56–59, 82, 99
    mine waste dispersion, Spain, 110
    NextMap UK, 94, 97, 102, 103
    'off-the-shelf' datasets, 137
    quality controls, 57, 58–59
    urbanization, Bukavu (R.D. Congo), 71, 73
    see also ASTER; CORONA; DSMs; DTMs; InSAR; Landsat; SPOT; SRTM; TOPSAR
deposition see sedimentation
Derbyshire (UK), 54–61
desertification, 159
differential GPS, 14, 36, 55
DifSAR (differential synthetic radar interferometry), 2, 13, 47–49
digital aerial photography, 94, 101
Digital Airborne Imaging Spectrometer (DAIS), 128–129
Digital Elevation Models see DEMs
digital photogrammetry, 6, 17
    coastal and riverine environments, 94, 97, 101
    for landslide monitoring, 56–61
    see also photogrammetry
Digital Surface Models see DSMs
Digital Terrain Models see DTMs
digital videography, 101
DigitalGlobe, 7
disaster management, 1, 149–164
Disaster Monitoring Constellation (DMC), 28
displaced vectors, automatic extraction of, 60–61
drought, and disaster management, 153, 157, 159
DSMs (Digital Surface Models), 95, 102, 137
DTMs (Digital Terrain Models), 95, 102, 135, 136, 137, 138

Earth Observation (EO), 7, 46–47, 137–138
earthquakes
    disaster management, 1, 159, 161
    ground deformation mapping, 48, 49
Ebro platinum thermometer probe, 36
education, and disaster management, 160–161
electromagnetic spectrum, 93, 128
emissivity, 34
Enhanced Thematic Mapper see ETM and ETM+
ENVI software, 34, 117, 122
environmental impact, acid mine drainage discharge, 107–116
ENVISAT, 11, 12, 27, 46, 49, 50
EOWEB, 7
epidote, 160
ERDAS Imagine, 34
Eritrea, 149, 153, 159, 160
EROS Data Center, 152, 161
erosion, 2, 93, 100–102, 108
ERS imaging radar, 2, 96, 102
    ERS InSAR, 102
    ERS SAR, 9–50, 23–26
    ERS-1, 6, 7, 11, 46, 49
    ERS-2, 6, 7, 11, 46, 49

estuaries, 102
ethane, 126
Ethiopia, 153, 157
ETM+ (Enhanced Thematic Mapper Plus), 7, 8, 150, 153, 161
Etna, Mt., 14
Eurimage, 7
European Space Agency, 47, 96
    see also ERS
evaporites, 125

famine, 159
faulting
    and disaster management, 2, 153, 157
    Jamaica (St Thomas), 82
    R.D. Congo (Bukavu), 67–68
feldspar-rich sediments, 160
Fernandina, Volcán (Galápagos), 15
ferrihydrite, 108, 110, 113–115
field surveying
    and landslide susceptibility, 80
    tailings dump, 109
FLIR systems ThermoCAM, 39
flood defences, 93, 96, 97
flooding, 2, 93, 95, 96–97
    disaster management, 153, 157, 161
forest fires, 157
fracture zone mapping, 160, 161
France, 117–124

Galápagos volcanoes, 9, 11, 15
GAMBIT system, 9
gas seepages, 125, 126, 127, 130–132
Geo-Tiffs, 34
geodetic datum, 32–34
geohazards, 136, 139–142
    and disaster management, 149–164
geological maps and mapping, 81–82, 138
geology, urban, 65–68, 70–73
geomorphological anomalies, and hydrocarbon seepage, 126
geomorphological applications, 95, 97, 98–99, 138–139
geomorphological mapping, and aerial photograph interpretation, 56, 65–73, 80
Geophysical and Environmental Research Corporation (GER), 109
geothermal activity, Iceland, 32
German Aerospace Agency, 7
Gers (France), 117–124
GIS, 1, 2, 6, 71–74
    and digital aerial photography, 94
    for LiDAR-CASI data, 97
    for terrain analysis, 135, 138, 140, 142
    see also ArcGIS; ArcMap
glacier motion mapping, 49
Global Land Cover Facility, 7, 8, 17, 157
global warming, 93, 97, 125
GMES (Global Monitoring for Environment and Security), 47
goethite, 108, 110
gold mines, abandoned, 107–116
Google Earth, 150, 152

INDEX

GPS, 49, 50
   datum, 32–34
   see also differential GPS
Grimsvötn volcanic system (Iceland), 31–43
ground displacement mapping and monitoring, 1, 47–48, 136
   see also landslides
ground instability, 93, 97–100
'ground truth', 140

Hawai'i, 6, 13, 14–23
heavy metals, mine waste dispersion, 107, 108, 112–115
Heimaey (Iceland), 14
hematite, 108, 110, 114
HEXAGON system, 9, 11
High Resolution Stereoscopic (HRS) instrument, 7, 9
historical archives, 136, 139–140, 142
Hough transforms, 130
humidity, 39, 42
Hungary, 127, 129, 130
hydrocarbon seepage, 125–133
hydrogeological planning, and disaster management, 159–160
hydrological applications, 95, 97
HyMap, 94, 128
hyperspectral imagery, 1, 2, 94, 103, 107–116, 117–124, 125–133, 136, 137
   see also CASI; HyMap

ice cauldrons, 31–32, 41–42
ice motion mapping, 49
Iceland, 14, 31–43
IfSAR DEMs, 102
IGOS (Integrated Global Observing Stategy), 161
IKONOS, 2, 6, 7, 11, 66, 70–74, 95, 102, 136, 138, 139, 141, 153
illite, 120–121, 122–123
IMAGINE OrthoBASE Pro 8.6 software, 54
Indian Ocean tsunami disaster, 1, 153–157
Indonesia, 1, 11
infrastructure planning, and disaster management, 157–159
infrastructure projects, terrain analysis, 3, 135–142
InSAR, 1, 2, 94, 136
   coastal and riverine environments, 96, 100, 102, 103
   DEMS, 12–14, 94, 96
   for volcanic terrain mapping and monitoring, 5, 6, 12–14, 23–26, 27–28
   see also AirSAR; DifSAR
Integrated Global Observing Stategy (IGOS), 161
interferometry see InSAR
Intermap Technologies Inc., 94, 102
Internal Average Relative Reflectance algorithm, 117
International Committee of the Red Cross (ICRC), 160
Iran, 143–147
iron, secondary, 107, 108, 110, 113–115
IRS, 95
Italy, 60
ITC system of geomorphological mapping, 56
Izmit (Turkey), 49

Jamaica, 77–91
jarosite, 107, 108, 110

JERS SAR, 7, 11, 12
Jet Propulsion Laboratory (NASA), 13, 143, 149

Kamchatka, 11
kaolinite, 121, 123
KH satellite sensors, 9–11
Kiejo volcano (Tanzania), 23, 24
kriging, 36
KVR-100, 136

L-band SAR, 7, 11, 26, 27–28
Land Process Distributed Active Archive Center, 7, 8, 17
land use patterns, St. Thomas (Jamaica), 83
Landsat, 1, 2, 7, 97, 102, 128, 136, 137, 138, 140
   and disaster management, 149, 150, 152, 160, 161
   volcanic terrain mapping, 6, 8, 15, 17–26, 27
landscape evolution, 139–140, 142
landscape recognition, and disaster management, 153
landslides, 2, 3
   disaster management, 153, 157, 161
   monitoring, 53–63, 65–75
   susceptibility assessment, 77–91
   see also slope instability
laser altimetry see LiDAR
lava flows, 5, 6, 8, 11, 14–23, 67
   and disaster management, 157
Lee, river (N of London, UK), 97
Leica 300, 14
LiDAR (Light Direction And Ranging), 1, 2, 95, 97, 99, 101, 102, 103, 135–137, 142
   LiDAR-CASI data, 97, 102, 103
linear projects, terrain analysis, 135–142
Linear Spectral Unmixing technique, 109, 110
lithologocal mapping, St. Thomas (Jamaica), 81–82
London, 99–100

Mam Tor (Derbyshire, UK), 54–61
mapping, 2
   see also specific types of mapping
Matched Filtering techniques, 109, 110, 122–123
Mauna Kea volcano (Hawai'i), 6, 14–23
Mersey estuary (UK), 102
methane, 126, 128
Mexico City, 14
micas, 160
micro-graben, Bukavu (R.D. Congo), 65, 68, 69
mineralogical anomalies, and hydrocarbon seepage, 126, 128
mineralogy
   mine waste, 107–116
   shrink-swell clays, 120–121, 122, 123–124
'minimum distance to class means' algorithm, 128–130, 131–132
Minimum Noise Fraction (MNF) algorithm, 109
mining
   subsidence measurement, 49
   waste discharge, 107–116
Mississippi river (USA), 95
Mixture Tuned Matched Filtering (MTMF) algorithm, 122–123
modelling of slope instability, 85–90
MODIS, 1, 152, 161
moisture level mapping, 117, 135, 140

Mojave Desert (USA), 149
molasse deposits, mapping, 117–124
montmorillonite, 121, 123
mosaicing, 143–147
multi-temporal DEMs, 59
multi-temporal geomorphological maps, 56
multispectral imagery, 2, 136, 138
    coastal and riverine environments, 94, 95, 99, 102, 103
    volcanic terrain mapping and monitoring, 6–11, 23–26, 27–28
    see also hyperspectral imagery

NASA, 138, 143, 149, 152, 161
NASA Space Shuttle, 1, 138
    see also Shuttle Imaging Radar; SRTM
New Zealand, 13
NextMap UK, 94, 97, 102, 103
Ngozi volcano (Tanzania), 23, 24
Nicobar Islands, 157
NOAA, 152, 161
Norfolk, 101
Normalized Difference Vegetation Index (NDVI), 71, 117

oil seepages, 125, 126–127, 128, 130–132
OrbView, 138
OrbView-3, 136, 138
OrbViews-5, 27
Orthoengine (PCI), 144, 146, 147
orthophotos, 60–61, 135

PALSAR, 27, 46
Parádfürdo (Hungary), 127, 129, 130
pattern recognition, 130–131, 132
PCI Orthoengine, 144, 146, 147
pentane, 126
persistent scatterers interferometry see PS-InSAR; PSI
photogrammetry, 55, 98–99, 100–101, 102
    software, 135
    see also digital photogrammetry
PIMA portable spectrometer, 123
Pinatubo, Mt., 6
pipelines
    detection, 97
    leakage detection and monitoring, 130–131, 137
    routing, 140
Planck's law, 34
pollution
    disaster management, 2, 153
    see also acid mine drainage discharge; hydrocarbon seepage
PRISM, 28
propane, 126
Protected Area Archive, 149
PS-InSAR, 100, 102, 103
PSI (Persistent Scatterers Interferometry), 13, 25–26, 49–50, 96, 102
pyrite, 108, 109, 113

quartz-rich sediments, 160
Quickbird, 2, 6, 7, 95, 102, 136, 138, 139, 142, 153

radar, 138–139
radar see ASTER; ERS; SAR

RADARSAT, 7, 11, 12
    RADARSAT-1, 46, 49, 50
    RADARSAT-2, 27, 46
Radarsat International, 7
radiance, converting to temperature, 34–36
radon, 125
rainfall
    effects on mine waste, 107–108
    hazard, St Thomas (Jamaica), 77, 78, 83–84
RapidEye constellation, 28
Red Cross (ICRC), 160
redox environment, 126
refugees, 2, 157
relief models, 137
RGB images, 150, 160
Rhine river, 95, 102
Rhône river, 95
riverine environments, 3, 93–106
Rodalquilar (Spain), 107–116
Rungwe Volcanic Province (Tanzania), 6, 23–26
Russia, 49
Rwanda, 69

Sabancaya volcano (Chile), 9
St Peterburg (Russia), 49
St Thomas (Jamaica), 77–91
SAM (spectral angle mapper), 109, 121–122, 123, 128–130, 131–132
SAR (Synthetic Aperture Radar), 2, 45, 46–47, 95–96, 97, 143
    interferometry see InSAR
    volcanic terrain mapping, 6, 11, 24–25, 26, 27, 28
    C-band, 7, 11, 13–14, 46–47
    L-band, 7, 11, 26, 27–28
    X-band, 7, 11, 13–14, 94
    see also AirSAR; DifSAR; ENVISAT; ERS SAR; JERS SAR; RADARSAT; SRTM
Satellite Pour l'Observation de la Terre see SPOT
scoria cone morphometry, 20–23
sea-level rise, 93, 97
Seamless Data Distribution System, 7, 17
sedimentation, 2, 93, 100–102, 108
seepage haloes, 129, 130–132
seismic applications, 1, 2
    see also faulting; fracture zone mapping
Severn estuary (UK), 102
shrink-swell clays, 99–100, 117–124
Shuttle Radar Topography Mission see SRTM
Shuttle-Imaging Radar (SIR), 143
Shuttle-Imaging Radar-C (SIR-C), 11
slope instability, 153
    and human settlement, 65–75
    modelling, 85–90
    see also cliffs; landslides
Slumgullion landslide (USA), 60
snow cover, 123, 144
soil erosion, Bukavu (R.D. Congo), 65
soil moisture content, 117, 135, 140
Solid Terrain Modelling, 153
Space Imaging, 7
Space Shuttle see NASA
Spain, 107–116

INDEX

spectral analysis techniques, 109–111, 121–122, 123, 128–132
spectral angle mapper technique *see* SAM
spectrometry
    mine waste dispersion, 110
    shrink-swell clays, 99, 123
SPOT, 2, 95, 97, 102, 136, 137, 138, 144
    DEMs, 9, 95
    SPOT HRS, 7
    SPOT-5, 7, 9, 136
    volcanic terrain mapping, 6, 8, 9, 15
Spotimage, 7
spy-satellite data *see* CORONA
SRTM (Shuttle Radar Topography Mission), 2, 6, 7, 12–14, 136, 138, 150, 152
    DEMs, 2, 17–26, 102, 143–147, 150, 161
stereo orientation, 144–147
stereo photography, 56–57, 69–73, 102
stereoscopic viewing, 1, 53, 55, 100, 135, 140, 161
storms and storminess, 1, 93, 97, 151
subsidence measurement and monitoring, 49, 51
Sudan, 157
Sumatra, 1
susceptibility-hazard mapping, 77–91
Swiss Alps, 60
Synthetic Aperture Radar *see* SAR

Tanzania, 6, 23–26, 150
Tay river (Scotland), 101
tectonics, Bukavu (R.D. Congo), 65–68
temperatures, ground-based data compared with airborne thermal imagery, 31–43
TERRA Satellite, 8
Terrafirma Project, 47
terrain analysis, 135–142
terrain modelling, 153
    see also DTMs
TerraLook, 149, 152
TerraSAR, 27
TerraSar-X, 46
TerraWeb, 152
Thames floodplain (London, UK), 97
thermal imagery, 1, 31–43, 102, 125, 136, 137
tidal conditions, 95, 102
tilt and offset, 144–147
TM (Thematic Mapper), 8
topographic maps and mapping, 14, 22–23, 70, 135, 138
TOPSAR-derived DEM, 15
toxic mine drainage discharge, 107–116
Trannon river (Wales), 101

TRIMBLE Pathfinder, 36
TRIMBLE Pathfinder Pro XRS, 14
tsunamis, 1, 97, 153–157, 161
Turkey, 49

UK Centre for Observing and Modelling Earthquakes and Tectonics, 143
UK Environment Agency, 97
UK Natural Environmental Research Council (NERC), 32
UK Overseas Development Agency, 150
UN Development Programme, 161
UN Office for Outer Space Affairs (UNOOSA), 161
University of Maryland *see* Global Land Cover Facility
urban geology, 65–68, 70–73
US Department for International Affairs, 161
US Geological Survey, 7, 9, 11, 161
USA, 13, 60, 95, 101, 127, 128, 130, 143, 149
USAID, 150, 161

variograms, 36
Vatnajökull (Iceland), 31–43
Vegetation Index, 71, 117
vegetative cover, 70, 71, 83, 135, 137, 153, 157
    coastal and riverine environments, 94, 95, 96, 102
    shrink-swell clay detection and mapping, 117, 123
    volcanic terrain mapping, 13, 17, 18, 22, 27–28
videography, 101
Virtuozo digital photogrammetry software, 17
volcanic terrain, mapping and monitoring, 2, 5–30, 48
volcanoes
    disaster management, 153, 157, 159, 161
    subglacial, 31–43

Wales, South, landslide evolution, 61
walk-over surveys, 140
waste *see* acid mine drainage discharge
water supplies, 149, 157, 159–160, 161
wetland mapping, 94, 140
WGS 1984 geodetic datum, 32–34
World Bank, 27, 161
World-View, 138

X-band InSAR, 94
X-band SAR, 7, 11, 13–14, 94
X-ray diffraction analysis, 109
X-ray fluorescence analysis, 109
X-ray spectrometry, 110

Zagros Mountains (Iran), 143–147